FRESHWATER AQUACULTURE

A FUNCTIONAL APPROACH

(With Intricate Informations on Integration of Fish with other Crops, Diversifications of Suitable Alternatives of Indian Major Carp Culture Practices and Sewage-fed Aquaculture)

The book entitled 'FRESHWATER AQUACULTURE -A FUNCTIONAL APPROACH' is the outcome of more than thirty years of association and the day to day practical experience of the author with the grass root level fish farmers. The entire book is divided into six chapters excluding the References cited.

Freshwater Aquaculture – the study of breeding, rearing and commercialization of organisms, fish in particular, which inhabit in fresh water. Even though, there remain some fragmentary informations regarding the history of development of aquaculture in India but those seem to be far from being complete. In the present communication, the same has been given elaborately. The book concentrates on the culture technology of commercially important fresh water fishes.

Various types of culture techniques including Aquaponics, Bioflocs, Recirculatory Aquaculture Systems (RAS) apart from the conventional Cage culture, Pen culture, Integration of fish culture with other crops viz. paddy, vegetables, dairy, piggery, poultry etc. have been dispensed in detail.

Dr. Biplab Kumar Bandyopadhyay a well-known fishery scientist who obtained his Doctoral degree from University of Calcutta under the guidance of Prof. Naresh Chandra Datta, Department of Zoology, University of Calcutta. His remarkable work is on the use of different micro- and macronutrients, organic and bio fertilizers, nutritional supplements and more so the disease abatement of fishes, which deserve special mention. Proper aquaculture management practices are his special area of interest.

Dr. Bandyopadhyay is probably the first to use different types of agro-chemicals in aquaculture practices to promote sustainable growth and prevent various types of fish diseases which very often create tremendous financial losses to the farmers, at least in the states like West Bengal, Odisha, Jharkhand, Bihar, Assam and Tripura.

FRESHWATER AQUACULTURE
A FUNCTIONAL APPROACH

(With Intricate Informations on Integration of Fish with other Crops,
Diversifications of Suitable Alternatives of Indian Major Carp
Culture Practices and Sewage-fed Aquaculture)

By
BIPLAB KUMAR BANDYOPADHYAY
M.Sc, Ph.D, FZS (Cal)

NARENDRA PUBLISHING HOUSE
DELHI (INDIA)

First published 2022
by CRC Press
4 Park Square, Milton Park, Abingdon, Oxon, OX14 4RN

and by CRC Press
6000 Broken Sound Parkway NW, Suite 300, Boca Raton, FL 33487-2742

© 2022 Narendra Publishing House

CRC Press is an imprint of Informa UK Limited

The right of Biplab K. Bandyopadhyay to be identified as the author of this work has been asserted in accordance with sections 77 and 78 of the Copyright, Designs and Patents Act 1988.

Print edition not for sale in South Asia (India, Sri Lanka, Nepal, Bangladesh, Pakistan or Bhutan).

British Library Cataloguing-in-Publication Data
A catalogue record for this book is available from the British Library

Library of Congress Cataloging-in-Publication Data
A catalog record has been requested

ISBN: 978-1-032-29170-3 (hbk)
ISBN: 978-1-003-30033-5 (ebk)

DOI: 10.1201/9781003300335

Printed in the United Kingdom
by Henry Ling Limited

DEDICATED TO MY BELOVED TEACHER:
Late PROF. NARESH CHANDRA DATTA

(Who taught me aquatic ecology, fishery biology and aquaculture science)

Contents

Preamble

Aquaculture has a long history and the 'Art' of aquaculture is very old. The evidence that Egyptians were probably the first in the world to culture fish as far back as 2500 B.C. available from the pictorial engravings of an ancient Egyptian tomb showing tilapia being fished out from an artificial pond. The Romans are believed to have reared fish in circular ponds divided into breeding areas. Culture of Chinese carps was wide spread in China in 2000 B.C. Writings in India were available in 300 B.C. which suggest means of rendering fish poisons in the Indian sub-continent in times of war. This implies that fish culture prevailed in those time in some Indian reservoirs. Some historical documents compiled in 1127 A.D describe methods of fattening fish in ponds in India. Culture of Gangetic carps in Bengal in the Indian Sub-continent is of historical origin.

In the ancient times the human beings were residing in forests, used to hunt fishes from the aquatic systems and used to eat it raw or either by burning or smoking. As the humans gradually felt to reside collectively in some places, unknowingly giving a shape of a society, realized that hunting fishes from the nature may serve the need of the day, but may not be consumed in future. They started stocking fish in ditches, small ponds etc. gradually it became a hobby and this stocking gave the shape of 'rearing' and finally to Aquaculture.

[Hunting ⟶ Hobby ⟶ Stocking (gathering) ⟶ Rearing ⟶ Culture]

Aquaculture, the farming of aquatic organisms, has existed in some form for 4000 years. *There are various hypotheses of how aquaculture came to be including the Oxbow, Catch-and-Hold, Concentration, and Trap-and-Crop theories. It has evolved from extensive culturing into very intensive.* There are a variety of types of aquaculture systems including pond, rice and fish integration, cage culture, raceways and recirculating systems. The intensification of aquaculture occurred following the Blue Revolution of the 1950's when demand for fish products spiked. This increase in demand occurred due to the understanding of fish health benefits including Omega-3fatty acids and strives to eat locally.

Aquaculture, the farming of fish and aquatic plants has existed for over 4000 years (Rabanal.H.R.1988. Historyof Aquaculture, FAO). Over those 4000 years of its existence aquaculture has fluctuated in methodologies, species farmed, environmental effects and economics. The present circumstances in aquaculture are more technologically based than ever before. *The future of aquaculture is bright due to the trend of depleting of the oceans' fish resources.* By understanding the past and education in the present, the future of aquaculture can aid in decreasing the overuse of marine resources and help reduce food shortages for the world's poverty stricken populations.

There are multiple theories as to how this husbandry technique were developed, including the **oxbow, catch-and-hold, concentration, and trap-and-crop theories** (Rabanal, 1988).

The **Oxbow theory** reflects nature's natural system of developing oxbow lakes from meandering rivers. As the river changes its course, a section may no longer be attached to the main river system, thus a new lake is formed. Oxbows are natural locations for high concentrations of fish due to the yearly stocking by floods. Humans, being naturally entrepreneurial, took advantage of these by enhancing the habitat, making more oxbow areas and eventually stocking additional fish into existing lakes. According to Rabanal, Bangladesh may have been a good location for this to have developed.

The second one is the **Catch-and-Hold Theory** which focuses on the tendency for nobles to request fish all year round. In order to elicit this, workers would catch wild fish and transport them to neighboring water locations where they could be accessed easily. Many water areas that were artificially constructed for other purposes such as in gardens or defense around a dwelling were stocked with these wild fish. The common carp (*Cyprinus carpio*) is a fish species that would be able to survive and flourish in this type of environment.

The **Concentration Theory** focuses on environments that have seasonal monsoon periods that inundate the landscape with water, but when the water subsides during dry seasons and the areas where it stays become concentrated with large populations of fish due to the smaller quantity of water. As time progressed, the humans of the areas began differential harvesting focusing on the larger fish and allowing the smaller fish to grow and reproduce. The depressions that the water stays in, were improved for future water catchment and enhanced populations of fish.

The final theory is termed the **Trap-and-Crop Theory** which was developed in marine locations, especially tidal coastal locations that receive nutrients and flushing with each tidal movement. These areas have natural lagoons that fill with water and organisms during high tide, and maintain a level of water during low tide when the water recedes. This provides the optimal location for an entrepreneurial human to develop an aquaculture system. It is believed that fences and traps were created that allow fish and crustaceans in, but not out, allowing easy harvest. **Over time there was a switch to allowing the fish that entered the trap to grow to a larger size, thus aquaculture**. Areas of southeastern Asia, primarily Indonesia, the Philippines, Thailand, Malaysia, and India have the ideal environment for this theory to have developed.

Fisheries and aquaculture products are now globally important sources of much needed, high quality aquatic animal proteins, and invaluable providers of employment, cash income, and foreign exchange. Fisheries products are the world's most widely traded foods, with commerce dominated by the developing countries. Fisheries products are the primary protein sources for some 950 million people worldwide, and are an important part of the diet of many more. In comparison to other sectors of the world food economy however, the fisheries and aquaculture sectors are poorly planned, inadequately funded, and neglected by all levels of government. This neglect occurs in a paradoxical situation: fishing is the largest extractive use of wildlife in the world; and aquaculture is the most rapidly growing sector of the global agricultural economy.

Aquaculture is an ancient practice but until the 1950's depended less on innovations arising from applied science and directed management skills than accomplishments derived from trial and error. Of the 25,000 fish species, only a few are harvested for direct human consumption. Aquaculture science is still poorly developed and undestood. There are very few centers of excellence in aquaculture, and few aquaculture experiment stations. *Clearly, freshwater aquaculture is the "poor cousin" of agriculture and capture fisheries in the marine sectors.*

Aquaculture especially in freshwater became important in the late nineteen-forties, since the methods of aquaculture could be used to restock the waters as a complement to natural spawning. In recent times, aquaculture has taken a shape of lucrative industry (Timmons *et al.*, 2002 and Cressey, 2009).

However, the intensification of aquaculture practices are the need of the day which necessitate the cultivation at towering densities resulting significant damage to the environment principally due to the discharge of concentrated organic wastes, depletion of dissolved oxygen content in pond water, increased toxic metabolites like hydrogen sulfide, methane, ammonia, and nitrites etc., which are very often liable for mortalities of aquatic flora and fauna. Moreover, under these conditions of intensive production, aquatic species are subjected to high-stress conditions, increasing the incidence of diseases and causing a decrease in productivity (Bondad *et al.*, 2005).

Outbreaks of viral, bacterial, and fungal infections have caused devastating economic losses worldwide, and India is reported to bear $210 million losses during 1995 and 1996. Added to this, significant stock mortality has also been reported due to poor environmental conditions of farms, unbalanced nutrition, generation of toxins, and genetic factors (Kautsky *et al.*, 2000). In recent decades, prevention and control of animal diseases has attracted doubtful attention on the indiscriminate use of chemical additives and veterinary medicines, especially antibiotics, which generate significant risks to fishes as well as public health by promoting the selection, propagation, and persistence of bacterial-resistant strains WHO(2012).

Sustainable aquaculture evidently indicate the culture of commercially important fishes, prawns, molluscs and some aquatic plants ensuring better quality of their lives now and for their generations to come.

Like all the culture practices, whether it is Agriculture, Horticulture and Animal Husbandry which includes, Poultry, Duckery, Dairy, Piggery etc., **Aquaculture too is based on three Q's:**

1) Quality seed, 2) Quality feed and 3) Quality environment. It is the responsibility of the workers of Extension services in the respective fields to make understand to the ignorant grass root level farmers about the immense utility and importance of the three Q's.

Since independence Fisheries sector in India has been recognized as an important sector and contributed about 0.75million tones in 1950-51and gradually increased to more than 10million tones at present. This achievement placed the country in second largest global fish producer, only after China.

The sector has shown significant growth from traditional culture practices to commercial methods of culture, enhancing fish production from a mere 7.5 lakh tonne in 1950-51 to 107.95 lakh tonnes during 2015-2016, while earnings has of Rupees 33,441 crore in 2014-15 (US $ 5.51 billion) through export of fish to different countries. Fisheries and aquaculture sector has achieved a significant overall annual growth rate of about 4% during the 11th Five Year Plan period. As such fisheries and aquaculture has made contribution of about 0.91% to the National Gross Domestic Production (GDP) and 5.23% to the agricultural GDP (2014-2015), which is very significant. At present India contributes about 6.30% to the global fish basket and 5% of global fish trade (Misra *et al.*, 2017).

It is estimated that the contribution of capture fisheries to the total fish production of the country has shown a gradual declining trend over the years, from almost 100% during1950-51 to the present level of about 45%, aquaculture sector on the other hand have shown marked transformations contributing significant increase.

Fresh water aquaculture resource of the country largely confined to 2.41million hectares of ponds and tanks. All these water bodies varied from small backyard ponds of eastern Indian states to large deepwater tanks of south India. Till date utilization of only 60-70% of these resources where from the production as achieved, contribute a little more than 50% of the total fish production of the country.

State wise marine output and culture fisheries production during 2012-2013 has been worked out by Goswami and Zade (2015).

There is global need for food and nutritional security especially amongst developing and under-developed countries. Fisheries in India, is a progressively growing sector with varied resources. It has been estimated that more than 14.50 million people at the primary level are directly or indirectly dependent on this sector for their livelihood security (Misra et.al 2017).

Fish and fishery products represent a valuable source of nutrients of fundamental importance for diversified and healthy diets. With a few exceptions for selected species, fish is usually low in saturated fats, carbohydrates and cholesterol. Fish provides not only high-value protein, but also a wide range of essential micronutrients, including various vitamins (D, A and B), minerals (including calcium, iodine, zinc, iron and selenium) and polyunsaturated omega-3 fatty acids (docosahexaenoic acid and eicosapentaenoic acid). While average per capita fish consumption may be low, even small quantities of fish can have a significant positive nutritional impact by providing essential amino acids, fats and micronutrients that are scarce in vegetable-based diets.

According to the latest data, Indians on an average consume just 269 grams of fish per month in rural areas while in urban areas it's 238 gram (Approx.6.1kg/ year) Noticeably, just about 282 of 1000 households in rural areas consume fish, while the number is 209 households for urban areas. This apparently indicates a vast majority of the Indians prefer consuming vegetarian dishes.

India is the third largest producer of fish while 7th in shrimp aquaculture production. India has a long coastal line stretching about 8129 km covering 9 coastal states, 4 union territories (enriched with wide spectrum of flora and fauna), 3.9 million hectares of estuary, 2.54 million hectares of salt affected coastal soil and 5 million hectares of mangrove forest.

The projected demand for fish in the country (2012) is 9.74 mmt; demand for fish for the domestic market is 5.9 mmt. The projected supply of fish is 9.60 mmt (by 2012) with major share is from 2 are as viz. 5.34 mmt from inland aquaculture 3.10 mmt from marine fisheries The fisheries sector contributed Rs. 67,913 crores to the GDP (at current prices) during 2009-2010. When considering the fish production state wise:

West Bengal occupy the first position with 1615.313 tones fish during 2010-2012, Andhra Pradesh occupy second position 1349.940, Gujarat third position774.902 and Kerala have fourth position 681.613 followed by Maharashtra having of 576.987mt. (FAO, 2012). Recent data depicts that West Bengal exported about 1.04 lakh metric tones of fisheries products amounting Rs. 4456 crores in 2017-2018 (source: Department of Fisheries,Govt.of West Bengal).

It is no exaggeration that both the capture and culture sectors are subjected to manmade and nature driven calamities, yet the contribution of fish production of India to the global production is on the steady rise, from 32% in1950-51 to 58% as on date. However, considering the target posed, before the country to produce 16.0 million tons of fish by 2030, it is necessary to impose appropriate sustainable management strategies those are made to utilize the available aquatic resources of the country (Ayyappan. S. 2016. Key note address during"International conference on Aquatic Resources and sustainable Management" Kolkata).

In modern aquaculture practices the contributions of late Prof. Hiralal Chowdhury (1921-2014), the father of induced breeding of fishes and the forerunner of first blue revolution in India cannot be ruled out. He was the Fish breeder par excellence and could successfully breed *Esomas danricus* in 1955 and *Pseudoutropis atherinoides* in 1956, soon after on 10th July1957, he succeeded in induced breeding of the minor carp *Cirrihinus reba* in captivity. Subsequently, Prof.Chowdhury with the help of pituitary extracts revolutionized in successful breeding of *Cirrhinus mrigala, Labeo rohita* and *Catla catla* in captivity. Successful production of millions of fertilized eggs bring forth miraculous development and the outcome of which is nearly more than about 1800 fish hatcheries today, through both private and public participation came into existence. Albeit, the availability of fish seed in terms of quantity has not remained an issue, but the availability of scientific protocol for nursery rearing, fingerling production, technology of carp poly culture, integrated farming in the country, eastern India in particular is still in a state of infancy. In the recent past the large scale adoption of cat fishes (*Pangassius* spp.) farming, also a suitable alternative of carp culture, at least in southern part of India might reach the production target by 2030, but the sustainable culture technology still remain in a question (Ayappan, 2016).

Besides, in India 2,319 species of fin fish have been recorded of which 838 from freshwater, 113 brackish water and 1,368 from marine environment. Small indigenous freshwater fish species (SIF) are defined as fishes which grow to the size of 25-30 cm in mature or adult stage during their life cycle. Majority of them resides in rivers and tributaries, floodplains, ponds and tanks, lakes, beels, streams, lowland areas, wetlands and paddy fields. In general these small indigenous fishes are highly acclaimed because of their taste and high nutritive value. However, very little attention has been paid on their role in aquaculture enhancement, nutrition, captive breeding and conservation needs. Subsequently, many small indigenous fishes (SIF) have become threatened and endangered due to pollution, over exploitation coupled with habitat destruction, diseases and introduction of exotic varieties.

Among small indigenous fishes, many species are cultivable with high demand, and can be introduced as a candidate species in freshwater aquaculture system. These are *Amblypharyngdon mola, Notopterus notopterus, Puntius sarana, Labeo bata, Puntius ticto, Cirrhinus reba, Nandus nandus, Anabas testudineus, P. sarana, Mystus vittatus* and air breathing fishes like *Clarias batrachus, Notopterus notopterus, Osteobrama belangeri* the– *Pengba* of Manipur Loktak lake etc. Some of these species are being cultured at minimum scale, mostly based on wild seed collection. With the technology available for seed production, culture and expertise of disease remedial treatments now available, large scale farming need to be popularized and expanded.

The technology pertaining to breeding, seed production and farming at grow out state of magur (*Clarias batrachus*) and singhi (*Heteropneustes fossilis*) though have been developed but their large scale seed production and scientific farming has not yet been taken up at commercial scale. In the present communication however, an attempt has been made to throw some light on this aspect along with food and feeding habits, artificial propagation, larval rearing, expected diseases along with their remedial measures and the grow out technology of some of the fresh water cultivable fishes like, the air breathing fish *Clarias batrachus*, climbing perch *Anabas testudineous'* (the koi) fresh water stripped cat fish like *Mystus vittatus*, butter cat fish *Ompokn bimaculatus*, the milk fish *Chanos chanos* and the slim-bellied Amur carp *Cyprinus carpio haematopterus*. An attempt has also been made to discuss on the biology and culture possibilities of some of the endangered fresh water fishes.

Acknowledgements

The inspiration injected in my blood by Late Rathin Chandra Sengupta, the then Deputy Director, Directorate of Fisheries, Government of West Bengal, referring a Chinese proverb which says *"If you give a man a fish he will eat for a day, but if you teach him the technology of fish farming he will eat for his entire life"*. This particular sentence awakened me since my university days to recent.

Since last thirty five years I had the opportunity to work in different types of Fishery environments. In the present communication, I have tried my level best to share my experiences with grass root level fish farmers. In this connection, I do not have any hesitation to convey my sincere gratitude to my teacher, path finder and guide Late Prof. Naresh Chandra Datta, Former Head, Department of Zoology, University of Calcutta, who taught me Fisheries Science and extended his blessings and helps as and when asked for and supported me to solve various problems related to fresh and brackish water aquaculture. It is quite unfortunate that Prof. Datta is nor more with us and it is also heart breaking that this publication could not be sent to the press before his heavenly aboard.

Initially I would like to convey my heartiest gratitude to Dr. Mrinal Kanti Das, retired faculty of the Department of Zoology, Behala College Kolkata, West Bengal, happened to be one of my colleagues since a few decades who inspired me to write few lines for the grassroot level aquafarmers. The present communication is the outcome of his valued request and inspirations.

I would like to thank Dr. Niranjan Sarangi, Former Director, Central Inland Fresh water Aquaculture (CIFA),ICAR, Kausalyaganga, Bhubaneswar,Odisha,Dr. Bijay Kali Mahapatra, Principal Scientist, CIFE (ICAR), Salt Lake City, Sector-V, Kolkata-700 091, Prof. Sushanta Chakraborty, Prof. Bidhan Chandra Patra of Department of Zoology, Vidyasagar University, Midnapur, Prof. Sumit Home Choudhury, Former Head, Department of Zoology, University of Calcutta, Prof. T. J. Abraham of West Bengal University of Animal and Fisheries Sciences, Chakgaria, Kolkata-94 and especially Dr. Uttamn Laha of CADC (Comprehensive Area Development Corporation), Tamluk Centre, who has provided various information's and photographs related to the subject as an when required.

I specially extend my sincere gratitude to aquaculture professional's viz. Bijoy Roy of Kalna, Dist.-Burdwan, Sri Samir Das, Subrata Das, Sanjay Das, Jayanta Das (Bapi) and many more aqua farmers of Risra, District-Hoogly , Sri Bablu Ghosh, Sri Bikas Roy, the expert in fish breeding and fish seed marketer of Naihati, 24-parganas (N), Sri Soumitra Mondal of Baharampur ,Dist-Murshidabad and Sri Dilip Dutta of Uluberia Industrial Growth Centre, Howrah for their co-operations on trust to spare their grow out ponds for experimental purposes of various aquaculture products of repute.

While composing this book I took the opportunity to consult various communications of the aquaculture scientists and professionals viz. Md. Tarique Sarkar of FISHTECH (Bangla Desh) Ltd., Dr. Md. S.I. Bhuiya of BAU, Mymensingh, Bangla Desh, Dr. Subhendu Datta and Dr. P. Kumar of Central Institute of Fisheries Education (CIFE), Salt Lake, Sector-V, Kolkata Chapter, Dr. Pratap Mukhopadhyay, Retired Scientist, CIFA, Kausalyagang, Bhubaneswar, Odisha, Dr. Manas Kr. Das, Dr. Utpal Bhowmik and Dr. Sandipan Gupta of CIFRI (ICAR), Barrackpore, West Bengal, Dr. Dipankar Saha, former Head, Nimpith Krishi Vigyan Kendra, 24-Parganas (South), M. Mukherjee, Aloke Praharaj, Shamik Das and Tuhin Baguli, Deputy Directors, Department of Fisheries, Government of West Bengal, Agro-fisheries input seller's and a number of students of Aquaculture and Fisheries Science, Mr. Arijit Mazumder, Principal consultant of Creative Research Group, Kolkata especially for providing informations on sewage fed aquaculture in EKW (Section-V), and many others, for their kind help and valuable information.

While writing this book I have received help and encouragement from so many people related to this field, which is not possible to name them all here. Those who have freely shared with me and their outcome of many years' of experience and study represent from a wide variety of research institutions, and of many professions. To all of them I express my deepest thanks for time and thought so generously given.

I have taken numerous information's from various electronic media including subject related websites and internet time to time. The author expresses his sincere gratitude to both of these two agencies.

I wish to thank Mrs. Tota Bandyopadhyay, my wife who motivated me to write this book for the interest of fish farmers in spite of several difficulties in our day to day life schedule. I extend my heartiest thanks to Dr. Jharna Nath, retired teacher Shyama Sundari Balika Vidyapith, Behala, Kolkata-700 0034, and Dr. Subrata Kar retired scientist of Zoological Survey of India who took all pains to share their valuable experiences in aquaculture time to time and scrutinize this manuscript meticulously and sincerely.

I personally indebted to Mr.Sekhar Kumar Karmakar and Miss.Sarbani Chakraborty of **DURON AGRO INDUSTRIES, Kolkata** for their positive support and encouragement to write this book for the interest of entire fresh water fish growers, related fishery officials and agro-fisheries input sellers.

It will be unjustified if I do not express sincere gratitude to all of my office colleagues including the marketing professionals of **DURON AGRO INDUSTRIES,** Kolkata, for their all out inspiration while preparing this manuscript.

Last but not the least; I have tried to share almost all the minute details of aquaculture aspects, problems and their possible rectifications. A constructive suggestion and criticism if any, is welcome with gratitude for further upgradations.

Biplab Kumar Bandyopadhyay

Kolkata,

May' 2018

CHAPTER

1

AQUACULTURE: TYPES

1. INTRODUCTION

Aquatic farming is one of the oldest farming practices in the world. It is believed that the Egyptians were probably the first in the world to initiate culture practices of the aquatic fauna (Fish) as far back as 2500 B.C. which are available in the pictorial engravings of an Egyptian tomb where in tilapia fish (?) was shown to been fished out from a pond.

In India, the aquaculture practice is perhaps much older than those of the Egyptians since the documentations on the subject are also mentioned in *Vedas & Upanishadas* which are existed between 4000-2500 BC. There are references to fish culture in *Kautilya's Arthasahastra* (321-300B.C) and king *Someswara's Manasoltara* (1127A.D) (Kumar and Sharma, 2012).In the "Kishkinda Khand" section of the Hindu epic "**Ramayana**" there are interesting references to fish in water in the lines: "*Jal Sankoch Bikal Bhai Meena; Abodh Kutambi Jimi Dhanabeena,* " (which means, due to paucity of water in the pond, especially in summer, the fishes are in distress like a foolish householder who is in distress for want of money). And further, "*Sukhi meena je neer to Agodha Jimi Hari Saran Na Akakum Badha*" that means, fish which are in deep waters are as happy as a man under God's sheltering care, (Jhingran, 1983).

The traditional practice of fish culture in small ponds of eastern India is known to have existed for even more than hundred years.

Fishes are defined to be the first vertebrates in the evolution of animal kingdom. Fishes are cold blooded aquatic vertebrates which breathe by means of pharyngeal gills, propelling and balancing themselves by means of fins (Jhingran, 2000).

Fish is an essential protein rich item in every one's day to day food dish in majority of the states. The supply of cultured fresh water fish though is at its increasing trend, but the demand perhaps seem to be much more. Aquaculture is the farming of aquatic organisms by intervention in the rearing process to enhance production. Aquaculture has a long history, originating at least in the year 475 B.C. in China, but became important in the late nineteen-forties, since the methods of aquaculture could be used to restock the waters as a complement to natural spawning (Boyd and Tucker, 1998).

Definition of Fisheries: *Fisheries may be defined as the culture, capture and conservation of commercially important aquatic organisms. This includes both aquatic animals like fishes, molluscs (clams, oysters etc.), arthropods (crabs, prawn and shrimps etc.), plants (aquatic weeds, macrophytes etc.).*

In India, the excessive price of this protein rich item coupled with the gap of demand and supply vis-à-vis comparatively less production, poses a great concern to the state governments. It may be inferred in this context that the less production from the local water bodies West Bengal in particular is subject to two possible reasons:

(1) *There is a shortage of technological knowledge pertaining to the various management practices involved in culture,*

(2) *Socio-economic and socio-political problems & interventions.*

Aquaculture is the fastest growing food-producing sector in the world, with an average annual growth rate of 8.9% since 1970, compared to only 1.2% for capture fisheries and 2.8% for terrestrial farmed meat production systems over the same period.World aquaculture has grown tremendously during the last fifty years from a production of less than a million tonnes in the early 1950s to 59.4 million metric tons (mmt)by 2004. This level of production had a value of US$70.3 billion or more. The diseases and deterioration of environmental conditions often occur and result in serious economic losses.

Global production of fish from aquaculture, grew more than 30 percent between 2006 and 2011, from 47.3 million tons to 63.6 million tons. It is estimated that by 2012 more than 50 percent of the world's food fish consumption will come from aquaculture, and is expected to overtake capture fisheries as a source of edible fish. This growth rate may be attributed to several factors:

(1) Many fisheries have reached their maximum sustainable exploitation,

(2) Consumer concerns about security and safety of their food,

(3) The market demand for high-quality, healthy, low-calorie, and high-protein aquatic productsand,

(4) Aquatic breeding makes only a minimum contribution to carbon dioxide emission (Cruz et al., 2012).

Indian aquaculture is an important economic enterprise and contributes to about 1.3% of national inland production (GDP) and the annual growth rate is 8%.

In India 2,319 species of fin fish have been recorded of which 838from freshwater, 113 brackish water and 1,368 from marine environment.

West Bengal contains 239 species belonging to 147 genera, 49 families and 15 orders of freshwater fishes. Since West Bengal is one of the eight maritime states lying on the north eastern coast of our country, some marine or estuarine fishes are also found along with the freshwater fishes in this state.

Definition of Aquaculture: *Aquaculture refers to the breeding, rearing, and harvesting of plants and animals in all types of aquatic environments (i.e. aquatic both lotic [flowing] &lentic [standing]) environments including rivers, ocean as well asponds and lakes.*

Aquaculture-produces food fish, sport fish, bait fish (Bait fish are small fish caught for use as bait to attract large predatory fish) particularly game fish e.g., marinebait fish are anchovies gudgeon. Freshwater bait fish include any fish of the minnow or carp family (Cyprinid etc.), ornamental fish, crustaceans, mollusks, algae, sea vegetables, and fish eggs.

The principal target of aquaculture is:

(1) To produce easily digestible animal protein for the mankind,

(2) Economic upliftment of the concerned family, and

(3) To keep the aquatic eco-system pollution free.

Social values of Aquaculture

Water farming in contrast to soil farming: To restore the aquatic biodiversity and natural balance, replenishment of underground water, flood prevention, purification of waste water, small irrigation, cooling of environment, disaster management, easily digestible supply of animal protein, economic upliftment, perfect utilization of leisure of house wives and children. Besides oxygen supply, absorption of carbon-di-oxide and atmospheric dust particles, water farming supply enormous pleasure to the farmer members behind the scene. Some of the fishes very often also help in mosquito control.

Importance of Fish

(1) Fish is highly nutritious, easily digestible, protein rich food.

(2) Besides protein, the fish also contain various other necessary components.

(3) Abundant quantity of vitamins, minerals, calcium, and iron are available in fish flesh.

(4) There are plentiful marine weed fishes, which the mankind do not consume, are dried and converted to fish meal which is used as major component of the animal feed.

(5) There are some larvivorous fish available in fresh waters which are cultured for prevention those are also considered as vectors of various diseases of mosquito larvae of various diseases. These fishes also rectify the water column by controlling various infectious diseases.

(6) Fish is an easily digestible mineral and protein enriched aquatic animal.Fish also alleviate mental anxiety and maintain mental peace, entertainment and relief.

(7) On average, fish provides only about 33 calories per capita per day. The dietary contribution of fish is more significant in terms of animal proteins, as a portion of 150 g of fish provides about 50–60 percent of the daily protein requirements for an adult.

Distribution and Habitats of fishes

Distribution

1. Bony fishes inhabit almost every body of water. They are found in tropical, temperate, and polar seas as well as virtually all fresh water environments (Bond, 1979).

2. Some species of bony fishes live as deep as 11 km. in the deep sea. Other species inhabit lakes as high as 5 km. above sea level.

3. About 58% of all species of bony fishes (more than 13, 000 species) live in marine environments. Although only 0.01% of the earth's water is fresh water, freshwater fishes make up about 42% of fish species (more than 9, 000 species).

Habitat

Fish habitat has been defined as those waters and substrate necessary to fish for spawning, breeding, feeding or growth to maturity. This includes all types of aquatic habitat, such as wetlands, coral reefs, sand, sea grasses, and rivers (Rosenberg *et al.*, 2000).

1. Bony fishes live in fresh water, sea water, and brackish (a combination of fresh water and salt water) environments. The salinity of sea water is about 35 ppt (parts per thousand). Some species can tolerate higher-salinity environments. Some species of gobies can tolerate salinity levels as high as 60 ppt.

2. Fishes live in virtually all aquatic habitats. Different species of fish are adapted for different habitats: rocky shores, coral reefs, kelp forests, rivers and streams, lakes and ponds, under sea ice, the deep sea, and other environments of fresh, salt, and brackish water.

3. Some fish are pelagic: they live in the open ocean. For example, tunas (several species in the family Scombridae, subfamily Thunninae) are pelagic fishes.

4. Some species, such as the flatfishes (order: Pleuronectiformes) are adapted for living along the bottom. Certain fishes, such as gobies (family: Gobiidae) even burrow into the substrate or bury themselves in sand.

5. Ocean sunfish (family: Molidae) are most often spotted at the ocean's surface.

6. Some lungfishes "hibernate" throughout a summer drought season, buried under the mud of a dried-up pond.

7. Several fish species live in freshwater habitats and even in the darkness of caves.

Carps in general are the inhabitants of fresh water ponds, lakes, reservoirs and also in rivers. In the present book, the carps found in warm water fish ponds and lakes are given importance.

Forbes (1887) for the first time characterized a lake as a microcosm. Later, Forel (1892, 1895, and 1904) supported the idea of Forbes and developed the study of fresh water impoundments which is now globally known as 'Limnology' in contrast to 'Oceanography' which means the study of marine ecosystem.

Elton and Miller (1954) commenting on the importance of the studies on ponds and tanks had suggested that "There is a bewildering variation among the ponds, even if they are situated in the same geographical area and shows great internal complexity within each pond."

Before we enter in the subject, let us clarify the definition of a pond. Albeit there is no set rules. However, most of the limnologists agree that ponds are small, completely enclosed body of water, shallow enough to encourage aquatic vegetation. As per Forel's (1892) classification of lakes, later modified by Whipple

(1927), ponds represent lakes of third order. No exact limit of areas and depth has been laid down for a pond.

Lake or Pond – What is the Difference?

From a regulatory viewpoint there is no distinction between a lake and a pond. Both are surface waters of the state and subject to the same water quality standards. Water bodies named "lakes" are generally larger and/or deeper than water bodies named "ponds." From an ecological or limnological perspective, there is a difference between the two: The term "lake" or "pond" as part of a water body is arbitrary and not based on any specific naming convention. In general, lakes tend to be larger and/or deeper than ponds although exceptions are there.

From limnological (the study of inland waters) view point pond surface waters are divided into lotic (waters that flow in a continuous and definite direction) and lentic (waters that do not flow in a continuous and definite direction) environments. Waters within the lentic category gradually fill in over geologic time and slowly, the evolution from lake to pond to wetland came into existence. This evolution is slow and gradual, and there is no precise definition of the transition from one to the next.

Early limnologists in the late 18th, early 19th centuries attempted to define the transition from a lake to a pond in various ways. Area, depth or both were an essential part of most definitions, but what area or what depth that differed. Some used thermal stratification – a lake is a body of water that is deep enough to thermally stratify into two or three layers during the summer in temperate regions, others used plant growth – a pond is shallow enough that sunlight can penetrate to the bottom and support rooted plant growth across its entire width. Some included all plant growth (including submerged plants) while others said a pond was shallow enough to support emergent or floating-leafed rooted plants throughout.

Limnologists today recognize that nature can't be divided into precise, neat categories and accept the fact that there will never be a precise definition. However, they also recognize that "deep" lakes and ponds function differently than "shallow" lakes and ponds, and modern limnology texts often discuss the two separately. The generally accepted definition of a "shallow lake or pond" is that class of shallow standing water in which light penetrates to the bottom sediments to potentially support rooted plant growth throughout the water body. Lack of thermal stratification and the presence of muddy sediments are also common characteristics of this class of water. In contrast, a "deep lake or pond" has both a shallow shoreline area that may potentially support rooted plant growth and a deeper portion where sunlight does not penetrate to the bottom. These water bodies frequently stratify into distinct thermal layers during the summer.

Ecological status of a pond

The understanding of the wellbeing of the soil beneath the water and the water itself is the key factor for success of sustainable aquaculture enterprise, whether it is fresh or sewage-fed fresh water, brackish or sewage-fed brackish water. The key factor lies on the knowledge of the physical, chemical and biological aspects of the water body where the fish farmer intends to start the enterprise. In fact, the success of fish farming largely depends on the physico-chemical and biological characteristics of the water and soil. The status of soil, its water retention capacity, alkalinity and acidity coupled with organo-chemicals as well as physical aspects are largely considered in this enterprise.

Feeding habits of fish

Fishes may be grouped into four principal categories according to the type of food which they prefer under natural conditions, and these are:

(i) **Herbivorus fishes**: The fishes those feed upon the green plants available in the medium i.e. water which is affluent in phytoplankton and some larger aquatic plants. The important fishes are: *Catla catla, Hypopthalmichthyes molitrix, Aristichthyes nobilis* etc.

(ii) **Detritivorous fishes**: These fishes feed efficiently on the dead organic matter associated with live organisms at the bottom of the pond. The gut content of the detritivore fishes indicate that the content mainly consists of fungal and bacterial, remainings of dead plant and animal matters.

(iii) **Carnivorous fishes**: The fishes which feed on comparatively smaller animals including insects and their larvae, frogs and tadpoles, snails, mollusks and smaller fishes (piscivore). Example may be cited to those of *Lates calcarifer, Wallago attu* etc.

(iv) **Omnivorous fishes**: Fishes those feed fairly unchoosingly on both the plants, detritus and animal sources of food depending on "what ever is available".

2. TYPES OF AQUACULTURE

Different types of fish culture techniques include monoculture, polyculture, selected breeding, intensive and extensive culture, inland and brackish water, in rice field, in floating cages and rafts.

Different types of fish culture techniques

Fish culture techniques can be described from different perspectives. They can be described on the basis of biological, economic and social factors. The type of rearing facilities, technology of production, number of species cultured, type of fish, cultivation stage, geographical origin of cultured species and target market.

A. **Fish culture technique based on rearing facilities**

  1. Fish culture in earthen ponds.
  2. Fish culture in concrete tanks.
  3. Fish culture in reinforced plastic tanks.
  4. Fish culture in plastic tanks.
  5. Fish culture in wooden troughs or vats and
  6. Fish culture in rearing cages.

B. **Fish culture techniques based on technology of production**

  1. **Extensive culture system** – Characterized by low stocking density, no input and output of less than 1000 kg/ha/year.

  2. **Semi-intensive culture system** – Characterized by moderately high density, addition of input (fertilizer and feed), yield around10, 000 kg/ha/year.

  3. **Intensive culture system** – Characterized by high stocking density, addition of high input (fertilizer and nutritionally balanced diet), and cost of production is generally high with yield above 10, 000 kg/ha/year.

  4. **Surface Based Aquaculture system-** Characterized by the proper utilization of periphytic community which grows on submerged artificial substrates especially during fry and fingerling stages of fishes and fresh water prawn *Macrobrachium rosenbergii.*

C. **Fish culture technique based on number of species cultured**

  1. *Monoculture* – This is the culture of single species of fish in a pond or tank. The culture of *Clarias* spp. only or *Oreochromis niloticus* are typical examples, either of which can be selected during monoculture. The advantage of this method of culture is that it enables the farmer to make feed that will meet the requirement of fish especially in intensive culture system. Fish of different ages can be stocked thereby enhancing selective harvesting.

Common practices around the world:

- Common carp in East Germany.

- Common carp in Japan.

- *Tilapia nilotica* in several countries of Africa and Asia.

- Rainbow trout (*Salmon gairdneri*) culture in several countries.

- Channel catfish (*Ictalurus punctatus*) in U.S.A.

- Catfish, *Clarias gariepinus* in Africa.

A. Advantages, Disadvantages and Objectives of monoculture of fish farming

Advantages: Comparatively much easier to monitor the performance of growth, food intake, and breeding of the cultured fish. There exist no competition for space and food with other varieties of fishes.

Disadvantages: One of the major disadvantages of monoculture is the deterioration of water quality followed by cannibalistic nature if any among the fishes, various diseases due to overstocking of the concerned fishes.

B. Objectives of monoculture

(1) To maximize the production.

(2) To supply high quality animal protein and vitamin rich oils to pharmaceutical, solvent factories and soap industries.

C. Monoculture of monosex (all male) tilapia

The fish farmer's at present focusing on Nile Tilapia (*O. niloticus*) because of its ability to breed and produce new generations rapidly, its tolerance for shallow and turbid waters, its high level of disease resistance and its flexibility for culture under many different farming systems.However, the major drawback of pond culture is the higher levels of uncontrolled reproduction which may occur in grow-out ponds. Hence, monosex culture is one the basic methods of controlling Tilapia populations.Among the various techniques viz. manual separation of sexes, environmental manipulation, hybridization, hormone augmentation (sex reversal) and genetic manipulation methods such as androgenesis, gynogenesis, polyploidy and transgenesis which were carried out in various countries including India, none of these methods is consistently 100% effective, and thus a combination of methods is suggested. Males are preferred because they grow almost twice as fast as females, which may be caused by a sex-specific physiological growth capacity, female mouth-brooding or the more aggressive feeding behavior of males. Expected survival for all-male culture is 90% or greater (Silva *et al.*, 2013).

D. Disadvantages of monosex (All male) Tilapia culture:

(1) The major problem of male monosex culture is that fingerlings have to be grown until it is possible to distinguish the female and male juveniles (at least up to 50 g) and then the female juveniles are discarded.

(2) Accidental occurrence of even a single female tilapia mistakenly included in a population of mostly male Tilapia affects the maximum attainable size of the original stock in grow-out phase.

Male Tilapia production has an economic importance both to its producers and sellers. The increase in employment in the sector out pacing world population growth and employment in traditional agriculture is a crucial source of income and livelihood for hundreds of millions of people around the world (Soto-Zarazúa *et al.*, 2010b). More so, it also plays an important role to provide food security for the general population as an excellent source of high-quality protein (Ghosh, 2017).

E. Sex Reversal Or Hormone Augmentation

This method can be performed by oral administration of feed incorporated with androgen and eggs or fry immersion in different concentrations of the male hormone. The principle behind this method lies on the fact that at the stage when the Tilapia larvae are said to be sexually undifferentiated (right after hatching up to about 2 weeks or up to the swim-up stage), the extent of the androgen (male hormone) and the estrogen (female hormone) present in a fish is equal. Thus, augmenting one of the hormones that is originally present in the fish will direct the fish to either male or female depending upon the hormone introduced. Accordingly, if the Tilapia larvae are fed with feeds that are incorporated with male hormone as example 17α-methyl testosterone, the fish will develop into phenotypic male physically and function as male but possess the female genotype (XX). This is commonly referred to as "sex reversal". Production of male tilapia through the use of androgens is very effective. Sex reversed "male" reached similar average weights as genetically male tilapia (Mair *et al.*, 1995). Sex reversal by oral administration of feed incorporated with methyl testosterone is probably the most effective and practical method for the production of all male Tilapia, However, the technique has some limitations such as the uniform age of fish that should be used at the first feeding stage to ensure high reversal rate and less control of reversal efficiency especially when done in the natural environment where natural feed is present.This technique has achieved successful results up to 100% and feed with the male hormone is commercially available or can be prepared.

Culture of monosex tilapia is a recent concept in aquaculture at least in India. *Oreochromis niloticus,* the fish sometime is designated as the "food of the future" at least in a country like India.Filter-feeding tilapia *Oreochromis niloticus* is rightly designated as a core whitefish, "every-mans fish", "the aquatic chicken" and "not unlike the culinary versatility of chicken" and is able to convert plant proteins, diatoms, algae, heterotrophic plankton (biofloc), bacteria, zooplankton and low trophic level organisms into a valuable, succulent and mild tasting fish being enjoyed universally. Indeed, it has been argued that tilapia are perhaps the only truly herbivorous fishes since they possess not only a highly efficient digestive system (3.5 times the length of the fish) but also a highly specialised one, capable of producing pH values < 1 facilitating the digestion of highly refractory plant carbohydrate compounds truely commented by Ramon Kourie, Chief Technical Officer, Sust. Aqua Fish Farms (Pvt) Ltd. (Engormix/AquacultureTechnical Articles/ Article published the April 25, 2018).At present the expected growth of 2.6 percent in 2018 with average production of 6.5 million metric tonnes (MMT).

Most of the global production of tilapia is produced in freshwater pond systems and consumed in producing countries contributing to food security in the developing world where the sector is concentrated. China is the leading producer country followed by Egypt and Indonesia. Production estimates in 2017 have been pegged at 1.7 MMT for China, almost 900, 000 metric tonnes (MT) for Egypt and 800, 000 MT for Indonesia.

Tilapia is the most widely cultivated of all species with more than 120 countries reporting some commercial activity. In addition, tilapia are cultivated in the highest number of production environments from rice paddies and simple fertilized earth ponds to cages in lakes, aquaponics systems, biofloc technology (BFT)

Fig. 1: *Oreochromis niloticus*

tanks and Recirculation Aquaculture Systems (RAS) and are considered easy to cultivate.The Genetically Improved Farmed Tilapia (GIFT), a faster-growing strain of Nile tilapia (*Oreochromis niloticus*) is suitable for both small-scale and commercial aquaculture. Nile tilapia was selected due to its popularity in aquaculture, short generation time of approximately 6 months, and naturally hardy and high tolerance to variable water quality, good disease resistance, and ability to adapt many different farming systems.

Sex Reversal of Tilapia: (17 α-Methyl testosterone [MT]: Dose rate)

Early sexual maturity in tilapia culture is a well-recognized problem. There are a number of ways to control reproduction in mixed sex population. One of these is the culture of all-male tilapia. Sex reversal by oral administration of feed incorporated with 17α-methyl testosterone (MT) is probably the most effective and practical method for the production of all male tilapia. This is the most common method of sex reversal.

Sex reversal of newly hatched tilapia generally is accomplished via oral administration of 17 α–methyl testosterone (MT), which has been incorporated into a starter fish feed @ 5-6 mg/kg feed. Although the use of the 5-6 mg MT/kg feed dose consistently yields populations comprised of less than 5% females (i.e.> 95% males). Other investigators have reported sex reversal of tilapia at dose rates less than 5-6 mg/kg feed; however, results from these studies are inconsistent, and it is difficult to separate treatment environment effects. Thus, it is necessary to identify the optimal dose of MT for consistent, successful sex reversal in a variety of treatment environments.

Although a variety of hormones have been used for sex reversal, 17α - methyl testosterone is the most commonly used androgen. Dose rate and treatment durations vary depending on the environment and the experience of the producer. Tilapia fry <14 mm should be treated for at least 14 days before reaching 18 mm, if growth is slower the duration of treatment should be extended until all fish reach this size or a total treatment period of 28 days are exceeded. A dose rate of 30-60 mg of Methyl testosterone/kg of diet fed at an initial rate of 20%body weight/day should result in successful treatment. The efficacy of treatment should be based on gonadal development.

F. Food and feeding habit of *Oreochromis niloticus*

Nile tilapia, *Oreochromis niloticus* is one of the most known members of the tropical and subtropical freshwater fishes. It is recommended by the FAO as a culture fish species because of its importance in aquaculture and its capability in contributing to the increased production of animal protein in the world. Therefore, it is now globally distributed and has become very popular through the advances in the cultivation techniques. Juvenile and adult Nile tilapias are reported to filter phytoplankton hence indicate the herbivorous behavior. Since Nile tilapia use algal protein as food at lower trophic level can be a cost-efficient culture method.

G. Monoculture of monosex tilapia: a case study (Ghosh, 2017):

In one of the village named Sonatikari, adjacent to Sundarbans (South 24-Parganas, West Bengal, India), an attempt has been made to culture all male tilapia.

From d-1to d-30 the (hatchling to early fry stage) these were fed synthetic androgen (17∞methyl testosterone) mixed diets (incorporated with 0, 25, 50, 75and100 mg/kg diet) for 30 days.

The 30-days old early fry were brought to the farm area from the fish hatchery and released in a nursery pond measuring 0.5 hectare area with a depth of 1.2 meters. The stocking density in the nursery pond was restricted to 2, 00, 000/hectare. The fishes were fed with floating feed (average protein content 34.0% and fat content 5.0%), twice a day. After 40-days in the nursery phase when the fishes attained 10gms (average) weight were transferred to the adjacent grow-out pond measuring about 1.2 hectares with an average depth of 2.6ft. The average stocking density was 60, 000/hectare. Broadcasting of floating feed within the nylon net enclosure (Fig.2) was made twice a day. The protein content at this stage maintained 32% and the fat content was 4%. The application of feeding was @ 3% of fish weight till harvest.

Fig. 2: Feeding within an enclosure.

The harvesting was done on 120th.day of rearing, when the fishes attained approximately 400-500gms.average.The production achieved after 120th days of grow out phase is approximately 10metric tones /Ha/crop.

2. *Polyculture-* This is the culture of two or more species of fish in a pond or tank. It is also a system that grows more than one species of fish from the same trophic level.

3. INTEGRATED MULTI-TROPHIC AQUACULTURE (IMTA)

Integrated multi-trophic aquaculture systems mimic natural ecosystems. Multi-trophic refers to the combination of species from different trophic levels in the same system. The multi-trophic sub-systems are integrated in IMTA that refers to the more intensive cultivation of the different species in proximity of each other, linked by nutrient and energy transfer through water. Fishes are present in more than one trophic level and waste products thus produced are recycled. An example is cultivating sea weed near mariculture fish pens. Nutrients in the fish waste fertilize algae, which in turn improve water quality for fish.

4. MARICULTURE

In Mariculture, fish are grown in pens or meshed cages in harbors or sea. (Both the sections 3&4 are out of the scope of discussing in detail in this book, hence ignored).

5. AQUAPONICS: The Future of Sustainable Food

i) What is Aquaponics?

The simplest definition of Aquaponics is the combination of aquaculture (raising fish) and hydroponics (the soil-less growing of plants) that grows fish and plants together in one integrated system. The fish waste provides an organic food source for the plants, and the plants naturally filter the water for the fish. The third participants are microbes (nitrifying bacteria). These bacteria convert ammonia from the fish waste first into nitrites, and then into nitrates. Nitrates are the form of nitrogen that plants can uptake and use to grow. Solid fish waste is turned into vermicompost that also acts as food for the plants. In combining both hydroponic and aquaculture systems, aquaponics capitalizes on their benefits, and eliminates the drawbacks of each.

Aquaculture and hydroponics are systems that complement each other as a single unit, not as separate units. The fish water is pumped to a greenhouse, which is evenly distributed. The fish water feeds the plants, such as lettuce and tomatoes, then filters through a porous material (volcanic cinders in Hawaii) and returns to the fish tank by gravity. Both systems are in a controlled environment, meaning light and temperature are controlled. The primary crop is the vegetable and the fish are secondary, meaning commercially, there is more money and turn around with the vegetables. The aquaculture is suitable for ornamental fish, prawns, tilapia, catfish, bass and escargot snails. Another example is those of the Hawaiian taro plants in cement pots that are placed in with the fish. Watercress can also be used. What is amazing is the stable and balanced pH and crystal clear water. The water is basically recycled, with a small amount of water added weekly for evaporation. This is a balanced, self-contained ecosystem that works. Solar powers, the water pump and no chemical are added what so ever, it is totally organic. Chameleons control any insects that might get into the house. Earthworms are raised to feed the fish and the earthworm compost is used in the garden or planter box gardens.

ii) The problems with traditional soil based gardening:

* Pesticide and artificial nutrient usage.

* Weeds, pests, and soil-borne insects.

The amount of water requires:

* The heavy digging, the bending, the back strain

* Knowledge regarding water and its quality.When and how to fertilize, and what is the composition of the soil.

* Location – traditional farms are often located a few miles away from where the food is consumed.

* *Although all these issues have already been solved with modern hydroponic system.*

● **The own problems of hydroponics**:

 – Traditional hydroponic systems rely on the careful application of expensive, man-made nutrients made from mixing together a blend of chemicals, salts and trace elements. In aquaponics, one may feed the fish inexpensive fish feed, food scraps, and food that is grown at the farm itself.

 – The strength of this hydroponic mixture needs to be carefully monitored, along with pH and total dissolved solids (TDS). In aquaponics a careful monitoring of the system during the first month, but once the system is established surveillance on pH and ammonia levels weekly or if the plants or fish seem stressed.

 – Water in hydroponic systems needs to be discharged periodically, as the salts and chemicals build up in the water, becoming toxic to the plants. This is both inconvenient and problematic as the disposal location of this waste water needs to be carefully considered. In aquaponics, replacement of water regularly is not essential; simply top it off as it evaporates.

 – Hydroponic systems are prone to a disease called "Pythium" or root rot. This disease is virtually non-existent in aquaponics.

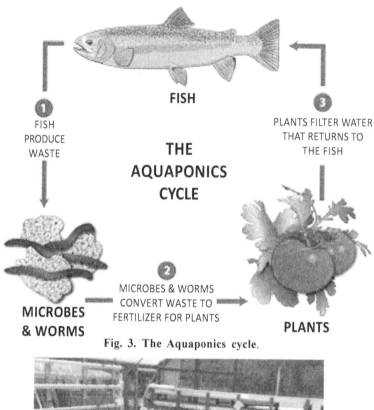

Fig. 3. The Aquaponics cycle.

Fig. 4: Newly constructed Aquaponics farm

Aquaponics combines aquaculture with traditional fruit, vegetable and herb farming using ecofriendly and organic biofiltration from the plants to recycle water for the fish. The fish produce waste that help the plants grow. The entire system uses 85-90% less water than traditional dirt farming. In aquaponics, almost anything may be grown, year round, and crop rotation is never a problem.

iii) Benefits of Aquaponics

In this system there is no waste and there remains no by-product.

Unlike hydroponics which produces waste water and requires monthly chemical doses, aquaponics allows for the harvesting of fish and uses the fish waste as the chemicals to feed the plants. There is no waste except for fish emulsion which can be used for raised grow beds and potted plants or used to feed additional growth in beds aquaponics system.The filtration of the waste is done since enough plants are not grown.

Expensive nutrients are replaced by less expensive fish feed. *Aquaponics is actually more productive than hydroponics. Aquaponics relies on nature-hence completely organic.*

a) **Resource:** The energy cost to nourish a family of four is about Rs.7000 to 10, 000 per year. Heating is done via natural solar methods (not electricity). Fish food can be home grown. Various compost methods produce fish food and get rid of human trash at the same time.

b) **Cost effectiveness:** The investment of even a small, family sized green house of 2500 liters, backup pumps, solar hot water heater, starter fish, starter plants, gravel, plumbing, etc is less than two lakhs.Small herb gardens can be created for only a few thousand rupees. This investment pays for itself almost within the first year.

c) **Low-maintenance:** Aquaponics is virtually maintenance free. Feed cost of fish 1 or 2 times a day, monitor temperature, planning of upcoming crops and harvest a meal at a time. There are no weeds, no tilling, and the watering is automatic too.

d) **Organic Aquaponics:** This system is completely organic. It must be organic because pesticides will kill the fish.

e) **Fresh:** One may walk into the backyard and harvest food year round, fresh – instead of eating something that is picked before its fully ripe, and then served to dish a month later.

f) **Water-Conservation:** Aquaponics a closed loop system (water isn't just poured on the ground, it is recirculated). Plants evaporate some water, but overall the same volume of vegetation can be harvested with very little additional water.

g) **Crop-Rotation:** In soil crops the amount of nutrients in the ground are limited. Aquaponics pumps fresh, natural nutrients every few hours into the growth medium and crops can be harvested repeatedly.

h) **Aquaculture:** Various fish can be easily farmed, such as cyprinid fishes belong to minor carps, catfish, tilapia, prawns, etc.

i) **Self-Sustainability:** In the present unsustainable world's food system one can get fresh food.

Fig. 5: Model of an Aquaponics

6. FISH REARING IN CAGES (CAGE CULTURE)

A fish cage is an enclosure made with net where fish are raised in free flowing water especially in lagoon waters, rivers, lakes and large reservoirs. Because of several advantages over stagnant pond culture, the cage and pen farming practices although gained wide acceptance in many countries but in India excepting some of the southern states (Natarajan *et al.,* 1983).Various biological factors like trophic efficiency, gregarious nature, faster rate of growth or biomass production, rate of survival, higher fecundity, reproductive nature especially in captivity, shorter reproduction cycle, disease resistance etc. are determined as bioeconomic matrix (Nash, 1974).

Cage culture of fishes in large waterbodies however, is considered to be the 'sleeping giant' in Indian aquaculture scenario.Although the practice in India till date is at its infancy, but some of the current achievements in this sector have established its commercial potential. Cage culture operations have succeeded with high - end varieties such as pearlspot, seabass, grey mullet, red snapper, cobia, Nile tilapia etc.

What is Cage Culture?

Cage culture is an aquaculture production system where fish are held in floating nets. Cage culture of fish utilizes existing water resources but encloses the fish in a cage or basket which allows water to pass freely between the fish and the pond permitting water exchange and waste removal into the surrounding water. Cages are used to culture several types of shell fish and finfish species in fresh, brackish and marine waters. Cages in freshwaters are used for food fish culture and for fry to fingerling rearing (Soltan, 2016).

Origin of cage culture is bit ambiguous.It is assumed that at the beginning fishermen might have used the cages as holding structures to store the captured fish until they are sent to the market. The first cages which were used for producing fish were developed in Southeast Asia around the end of the 19th century. Wood or bamboos were used to construct these ancient cages and the fish were fed by trash fish and food scraps. In 1950s modern cage culture began with the initiation of production of synthetic materials for cage construction. Fish production in cages became highly popular among the small or limited resource farmers who are looking for alternatives to traditional agricultural crops.

In India, cage culture has been attempted for the first time in 1970 in three different types of environments:

1) Swamps marked by low dissolved oxygen concentration, using air-breathing fishes,

2) Running waters of the Yamuna and Ganga Rivers at Allahabad, using major carps and

3) A static water body in Karnataka, using common carp, catla, silver carp, rohu, snakeheads and tilapia.

CIFE Mumbai though in an isolated manner started cage culture for raising fingerlings as well as table-fish in reservoirs such as Powai (Maharashtra), Govindsagar (H.P.), Halali (M.P.), Tandula (Chattishgarh) and Dimbe (Maharashtra). Later, a number of attempts were made to produce cage-stocked fish, especially fry to fingerlings (Banerjee, 1979).During 2005- 2006, floating cage culture experiments were conducted in Kabini reservoir, Karnataka for raising fingerlings for stocking, with moderate success, defined by a survival rate of around 40percent. Some trials on cage aquaculture were conducted during 1998 onward for production of fingerlings but results were not optimal. CIFRI has banked on the success of producing fingerlings for stocking reservoirs in floating cages installed in different agro climatic zones of India (Madhya Pradesh, Uttar Pradesh, Bihar, Jharkhand and Northeast states).With the lessons learned, cages were subsequently used for raising catla fingerlings in Karnataka (Govind *et al.,* 1988), also with little success. Production of fingerlings for stocking reservoirs was tried in Govindsagar (H.P), Getalsud (Ranchi), Gularia (U.P.) reservoirs, with poor success because proper monitoring was not initiated.More recently cage aquaculture experiments were conducted in Walvan Reservoir (Maharashtra), with fry to fingerling culture of *Tor putitora* and *T. khudree* (Kohli, *et al.*, 2002; Das *et al.*, 2014).

Why cage culture?

The reservoirs of India have a combined surface area of 3.25 million hectares (ha), mostly in the tropical zone, which makes them the country's most important inland water resource, with huge untapped potential.Truly, the scope of cage culture in Indian reservoirs, if properly exploited can revolutionerize the Indian aquaculture scenario and honestly being said that the reservoirs are the sleeping giants favouring aquaculture. Fish yields of 50 kg/ha/year from small reservoirs, 20 kg/ha/year from medium-sized reservoirs and 8 kg/ha/year from large reservoirs have been realized while still leaving scope for enhancing fish yield through capture fisheries, including culture-based fisheries. The success rate of auto-stocking is very low in Indian reservoirs, especially in smaller ones. Many of the smaller reservoirs dry up during the summer, partly or completely, with no stock surviving. A policy of regular, comprehensive and sustained stocking would greatly augment fisheries in such water bodies. The prime objective of cage culture discussed here is to rear fingerlings measuring >100 millimetres (mm) in length, especially carp, for stocking reservoirs.

Stocking with the right fish species, using seed of appropriate size and introducing it at the right time are essential to optimize fish yield from reservoirs. Though 22 billion fish fry are produced every year in India, there is an acute shortage of fish fingerlings available for stocking reservoirs. Where fingerings are available, transporting them to reservoirs usually incurs high fingerling mortality. In this context, producing fingerlings in situ in cages offers opportunity for supplying stocking materials, which are vital inputs towards a programme of enhancing fish production from Indian reservoirs.As many as 110 fish species from fresh, estuarine, brackish and marine systems have been found to be suitable for culture in pens and cages in India. Among fresh water species, the species like *Catla catla, Labeo rohita, Cirrhinus mrigala, Ctenopharyngodon idella, Hypopthalmichthys molitrix, cyprinous carpio, Channa* etc. are important and found to grow to a measurably marketable size in cages of large reservoirs and rivulets and tributaries. While in estuaries(brackish and coastal waters) of Kerala, Tuticorin, Mandapam, Chilka fishes like *Chanos chanos, Etroplus suratensis, Valamugil seheli, Liza macrolepis, Pincada fucata, Meretrix meretrix, Penaeus monodon, P.indicus* and particularly *Lates calcarifer* has given much more importance due to their slow moving and fast growing characteristics.

Advantages and disadvantages of Cage culture

As with any production system, fish culture in cages has advantages and disadvantages that should be considered carefully before choosing it as the production model.

Advantages

Following are the advantages of cage culture when compared to the other fish farming systems:

- Its installation is easy.
- Flexibility of management.
- Effective use of fish feeds.
- Less manpower requirement.
- Better control of fish population.
- In emergencies it can be removed from one place to another.
- Treatment of disease is much simple than that of pond culture.
- It requires less investment, because it use existing water bodies,
- Simple technology and swift return of investment
- Mixed sex populations can be reared in cages.
- Close observation and sampling of fish is simple and therefore only minimum supervision is needed.
- Many types of water resources can be used, including lakes, reservoirs, ponds and rivers.
- Fish handling and harvesting are very simple and helps to maintain the non-seasonal supply of the fish.
- Since the cage is meshed, fish inside have less chances of being attacked by predators.
- They can be used to clean up eutrophicated waters through culture of caged planktivorous species such as silver carp.

Disadvantages

- Feed must be nutritionally complete and kept fresh.
- Stocked fish simply affected by the external water quality problems e.g., low oxygen levels.
- Diseases are a common problem in cage culture.

- The crowding in cages promotes stress and allows disease organisms spread rapidly. Also, wild fish around the cage can transmit diseases to the caged fish.

- Caged fish are unable to get the natural food of their choice, whereas it is readily available to the free fish.

- During feeding a significant amounts of fish feed passes out through the mesh therefore, fish require feeding many times a day.

- The high fish density with the high feeding rates, often reduce dissolved oxygen and increase ammonia concentration in and around the cage, especially if there is no water movement through the cage.

- In public waters, cage culture faces many competing interests and its legal status is not well defined.

- There is usually a high mortality rate because of bacterial and fungal diseases.

- Water pollution

- Poaching

- Conflicts in the use of water with other users.

- Management includes routine check, provision of adequate security and good supply of quality feed to ensure faster growth.

Species Selection for Cage Culture

Many species are suitable to grow in cages. The production of fingerlings of Indian major carps, as well as some of the exotic species like common carp (*Cyprinus carpio*) and grass carp (*Ctenopharyngodon idella*) are being encouraged due to their market acceptability subject to the food niches are sufficient for their further growth after stocking in the reservoir. First, the locality should be surveyed to verify the availability of the stocking species.Oxygen packed fry can withstand a journey of 40 hours. The fry should be healthy when purchased and come from a clean water body with low eutrophication.Species selection depends on local interest and market value. For fingerling production, the species selected were primarily Indian major carps and, to some extent, common carp and grass carp. During winter, fish had access to natural food sufficient for further growth after stocking in the reservoir. For growing table fish, the species must have high market value, such as freshwater prawns, air-breathing species, sea bass and, to some extent, carps and Pangasius sp. Small indigenous fish species (SIFS) with high market demand are also suitable (Govind *et al.*, 1988; Kohli *et al.*, 2002).

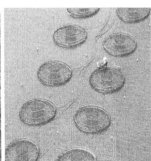

Fig. 6: Open cage farming in Kerala and USA

Cages are made of four parts: 1. Solid frame 2. Nets 3. Floats and 4. Anchor.

A cage can be floating type or fixed type. Floating type if the cage is submerged fixed and is attached to a stake fixed to the bottom. Cages have high potential to improve the status of the local fish supply.

7. FISH CULTURE IN PENS

Pens are constructed from bamboo or wooden poles that are forced down into the lake or shore bottom. Nets are strung from pole to pole to form an enclosure. The nets are anchored into the lake bottom with weights or sinkers. They are stocked with fish. Pens are often placed in the fertile lakes to yield high biomass. No feeding, no extra feed input but if lakes are not fertile feeding may be required. A Pen is defined as a*"fixed enclosure in which the bottom is the bed of the water body"* and is synonymous to 'enclosure'. By the very nature of the fixed enclosure walls of pens it is obvious that they cannot be moved about as in the case of cage. There is economy of material in the pen for the bottom material used is saved and therefore and for other reasons the pen can be and are much bigger. As we go by further differences and similarities will become obvious, especially when the different types of pens are described. Of the six possible zones:

i)   shore,

ii)   intertidal,

iii)   sublittoral,

iv)   surface floating

v)   mid-water and

vi) Sea/low tidal river or lake beds in the marginal waters (Milne, 1979).

Pen culture is possible only in the three zones, namely, intertidal, sublittoral and seabed -all having natural bottom as the limit of the lower side of the enclosure.In the case of freshwater except for the very large lakes – even here tidal influence is little compared with the sea, the intertidal zone is non-existent. Largely the enclosure of a pen is restricted to shallow area adjacent to the shore. The pen or enclosure may be:

(a) Completely enclosed on all four sides in the middle of a bay, with no foreshore or

(b) A shore enclosure with a foreshore extending to deep water surrounded by a net structure or

(c) A bay or loch enclosure with an embankment or net structure only at the entrance.

The pen culture of milkfish (*Chanos chanos*) is the most important fresh water pen culture in the world. The pen culture of milkfish has yielded the production much more than 4, 000kg/ha (average), without supplementary feeding. The Laguna Lake Development Authority has claimed that the lake fish pen industry can grow to 15, 000 hectares and a yield projection more than one lakh metric tons annually.

The origins of pen culture are more obscure, but it also seems to have begun in Asia. Pen culture originated in the Inland Sea area of Japan in the early 1920s. It was adopted by the People's Republic of China in the early 1950s for rearing carps in freshwater lakes, and was introduced to Laguna de Bay and the San Pablo Lakes in the Philippines by the Bureau of Fisheries and Aquatic Resources (BFAR) and the Laguna Lake Development Authority (LLDA) between 1968 and 1970 in order to rear milkfish (*Chanos chanos*) (Alfarez, 1977; PCARRD, 1981).

Pens are still constructed in much the same way as they always were, except that nylon or polyethylene mesh nets have replaced the traditional split bamboo fences. The nets are attached to posts set every few meters, and the bottom of the net is pinned to the substrate with long wooden pegs. But stressing may be used to strengthen the structures in exposed areas. Pens are usually built in shallow (<10m) waters, are 3–5m deep, and 1–50 ha in size (IDRC/SEAFDEC, 1979). Soft substrates are preferable.

The development and adoption of inland water pen culture has been much less dramatic than that of cage culture, and at present it is only practiced on a commercial basis in the Philippines, Indonesia and China (Dela Cruz, 1980, 1982; Lam, 1982). The principal species being cultured in these countries are milkfish and carps (e.g. grass carp, *Ctenopharyngdon idella*; bighead carp, *Aristichthys nobilis*; silver carp *Hypophthalmichthys molitrix*). Some experimental pen culture of carps has also been carried out in pens in oxbow lakes in Hungary (Muller & Varadi, 1980), and other countries such as Bangladesh and Egypt (Karim and Haroud-al-Rashid Khan, 1982). The production of tilapias in net pens is also currently being evaluated in the Philippines (Guerrero, 1983).

Because of their smaller size (generally $100m^2$ surface area) and because they are easier to manage, cages are more adaptable than pens and can be used not only for grow-out of fish to market size, but also for breeding and fry production of fishes such as the tilapias (Beveridge, 1984) and for nursing of the planktivorous juvenile stages of carps. Pens are largely restricted to lentic water bodies, whereas fixed and floating cages are also used in rivers and streams. However, in most cases both systems are used for monoculture.

Murugesan *et al.* (2005) while discussing the prospects and possibilities of pen culture in India emphasized that, till date experiments have been carried out to raise carp seed using pen culture in oxbow lakes, swampy tanks, beels and reservoirs, but the technology has not yet been standardized.

Advantages and Disadvantages of Pen Culture

The advantages and disadvantages of pen culture are in some cases common as those for cage culture. Obviously the pens are much larger and are stationery as their walls are fixed. It also appears that in large pens the culture will be less intensive. The mobility of the cage is its most definite advantage over the pen, but the later has the benefit that there can be interchange between the organisms within, with the natural bottom - at times of an inclement condition in the bottom the pen is decidedly difficult. Let us now enumerate the advantages and disadvantages of pen culture.

Advantages

i.   **Intensive utilization of space:** As it is already been mentioned that the requirement of a pen can be small (a few square meters) or large (over 100ha as in the case of the largest milkfish pen in Philippines), but in all cases the space given is intensively utilized. Even in the large milk fish pens, space is utilized intensively and their production as achieved is 4 – 10 times higher than the natural production.

ii. **Safety from predators:** Within the enclosure the predators can be excluded. Before stocking the predators will have to be removed; in the larger pens this would be more difficult, but in smaller pens this can be done as efficiently as in the cages.

iii. **Suitability for culturing many varied species:** Under artificial culture provided suitable environmental conditions are maintained, with artificial feeds, many varieties of species can be cultured as in the cage.

iv. **Ease of harvest:** Even though in the large pens the harvest may not be as easy as in the cages, it would be definitely more controllable and easier than in the natural waters.

v. **The flexibility of size and economy:** When compared with the cage, pens can be made much larger and construction costs will be cheaper than those of the cages.

vi. **Availability of natural food and exchange of materials with the bottom:** Since, the bottom of the pen is the natural bottom, unlike the cage which kept either on the bottom or floating, has always a netting/ screen separating the cage from bottom; the pen culture organisms are at an advantage that while enclosed they can procure food/exchange materials.

Pen culture, as cage culture is economical and is attributed to the multiple use of the same water body (e.g. pen culture in an irrigation reservoir).

Disadvantages

i. **High demand of oxygen and water flow:** Since the fish cultured are stocked in high density they deplete oxygen very fast and a good flow of water through the pen either by natural means or artificially by pumping is demanded for healthy and fast growing fishes.

ii. **Dependence on artificial feed:** Since high density (biomass) is to be sustained in a restricted area, for high production artificial feeding is necessary, increasing the cost of production.

iii. **Food losses:** Part of the feed is likely to be lost uneaten, and drifted away in the current, but the loss here would be less than in floating cages.

iv. **Pollution:** Since a large biomass of fish are cultured intensively a large quantity of excrements accumulate in the area and cause a high BOD - also substances such as ammonia and other excreted materials, if not immediately removed/ recycled. They pollute the water and cause damages.

v. **Rapid spread of diseases:** For the same reason of high stocking density in an enclosed area, beginning of any disease will spread very quickly and can cause immense mortality of stock and production decline.

vi. **Risk of theft:** Since the fish are kept in an enclosed area, 'poaching' and thefts can take place more frequently than in natural waters, but perhaps less than those from cages.

vii. Conflict with multiple uses of natural waters: In locations where a pen is constructed to the requirement of higher water level e.g. in a lake/reservoir, would be against the interest for irrigation water supply; enclosures can interfere with navigational routes and also with recreational activities, such as swimming, boating etc.

8. BIOFLOC TECHNOLOGY IN AQUACULTURE

Biofloc technology is a technique of enhancing water quality in aquaculture through balancing carbon and nitrogen in the system.The technology has attracted global attention as a sustainable method due to proven enhancement of production of the target species, quality, sustainability, biosecurity with an added value of protein rich feed *in situ* following market demand.

The biofloc is a protein rich macroaggregate assemblage of organic matter and microorganisms including the diatoms, bacteria, protozoa, algae, fecal pellets, remains of dead organisms with various types of invertebrates otherwise called periphytic organisms.

The basic technology was developed by Dr.Yoram Avnimelech in Israel and the technology was initially implemented commercially in the Belize province by Belize Aquaculture.The technology has been found to be of profound acceptance in farming of the Pacific White shrimp *Litopeneus vannamei* although exhibited encouraging results in fish farming especially *Oreochromis niloticus* farming.

Biofloc technology is mainly based on the principle of waste nutrients recycling, in particular nitrogen, into microbial biomass that can be used *in situ* by the cultured animals or be harvested and processed into feed ingredients (Avnimelech, 2009; Kuhn *et al*., 2010). Heterotrophic microbiota is stimulated to grow by steering the C/N ratio in the water through the modification of the carbohydrate content in the feed or by the addition of an external carbon source in the water (Avnimelech, 1999), so that the bacteria can assimilate the waste ammonium for new biomass production. Hence, ammonium/ammonia can be maintained at a low and non toxic concentration so that water replacement is no longer required (Bossier and Eksari, 2017).

It is no denying that that the biofloc system certainly provides nutritious food source as well as improves feed utilization efficiency since, the concentrations of free amino acids viz. alamine, glutamate, arginine, glycine etc.are all known attractants those present in the biofloc (Nunes *et al.*, 2006; Ju *et al.,* 2008). Furthermore , biofloc technology if adopted in the larval culture system provide easily accssible food source and thus minimize possible negative social interaction during feeding (Eksari *et al.,* 2015).

Introduction of biofloc system during culture also significantly reduce water utilization which is the principal resource in aquaculture and waste generation. It is calculated that intensive tilapia culture in a biofloc system requires 40% less water than those of recirculation system (RAS), while in an intensive zero exchange lined shrimp pond it requires only 1-2.26m^{3}per kg of shrimps compared to the conventional system with regular water exchange which require water to the tune of 80m^{3}per kg of shrimps (Luo *et al.,* 2014; Hargreaves, 2006).Luo *et al.,* (2014) emphasized that Nitrogen and Phosphorus waste in a biofloc system also significantly reduced.The heterotropic bacteria which plays a major role in the conversion, also facilitate other conversion mechanisms like nitrification (Eksari, 2014), phototrophic N uptake (Emerciano *et al.,* 2013) and denitrification (Hu *et al.,* 2014).The nutrient recycling involves the uptake of inorganic Phosphorus by heterotrophic bacteria also significantly reduce discharged Phosphorus and simultaneously enhance the bioavailability of this nutrient to the stocked shrimps and fishes as well (Kirchman, 1994 and Bossiewr and Eksari, 2017).The biofloc based integrated aquaculture system also directed to higher productivity through higher nutrient utilization thereby least aquaculture pollution is generated (Bossier and Eksari, 2017).

Maximization in nutrient utilization through nutrient recycle principle in an integrated aquaculture system has also been emphasized.The faster conversion of nutrients by the microbes associated in bioflocs or periphytic community (Awfuchs) may provide more digestible and nutritionally balanced other food source for both target species of culture or any other species which are present during biofloc culture system.In this way, utilization of the wasted nutrients is expected to be more efficiently being utilized there by minimum pollution is generated.

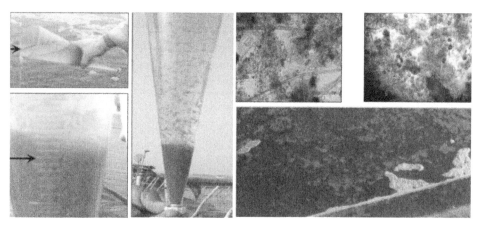

Figs. 7: Bioflocs

Table 1: Advantages and disadvantages of Biofloc technology in Aquaculture.

Advantages	Disadvantages
01. Biosecurity is appreciably good. Till date White Spot Syndrome Virus (WSSV) has not been detected using this system.	High energy input. Paddle wheels required: 28-30HP/Ha.
02. Zero water exchange-less than 100% exchange for whole culture period.	In case of failure of power opportunity of increased rate of mortality. Zero-hour failure is preferred.
03. FCR low- between 1.0-1.3	Entire pond should be lined with HDPE sheets. Initial expenditure is too much.
04. Production (Carrying capacity): 5-10% better than normal system.	Need special training to the technicians.
05. Shrimp size bigger than 2.0 gm than normal system.	
06. Low production cost (around 15-20%)	

9. POLYCULTURE-(utilization of all ecological niches of pond ecosystem)

Polyculture is the practice of culturing more than one species of aquatic organism in the same pond. The motivating principle is that fish production in ponds may be maximized by raising a combination of species having different food habits. *The concept of polyculture of fish is based on the concept of total utilization of different trophic and spatial niches of a pond in order to obtain maximum fish production per unit area.* The mixture of fish gives better utilization of available natural food produced in a pond. The compatible fish species having complimentary feeding habits are stocked so that all the ecological niches of pond ecosystem are effectively utilized. The habitat of an organism is "**the place where it lives**" i.e., the address whereas, the ecological niche, is its "**profession**. **Polyculture** also means cultivating various species of **fish** together by fully using of all water levels including available natural food in the pond to increase **fish** production in highest stage.

A. How does polyculture work?

Ponds that are generally enriched through fertilization, manuring or feeding practices contain abundant natural fish food organisms living at different depths and locations in the water column. Most fish, feed predominantly on selected groups of these organisms. Polyculture should combine fish having different feeding habits in proportions that effectively utilize these natural foods (Fig.8). As a result, higher yields are obtained. Efficient Polyculture systems in tropical climates may produce up to 8, 000 kg of fish per hectare per year (Das, 2017).

B. Fishes generally used in polyculture in India

Combinations of Indian Major Carps (Rohu, Catla and Mrigal) and Chinese carps (bighead, silver, grass carp, and the common carp) are most common in Polyculture. Other species may also be used.

C. Factors affecting species selection and stocking rates.

1. Water temperature.

2. Market value of fish.

3. Pond fertilization practices.

3. Feeding habits of fish.

4. Potential of uncontrolled spawning in grow-out ponds.

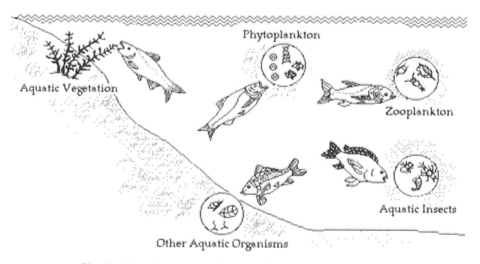

Fig. 8: Polyculture- for efficient utilization of natural foods

D. Problems in Polyculture

Polyculture is an effective way to maximize benefit from available natural food in a pond. But, pond management becomes more difficult when stocking fish species having specialized feeding habits in the same pond. Judicious application of fertilizers and nutritionally balanced feeding practices must have to be followed. Inadequate fingerling supply at the right time severely limits the choice of species available for polyculture; at least one species should have general rather than specialized feeding behaviour. This will allow more of the available natural food to be utilized.

An example of two species polyculture:

"Tilapia – Shrimp Polyculture in cages"

Methods of farming

Shrimps and Tilapia may be cultured simultaneously in the same pond. It is necessary to stock tilapia after 15 days of stocking shrimp @ 1000 and 1, 00, 000 respectively per hectare (1000 and 10,00,000 nos./Hectare). The average weight of the tilapia fish may be about 50gms each.Stocking more than the mentioned number may compete with the shrimp habitat. Only 1000 nos /ha of stocked fish will grow too fast and within 90 days of farming both fish and the shrimps may be harvested at one time. However, if the farmer wants much larger fish, may be transferred to another pond for continuation of farming. Subject to the provision of nutritionally balanced feed at daily basis, the average growth of Tilapia to the tune of 300gms may be achieved.

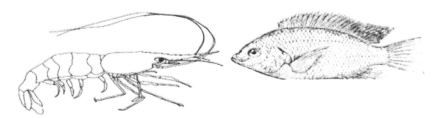

Fig. 9. Two species culture in cages (Tilapia–Prawn Polyculture)

Culture Tilapia in netted cages in shrimp ponds

This model is most applicable method and it is easy to harvest. In a 5000m^2 shrimp pond, cover 300m^2 areas with nets and stock 5000 tilapias in it: Raise tilapia in a separate pond, and then transfer water from tilapia pond to shrimp ponds for farming.

Farming of tilapia cages in shrimp ponds

- For intensive shrimp ponds with aerators installed, place tilapia cages in low-lying zones in the middle of the ponds. These cages and areas will account for 7-10% of each pond. Nets have sparse meshes (about 0.5-1cm). Only stock unisexual, large sized tilapia (15-20fish/kg). Stocking density should be at 6-8tilapia/m^2.

- In high-density shrimp pond, stock tilapia in cages or netted areas. There is no need of feeding. Ventilate the cages or nets once every 15 days.

- Regular change of water in ponds (once in every 1 month). Each time, replace 15-20% of water in ponds. Use of bio products (once in every 10-15 days).

10. RECIRCULATING AQUACULTURE SYSTEMS (RAS)

There is growing interest in recirculation aquaculture system (RAS) technology especially in intensive finfish culture in the world. This is due to the perceived advantages that RAS greatly reduces land and water requirements, offering a high degree of control of the culture environment that allows year round growth at optimal rates and fish biomass can be determined more accurately than in ponds (Masser *et al.*, 1999). A typical RAS consists of a water supply system, mechanical and biological filtration, pumps to maintain water flows, aeration and oxygenation system and other water treatment components that deliver optimal water quality for fish growth within the system (Hutchinson *et al.* 2004). With more stringent water pollution control, RAS provides greater environmental sustainability than traditional aquaculture in managing waste production and also a possibility to integrate it with agricultural activities such as using water effluent for hydroponics (Summerfelt *et al.*2004). Another key advantage is that, RAS technology is species-adaptable which allows operators to switch species to follow market preference for seafood products (Timmons *et al.*2002). "Eventhough RAS is capital intensive, claims impressive yields with year-round production is attracting growing interest from prospective aquaculturist" (Losordo *et al.*1998, p.1).

As the aquaculture industry continues to grow in response to the demand for increased fish products, the need for environmentally conscious operational practices and facility designs becomes more important (Peachey, 2008). The intensive aquaculture system culture employs intensive management of production system where culturist must provide for all the biological needs of the cultured organism. This method is often adopted in *Recirculating Aquaculture System*. Recirculating aquaculture system (RAS) is the newest form of the fish farming production system. RAS is typically an indoor system that allows farmers to control environmental conditions year round. The RAS is advantageous over other aquaculture systems in the reduction of incoming water volume (Verde gem *et al.*, 2006), reuse more water within the culture system, reduction in the amount of water released and the effluent quality for better hygiene and disease management and biological pollution control (Adamu *et al.*, 2014).

Recirculation aquaculture is essentially a technology for farming fish or other aquatic organisms by reusing the water in the production. The technology is based on the use of mechanical and biological filters, and the method can in principle be used for any species grown in aquaculture such as fish, shrimps, clams, etc.Recirculating aquaculture systems are indoor, tank-based systems in which fish are grown at high density under controlled environmental conditions. Generally, farmers adopt a more intensive approach (higher densities and more rigorous management) than other aquaculture production systems.Recirculating aquaculture systems can be used where suitable land or water is limited, or where environmental conditions are not ideal for the species being cultured. Hence, the operation of RAS which are mechanically sophisticated and biologically complex requires education, expertise and dedication (Duning et*al.* 1998).

In a recirculating aquaculture system the culture water is purified and reused continuously. A recirculating aquaculture system is an almost completely closed circuit. The produced waste products; non soluble waste, ammonium and CO_2, are either removed or converted into non-toxic products by the system components. The purified water is subsequently, saturated with oxygen, and returned to the fish tanks. By recirculating the culture water, the water and energy requirements are limited to an absolute minimum. It is however not possible to design a fully closed recirculating system. The non-degradable waste products must be removed and evaporated water must be replaced. Still our recirculating aquaculture systems are capable of reusing 90% or more of the culture water. To ensure good water purification, recirculating aquaculture systems consist of a number of components with specific functions. The fish culture systems must be able to:

● Remove solid particles (waste, spilled feed)

● Remove dissolved organic matter

● Convert ammonium into the less harmful nitrate

● Remove CO_2 and add O_2.

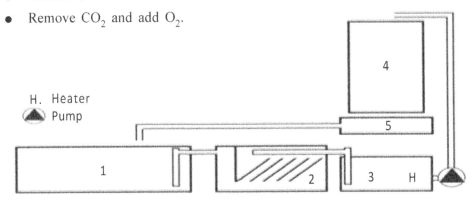

Fig.10:   The basic recirculating aquaculture system consists of the following components: 1. Fish tanks 2. Sedimentation tank 3. Pump tank 4. Biotower and 5. Reception tank

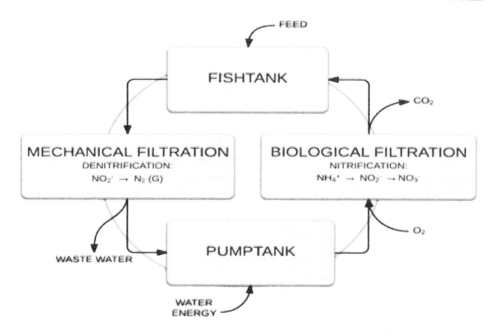

Fig. 11: The different processes within a recirculating aquaculture system

'RAS'= Re-defined

Recirculation aquaculture systems (RAS) represent a new and unique way to farm fish. Instead of the traditional method of growing fish outdoors in open ponds and raceways, this system rears fish at high densities, in indoor tanks with a "controlled" environment. Recirculating systems filter and clean the water and recycling back through fish culture tanks.

In a RAS, water flows from a fish tank through a treatment process and is then returned to the tank, hence the term recirculating aquaculture systems.

New water is added to the tanks only to make up for splash out evaporation and to flush out waste materials. In contrast, many raceway systems used to grow trout are termed "open" or "flow through" systems because all the water makes only one pass through the tank and then is discarded.

Fish grown in RAS must be supplied with all the conditions necessary to remain healthy and grow. They need a continuous supply of clean water at a temperature and dissolved oxygen content that is optimum for growth. A filtering (biofilter) system is necessary to purify the water and remove or detoxify harmful waste products and uneaten feed. The fish must be fed a nutritionally-complete feed on a daily basis to encourage faster growth and high survival.

Benefits of RAS

RAS offer fish producers a variety of important advantages over open pond culture. These include a method to maximize production on a limited supply of water and land, nearly complete environmental control to maximize fish growth year-round, the flexibility to locate production facilities near large markets, complete and convenient harvesting, quick and effective disease control.

RAS can be of various sizes ranging from large-scale production systems (more than 450 metric tonnes per year) to intermediate-sized systems (225metric tonnes per year), to small systems (more than 100metric tonnes per year). They can be used as grow-out systems to produce food fish or as hatcheries to produce eggs and fingerlings for sport fish stocking, growing out for minor carps and ornamental fish for residential aquariums.

This is an excellent alternative to open pond culture where low densities (extensive culture) of fish are reared free in large ponds and are subject to losses from diseases, parasites, predation, pollutants, stress, and seasonally suboptimal growing conditions.

RAS offers an unique opportunity of conversing both land area and water.Maximum production of fish may be achieved in a relatively small area of land and using a small volume of water.For example, using a RAS it is possible to produce more than 45 metric tonnes of fish in a 5, 000 square-foot building, whereas 20 acres of outdoor ponds would be necessary to produce an equal amount of fish with traditional open pond culture.

Similarly, since water is reused, the water volume requirements in RAS are only about 20% of what conventional open pond culture demands. They offer a promising solution to water use conflicts, water quality, and waste disposal. These concerns will continue to intensify in the future as water demand for a variety of uses increases.

Location flexibility of RAS

RAS are particularly useful in areas where land and water are expensive and not readily available. RAS requires relatively small amounts of land and water. In a country like India, due to urbanization and increasing trend of industrialization the facilities of gound water uses as open waters in the form of ponds, lakes, reserevoirs etc. is gradually coming to an end. Places which are geographically disadvantaged but the necessity of the availability of quality animal protein at a cheaper rate is a major concern, the development of recirculatory *aquaculture system* (RAS) reflects new light in the process. They can be located close to large markets (urban areas) and thereby may minimize hauling distances and transportation costs. There is an immense scope of the RAS near waste water disposals from municipal water supplies where dechlorination is of utmost importance for pisciculture. Nearly all species of food fish and sport fish that are commonly reared in ponds including major and minor carps, catfish, snake heads and tilapia etc. can readily be grown in high densities in confined tank systems having recirculation facilities.

Species and Harvest Flexibility

RAS are currently being used to grow Indian Major and Minor carps, Prawn and Shrimps, Catfish, Striped bass, Tilapia, Craw fish, Blue crabs, Oysters, Mussels, and Aquarium pets. Indoor fish culture systems offer considerable flexibility to:

(1) Grow a wide diversity of fish species,

(2) Rear a number of different species simultaneously in the same tank (polyculture) or different tanks (monoculture),

(3) Raise a variety of different sizes of one or several species to another depending on market demand and price.

RAS afford growers the opportunity to manipulate production to meet demand throughout the year and to harvest at the most profitable times during the year. This flexibility in the selection of species and harvest time allows the grower to rapidly respond to a changing marketplace in order to maximize production and profitability.

Disadvantages of RAS

RAS do have some disadvantages when compared to open pond culture. They are relatively expensive systems to develop (building, tanks, plumbing, and biofilters) and to operate (pumping, aerating, heating, lighting). Moreover, they are complex systems and require skilled technical assistance to manage successfully.

Constant supervision and skilled technical support are required to manage and maintain the relatively complex circulation, aeration, biofilter systems and to conduct water quality analysis. The danger of mechanical or electrical power failure and resulting fish loss is always a major concern when rearing fish in high densities in small water volumes.

Operating at or near maximum carrying capacity requires fail-safes in the form of emergency alarms and backup power and pump systems. The business and biological risk factors are correspondingly high. Continuous vigilance and quick reaction times (15 minutes or less) are needed to avert total mortality. However, the higher risk factor, capital investment, and operating costs can be offset by continuous production, reduced stress, improved growth, and production of a superior product in the RAS.

The'RAS' Design

The functional parts of a RAS include:

(1) Growing tank,

(2) Sump of particulate removal device,

(3) Biofilter,

(4) Oxygen injection with U-tube aeration and,

(5) Water circulation pump.

Depending on the water temperature and fish species selected, a water heating system may be necessary. Ozone and ultraviolet sterilization may also be advantageous to reduce organic and bacterial loads.

Biosecurity

Fish hatcheries with RAS facility are often fully closed and entirely controlled, making them mostly biosecure - diseases and parasites cannot often get in. Biosecurity means RAS can continusouly operate without any chemicals, drugs or antibiotics. Water supply is a regular route of pathogen entry, so RAS water is often first disinfected or the water is obtained from a source that does not contain fish or invertebrates that could be pathogen cariers.

Water quality and waste management

Water quality plays an important part in RAS systems which need continuous monitoring by skilled managers since these parameters are directly concerned with animal health, feed utilization, growth rates and carrying capacities. The critical water quality parameters that are taken care in RAS are *dissolved oxygen, temperature, pH, alkalinity, suspended solids, ammonia, nitrite and carbon dioxide* (CO_2). These parameters are interrelated in a complex series of physical, chemical and biological reactions. Monitoring and making adjustments in the system to keep the levels of these parameters within acceptable ranges is very important to maintain the viability of the total system.

A successful water reuse system should consist of tanks, filters, pumps and instrumentation (Helfrich.L.A and Libey.G 1991).

Fish tanks

Fish tanks may be round or octagonal or square design with rounded corners and the arrangement of in-and outlets of water treatment units should support the circular water flow. Aqua jets may be kept additionally for enhanced water flow and aeration.The circular flow promotes the performances of stocked fishes. Circular tanks are good culture vessels because they provide virtually complete mixing and a uniform culture environment. When properly designed, circular tanks are essentially self-cleaning. This minimizes the labour costs associated with tank cleaning. (Nazar *et al.*, 1990).

Aeration systems

Usage of air diffusers is preferred in RAS. These diffusers produce small air bubbles within the tank that rise through the water column. The smaller the bubbles and the deeper the tank, more oxygen is transferred.

Biofiltration

The biological filter (biofilter) is the heart of the RAS. As the name implies, it is a living filter composed of a media (corrugated plastic sheets or beads or sand grains) upon which a film of bacteria grows. The bacteria provide the waste treatment by removing pollutants. The two primary water pollutants that need to be removed are (1) fish waste (toxic ammonia compounds) excreted into the water and (2) uneaten fish feed particles. The biofilter is the site where beneficial bacteria remove (detoxify) fish excretory products, primarily ammonia.

Ammonia and Nitrite Toxicity

Ammonia and nitrite are toxic to fish. Ammonia in water occurs in two forms: ionized ammonium (NH_4^+) and unionized (free) ammonia (NH_3). The latter NH_3, is highly toxic to fish in small concentrations and should be kept at levels below 0.05 mg/l. The total amount of NH_3 and NH_4 remain in proportion to one another for a given temperature and pH, and a decrease in one form will be compensated by conversion of the other. The amount of unionized ammonia in the water is directly proportional to the temperature and pH. As the temperature of pH increases, the amount of NH_3 relative to NH_4 also increases.

In addition to ammonia, nitrite (NO_2) poisoning of fish also is an imminent danger in RAS. Nitrite levels should be kept below 0.5 mg/l. Brown blood disease (methemoglobinemia) occurs in cultured salmon and channel catfish when hemoglobin is oxidized by nitrite to form methemoglobin (a respiratory pigment of the blood that cannot transport oxygen). The disease can occur at nitrite concentrations of 0.5 mg/l or greater. As the name implies, the blood has a characteristic chocolate brown color. Adding salt (NaCl) at a rate of 1 pound per 120 gallons of water (a chloride to nitrite ratio of 16:1) will suppress this disease in soft water; a ratio of 3:1 is effective in hard water.

Calculating Ammonia Loading

The amount of ammonia excreted into a tank depends on a number of variables including the species, sizes, and densities of fish stocked and environmental conditions (temperature, pH). Ammonia loading can be roughly estimated from the biomass (weight) of fish in the tank or it can be based on the weight of feed fed each day.

On the average about 25 mg (milligrams) of ammonia per day is produced for every 100 grams (3.5 ounces) of fish in the tank. Therefore, in a tank containing 1,000 striped bass fingerlings each weighing 75 g (75,000 g total fish weight), the daily ammonia load produced by all the fish would be 18,750 mg (18.8 g). To remedy excessively high ammonia levels, add freshwater, eliminate feeding or reduce the density of fish in the tank.

Ammonia loading also can be estimated based on the total amount of feed fed. For manufactured fish feed with standard protein levels of 30 to 40 percent, simply multiply the total weight of the feed (in grams) 25 times. For example, if the fingerling stripers are fed 1 pound (454 grams) of pelleted feed per day, the amount of ammonia produced per tank would be about 11,350 mg per day.

Nitrification

Ammonia is a poisonous waste product excreted by fish. Since fish cannot tolerate this poison, detoxifying ammonia is fundamental to good water quality, healthy fish, and high production.

Detoxification of ammonia occurs on the biofilter through the process of nitrification. Nitrification refers to the bacterial conversion of ammonia nitrogen (NH_3) to less toxic NO_2, and finally to non-toxic NO_3. The process requires a suitable surface on which the bacteria can grow (biofilter media), pumping an continuous flow of tank water through the biofilter, and maintaining normal water temperatures and good water quality.

Two groups of aerobic (oxygen requiring), nitrifying bacteria are needed for this job. Nitrosomonas bacteria convert NH_3 to NO_2 (they oxidize toxic ammonia excreted by fish to less toxic nitrite), the Nitrobacter bacteria convert NO_2 to NO_3 (they oxidize toxic nitrite to largely nontoxic nitrate).

Nitrification is an aerobic process and requires oxygen. For every 1 milligram of ammonia converted about 5 milligrams of oxygen is consumed, and additional 5 milligrams of oxygen is required to satisfy the oxygen demand of the bacteria involved with this conversion. Therefore, tanks with large numbers of fish and heavy ammonia loads will require plenty of oxygen before and after the biofiltration process.

Nitrification is an acidifying process, but is most efficient when the pH is maintained between 7 and 8 and the water temperature is about 27-28⁰C. Acidic water (less than pH 6.5) inhibits nitrification and should be avoided. Soft, acidic waters may require the addition of carbonates (calcium carbonate, sodium bicarbonate) to buffer the water. The addition of a salt as a therapeutic in striped bass as freshwater bacteria temporarily adjust to alteration in salinity.

Biofilter Design and Materials

A biofilter, in its simplest form, is a wheel, barrel, or box that is filled with a media that provides a large surface area on which nitrifying bacteria can grow. The biofilter container can be constructed of a variety of materials, including plastic, wood, glass, metal, concrete, or any other nontoxic substance. In small-scale systems, some growers have used plastic garbage cans or septic tanks. The size of the biofilter directly determines the carrying capacity of fish in the system. Larger biofilters have a great ammonia assimilation capacity and can support greater fish production.

A biofilter must provide sufficient surface area for the colonization (attachment) of nitrifying bacteria. It needs to provide a large surface area to support bacterial populations at densities adequate to reduce the load of waste products (ammonia) excreted by the fish population in the tank. It is essential that the water flowing through the biofilter come into direct contact with the bacteria film growing on the surface media for a time period sufficient to allow the bacteria to convert toxic NH_3 and NO_2 to less toxic NO_3. Careful calculation of the flow-through rates (turnover or contact time) and size (volume and depth of the biofilter) is fundamental.

The biofilter media can be corrugated plastic, Styrofoam or glass beads, lava rock, sand, gravel, or similar material that supplies large surface area. The quality and quantity of surface area of the media provided for nitrifying bacteria are important determinants of the efficiency of the biofilter.

The ideal biofilter media has:

(1) High surface area for dense bacterial growth,

(2) Sufficient pore spaces for water movement,

(3) Clog resistance,

(4) Easy cleaning and maintenance characteristics.

Biofilter Sizing

The biofilter in any RAS design must be sized to correspond with the other system components.

Important factors that must be considered in designing a biofilter are:

1. Media surface area (square feet of surface for bacteria attachment),

2. Ammonia leading (ounces of ammonia that need to be converted per day per square foot of media area), and hydraulic loading (gallons of water per day per square foot media surface).

Types of Biofilters

Biofilters can be configured in many ways. The two general categories are:

(1) Submerged bed filters and

(2) Emerged bed filters.

Submerged bed filters can have fixed (immobile) media in which the water flow can be upward, downward or horizontally through the media.

The fluidized bed reactor (FBR) is a commonly used submerged bed filter. The FBR consists of fine particles (sand, dense plastic, glass beads, minerals, etc.) in a container through which upwelling water flows thereby "fluidizing" or suspending the media in the water column. FBRs offer large surface area per unit volume and theoretically greater nitrification. However, as with other submerged bed filters, all of the oxygen needed for conversion of ammonia to nitrate must be dissolved. Submerged bed filters often need supplemental aeration both before and after the water passes through the filter. If the inlet dissolved oxygen is low, the efficiency of ammonia conversion is reduced.

Emerged bed filters are of two basic types

(a) Trickling filter (TF) sometimes called packed columns, and

(b) Rotating biological contactors (RBC). These filters have the advantage of not requiring the addition of oxygen prior to water entering the filter. In fact, these filters frequently supply all the oxygen used to support fish respiration. For this reason these types of filters are often employed in RAS. The trickling filter is designed to have water slowly cascade down through the media column, which is suspended above the water. Water enters the column (which is filled with biofilter media) from an overhead spray pipe and trickles down through the biofilter media where nitrification occurs. The waterfall action of this filter adds oxygen to aerate the water.

Recirculation Rates (turnover times)

The recirculation rate (turnover time) is the amount of water exchanged per unit of time. This can easily be determined by dividing the volume of water in the tank by the capacity of the pump (in gallons per minute). For example, the turnover rate in a 2, 500 gallon tank system circulated with a water pump rated at 45 gallons (20, 052liters) per minute (63, 360 gallons per day or 2, 88, 03, 456 liters/ day) would be 25.3 tank volumes per day (a rate of slightly more than one volume per hour). Increasing the number of turnovers per day would provide increased biofiltration, greater nitrification (bacterial contact), and reduced ammonia levels. Most fish production recirculation systems are designed to provide at least one complete turnover per hour (24 cycles per day).

Compartmentalization

The ability of isolate the components of the system (biofilter, fish tank, and sump) is an important design feature, particularly critical when it becomes necessary to do filter maintenance or to treat the fish with chemicals and drugs. Cleaning and declogging static biofilters can pose difficult problems, particularly if there is no provision for shutting down the system for maintenance. Some therapeutic chemicals and drugs used to treat such fish may be harmful to nitrifying bacteria on the biofilter. A sudden drop in the efficiency of the bacteria can result in toxic NH_3 concentrations and fish kills. Other filters: Other types of filtration (mechanical and chemical) are available and can sometimes be used to supplement the efficiency of biofilters in removing ammonia in fish production systems. Most of these measures are useful only to temporarily control ammonia and nitrite in small systems. In chemical filtration, water is pumped through a chemical media of activated carbon, zeolite, or other substances. These chemicals have microscopic pores that trap ammonia ions and remove them from the water. The familiar activated charcoal filter, popular in aquaria, can be incorporated as an auxiliary filter to support biofiltration in fish production systems, but this form of filtration requires periodic replacement with large quantities of relatively costly activated charcoal. Zeolite filters are frequently used to remove NH_4 (and indirectly NH_3) at an estimated rate of 1 mg NH_4 per 1gm of zeolite. The use of zeolite requires regular and constant pumping of water through the filter and regular replacement of large quantities of expensive zeolite. Zeolite can be recharged with a salt solution (10%) and reused, but salt water disposal then becomes an environmental problem, particularly in inland waters.

Sump

A sump (clarifier tank) is used to prevent the excessive accumulation of fish excretory products and waste feed. Waste products increase the biological oxygen demand (BOD), decrease the dissolved oxygen content, lower the carrying capacity (density of fish) that can be reared, and may result in off-flavor in fish products. Accumulation and decomposition of waste material results in the production of toxic compounds such as ammonia (NH_4, NH_3, NO_2) and hydrogen sulfide (H_2S) that can be hazardous to fish health. The clarifier tank is designed as a settling basin (large volume tank with a slow flow rate to increase sedimentation). Its purpose is to concentrate and remove suspended solids (fish feces, uneaten feed particles) before they clog the biofilter or consume valuable oxygen supplies. The clarifier should be a separate tank, isolated from the fish tank and the biofilter, so that it can be cleaned periodically (daily) as needed. To increase the efficiency of the clarifier, various filters (plastic filters, sand filters, metal screens) can be inserted into the sump tank.The sump tank should have a v-shaped bottom to concentrate waste particles and facilitate cleaning.

Oxygen Management

Successful fish production depends on good oxygen management. The addition of oxygen in a pure form or as atmospheric air (aeration) is essential to:

(1) The survival (respiration) of fish held in high densities,

(2) The survival of aerobic, nitrifying bacteria on the biofilter and,

(3) For the decomposition (oxidation) of organic waste products.

Supplying sufficient oxygen to sustain healthy fish and bacterial populations and to meet the biochemical oxygen demand (BOD) for fish waste and unconsumed food is critical. Maintain oxygen levels, near saturation or even at slightly super-saturation at all times. Low oxygen levels will reduce growth, feed conversion rates, and overall fish production. The amount of oxygen needed in RAS depends on a number of factors. Oxygen demand is directly correlated with the density of fish in the tanks, feeding rates, water temperatures, flow rates, and nitrification. It is also a function of physical conditions such as water temperature and water volumes. Increasing dissolved oxygen concentrations through oxygen injection, aeration, and increasing water flow rates (turnover times) are ways to increase the density (carrying capacity) of fish that can be held in tanks of fixed size. Atmospheric oxygen can be added to the tanks by surface agitation with aerators or by large blowers. Surface aerators may not be cost effective or efficient in evenly distributing oxygen throughout large commercial-scale systems. Blowers can be effectively used to supply oxygen and also to mechanically rotate RBC.

Pure Oxygen

Pure oxygen injection systems are increasingly being used in aquaculture. They are particularly useful in maintaining oxygen-saturated conditions in recirculating systems with high densities of fish. Pure oxygen can be delivered and stored in a tank as liquid oxygen or it can be produced on-site by a oxygen generator. Bottled oxygen gas also is sometimes kept as an emergency backup system for RAS, but this alternative usually is too expensive and bulky to be practical. Liquid oxygen technology is relatively simple, efficient, and cost-effective; especially if purchased in bulk quantities and if the site is located near a reliable supplier. A liquid oxygen system consists of a storage tank for the liquid gas, vaporizers to turn liquid oxygen to gas, and supply lines to the fish tanks. It conveniently requires no external power supply and is therefore free of power failures and the consequent fish kills. Most growers rent or purchase a liquid oxygen storage tank of a size sufficient to provide two to four week supply of oxygen. The size of the tank corresponds with the fish production capacity of the system.

Oxygen Diffusion

Effective diffusion of pure oxygen gas into a liquid (water) can best be accomplished using a U-tube oxygenation, counter-current flow injectors, or micro-bubble devices (tubes or fine wetstones). The purpose is to dissolve much of the oxygen injected so that it is available to the fish, rather than wasted by bubbling out of solution to the atmosphere.

Ozone Sterilization

Ozone (O_3) is a naturally occurring gas (upper atmosphere) that consists of three atoms of oxygen. It is a powerful oxidizing agent that can be used to break down compounds. Ozone must be used with caution since it is directly toxic to aquatic life and may form harmful biproducts (hypochlorite, hypobromite). Careful redox potential measurements and special injection equipment apparatus are needed to determine and control ozone applications.

Carbon Dioxide (CO_2) Control and Removal

In addition to toxic ammonia, carbon dioxide tends to concentrate in intensive fish production systems. As carbon dioxide increases, the pH of the water decreases, and fish respiration is affected. Carbon dioxide levels should be maintained at levels less than 20 mg/l for good fish growth. Some carbon dioxide is beneficial since it reduces pH and mitigates ammonia toxicity. Carbon dioxide removal can be accomplished with any device (RBC, packed column) that increases air-water contact.CO_2 is usually removed through some form of gas exchange process either by exposing the water to air in a "waterfall" type of environment, or mixing air into the water to remove excess CO_2.

Disinfection of culture water

Installation of suitable UV sterilizers or ozonisers in the water flow would remove unwanted bacteria, algae and pathogens. The capacity and the flow rate of the UV sterilizer/ ozoniser should be calculated based the on quantity of water to be treated and effectiveness of treatment.

Fig.12: Temperature control unit in RAS

Fig.13: Drum filter in RAS

Fig.14: Sprinkler device for aeration

Fig.15: The RAS shed

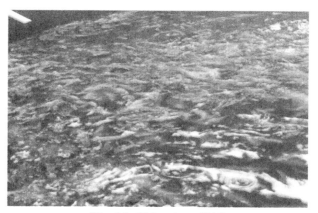

Fig. 16: Fishes in a RAS

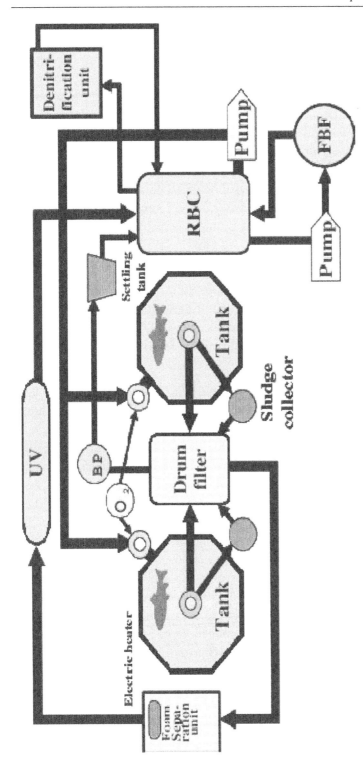

Fig. 17: Schematic view of closed recirculating aquaculture system (After Yoshino. H, D.E. Gruenberg, I. Watanabe, K. Miyajima, O. Satoh and K. Nashimoto. Changes in Water Quality and Performance of a Closed Recirculating Aquaculture System. https://www.Researchgate.Net/publication/237797132.

11. INTEGRATED FISH FARMING

Prof.Chandra Sekhar Chakraborty, the then Vice-Chancellor of West Bengal, University of Animal and Fisheries Sciences (**WBUAFS**) while delivering his address to a group of farmers engaged in agro-fisheries activities on the possibilities of agriculture developments vis-à-vis the growing requirement and demand of food stuff, expressed the idea *of integrating the farming systems, can only save the poverty at least in a country like India,* and*"Integration of fish farming is no exception."* Integrated Farming System is a resource management strategy to achieve economic and sustained agricultural production to meet diverse requirements of farm household while preserving the resource base and maintaining high environmental quality. This farming system seems to answer the problems of increasing food production, increasing net farm income, improving nutritional status, promoting natural resource management, sustainable use of land, water and biota.

The concept of an all-round development of agriculture, animal husbandry, fisheries and other sideline occupations are economically and environmentally sound, the motivation for integration would appear to be the national policy of diversification of production.

Under the gradual shrinking of land holding, it is necessary to integrate land based enterprises like fishery, poultry, duckery, apiary, field and horticultural crops etc. within the bio-physical and socio-economic environment of the farmers to make farming more profitable and dependable (Rana, 2015).Integrated farming involving aquaculture defined broadly is the concurrent or sequential linkage between two or more activities, of which **at least one is aquaculture**. These may occur directly on-site or indirectly through off-site needs and opportunities, or both. Benefits of integration are synergistic rather than additive; and the fish and livestock components may benefit to varying degrees. The term "waste" has not been omitted because of common usage but philosophically and practically it is better to consider wastes as "resources out of place" (Little and Edwards, 2003).Farming is a stochastic dynamic, biological and open system with human and social involvement. It specifically refers to crop combination or enterprise-mix in which the products and/or the by-products of one enterprise serve as the input for the production of other enterprise (Bhuyan, 2014).

A. Basic Principles

Integrated fish farming is based on the concept that 'there is no waste', and waste is only a misplaced resource which can become a valuable material for another product (Vincki, 1977). In integrated farming, the basic principles involve the utilization of the synergetic effects of inter-related farm activities and the conservation, including the full utilization of farm wastes.

Integrated fish farming is a system of producing fish in combination with other agricultural and livestock farming operations centered on the fish pond. The farming sub-systems e.g. fish, crop and livestock are linked to each other in such a way that the by products/and wastes from one sub-system become the valuable inputs to another sub-system and thus ensures total utilization of land and water resources of the farm resulting in maximum and diversified farm output with minimum financial and labour costs (FAO: Corporate Document Repository).*The integration of aquaculture with livestock or crop farming provides quality protein food, resource utilization, recycling of farm waste, employment generation and economic development*. Integrated fish farming is well developed culture practice in China followed by Hungary, Germany and Malaysia. Our country, India--the country is organic-based and derives inputs from agriculture and animal husbandry. The integrated fish farming is accepted as a sustainable form of aquaculture. For integration we can use recycled effluents from agro-based industries as well as food processing plants at minimum financial and labor costs. Integrated aquaculture is the concurrent or sequential linkage between two or more farm activities, of which at least one is **aquaculture.**

Integrated Farming System is a resource management strategy to achieve economic and sustained agricultural production to meet diverse requirements of farm household while preserving the resource base and maintaining high environmental quality. This farming system seems to answer the problems of increasing food production, increasing net farm income, improving nutritional status, promoting natural resource management, sustainable use of land, water and biota.

- Integrated fish farming offers tremendous potential for food security and poverty alleviation in urban and periurban areas.

- It is an efficient way of using the same land resource to produce carbohydrate as well as animal protein and important micronutrients concurrently or serially.

- Optimization of available natural resources use.

- Diversification of income generating activities.

- Improvement of soil fertility.

- Improved pest control with less use of chemicals (pesticides, fertilizers).

- Aquatic Biodiversity conservation and sustainable use could be enhanced.

- Efficient waste utilisation from different culture practice for fish production.

- It reduces the additional cost for supplementary feeding as well as fertilisation.

- It is an artificial balanced ecosystem where there is no waste.

- It provides more employment avenues.

- It reduces the input and increases output and economic efficiency.

- The integrated fish farming provides fish along with meat (chicken, duck, beef, pork etc.), milk, vegetables, fruits, eggs, grains, fodder, mushroom etc.

- This practice has potential to increase the production and socio-economic status of weaker section of our society.

Table. 2: Types of integrated fish farming

a) Fish cum Agriculture system	b) Fish cum Animal system
Fish cum paddy culture	Fish cum Dairy
Fish cum water chest nut	Fish cum Pig farming
Fish cum papaya	Fish cum rabbit farming
Fish cum mulberry	Fish cum poultry
Fish cum mushroom	Fish cum duck farming

B. Paddy-cum-Fish culture

Production of fish in paddy fields is almost as primitive as the practice of paddy culture itself. Paddy farming with fish culture is a type of dual-culture farming system in which paddy is the sole enterprise and fishes are taken to initiate additional for extra income.

Paddy-cum-fish culture is practiced in many rice-growing belts of the world mainly in the South Eastern Asian region. Rearing of fish along with paddy is a classic traditional farming practice adopted in India. It has largely been practiced in a primitive way in the coastal areas of the country.As a socio-economic activity, fisheries ranks second in the world to agriculture sector. India is an agro-based country more so both rice and fish are considered as the staple and most preferred food items and Fishery has been playing a significant and dynamic role in terms of nutrition, employment, foreign exchange earnings, and more importantly socio-economic stability in the rural areas.However, during the beginning of 'green revolution' at least in India paddy-cum-fish culture practices declined drastically as rice production intensified and chemical insecticides were heavily applied. From the late 70's on, however, less toxic chemicals, better varieties of paddy, and the increase in availability of fish seed as aquaculture developed, all led to renewed interest in fish production from rice fields. Though the paddy-cum-fish culture a large number of small rural producers are benefitting from aquaculture in their rice fields. The fish produced are either fingerlings for ongrowing in other aquaculture systems or table fish for domestic market and house hold consumtion (Williams, 1997).It has been reported that rice fish culture led to financial returns that were 41% higher than rice alone. Part of the increase was due to increasing rice yields as a result of the apparent mutualism of rice-fish production, i.e., adding fish to rice fields tends to increase, or at least not decrease, rice yields.India enjoys about 4% of world's freshwater resources ranking it among the top ten water rich countries.It is very rich in natural water resources in the form of rivers, reservoirs, ditches, lakes, ponds & tanks, flood plains and large areas of rice fields etc. and in monsoon it receives every year, precipitation in the form of rain and snowfall which provide over 4000 cu km of freshwater to India, of which 2047 cu km return to oceans or is precipitated.The paddy (rice)-fish farming culture (Fig.18) involves the simultaneous production of rice and fish within irrigated rice fields to obtain an added profit multiplier in terms of value and production. In paddy cum fish culture, initially strong attention is given on technically feasible paddy cum fish pond bed preparation, ecologically sustainable stocking of different species of fish along with cultivars of selected paddy variety, addition of inputs like feed, mineral and organic fertilizers and ancillary post stocking managements is undertaken. Fish harvesting is done after paddy is reaped from the agrarian wetland. As the water level increases with the onset of monsoon during June-July, the farmers release advanced fingerlings at a modest density of about 500-1000/ha.

Fig.18: Integration of fish with paddy

Fish stay in the paddy field up to December and start migrating to the actual fish pond as the water level goes down in the paddy field (Fig.18). Paddy is harvested during November-December followed by harvest of fish in January-February. Farmers use organic and inorganic fertilizers during paddy field preparation. Application of lime, to keep the alkaline state of soil and water.

Fig. 19: Releasing of fish seeds in paddy cum fish culture farm

The principal objective of sustainable paddy-cum fish culture:

i.   To assess the relative profitability of using paddy-cum-fish culture compare to monoculture of paddy crops.

ii.   To determine the effects of the paddy-cum-fish culture in changing yields, total costs, fish consumption and labour employment as compared to the monoculture.

iii. To identify the major problems in conducting integrated paddy-fish farming and

iv.   To determine the effect of different fish combinations on fish yield in Paddy-cum-Fish culture system, with a view to recommending the outcome to prospective rice/fish farmers.

Fig. 20: Culture of paddy and fish together

Benefits of Integrated paddy cum fish culture system

1. Improves the soil fertility & soil health.
2. Increasing economic yield per unit area.
3. Reduction in production costs.
4. Decreases farm input requirements.
5. Multiple income sources.
6. Family income support.
7. Efficient utilization of family labour.
8. Reduction in animal feeding requirements.
9. Minimize the use of chemical fertilizers.
10. Provides balanced nutritious food for the farmers.
11. Solves the energy problems with biogas.
12. Avoids degradation of forests.
13. Enhance employment generation.
14. Pollution free environment.
15. Recycling of resources.
16. Improves the status & livelihood of the farmer.

Practices / or technology engaged in paddy cum fish farming

Site selection

The site selection for rice cum fish farming is low lying area where water flows easily and available at any time in needs.

Soil quality

The soil of the paddy field should be fertile, rich in organic manure and has high water holding capacity. Usually medium textured soils like silty clay or silty clay loam are most suitable for paddy cum fish farming/ or shrimp culture.

Bund preparation

The plots selected for paddy cum fish culture are normally prepared in the month of February by raising their embankment all along the plots. The paddy fields are suitable for fish culture at the areas because of strong bund, which prevent leakage of water to retain water upto desired depth and also guarded the escape of cultivated fingerlings/ or fishes during the floods. The dykes should be built strong enough to make up the height due to geographical and topographic location of the field. The bamboo screen mating should be done at the base of the bund area for its support.

Field leveling

After the completion of bund construction the base of paddy fields are leveled with the help of spade and local made wooden plates. Manual weeding is done during the month of February followed by construction of irrigated channel for easy passage, storage and draining of water. There are 2-3 channels constructed at the middle of paddy field for water management. The channel divides the paddy field perpendicular and horizontally bisect at a point. It is important to note that almost all paddy field have one or two inlets and more outlets. The former serve as entry of water required for the field and the later as outlets, one which remains at the bottom side of the dykes is meant for draining out the water for harvesting paddy crops and fishes. The remaining outlet constructed at the middle height of the dykes is meant for maintaining desirable water depth. Once the dressing work is over, the paddy field is ready for transplantation of rice seedling and fish seed stocking. However, the stocking of fish seed is done after 10-15 days of transplantation of rice seedling from its nursery bed.

Pond construction

The paddy plots are renovated suitably for the purpose of paddy cum fish culture. These may be fresh water or, brackish water. Construction of an earthen dyke surrounding the paddy plot is essential for retaining water and also for holding the fish and shrimp during aquaculture. The height of the dyke is required to be maintained between 50 and 100 cm depending upon the topography of the plot and tidal amplitude at the site. A perimeter canal is necessary on the inner periphery of the plot. For a one ha paddy plot, the width and depth of the canal may be about 2 m and 1 m respectively. The earth removed from excavating the canal is generally utilized for constructing or strengthening the dyke. In addition to the perimeter canal, two cross trenches of about 1 m width are also constructed at both the directions. The bottom of the trenches generally kept above the perimeter canal so that during the course of desalination, entire water can easily be removed to the canal. The area covered by the perimeter canal and the trenches are about 12% of the total land area

Flooding and Weeding of Paddy

After transplantation of paddy seedlings the field area are flooded. Adequate care is taken to keep the water level at a minimum of 5-7cm to allow proper filtering therefore, the water level is kept at 30 -50 cm deep, until the rice matures. The weed control measures was adopted by manual method by uprooting the weeds on both side of the plots of cultivated rice and generallyit is carried out twice/ or thrice in a week. Chemical method of weed control is avoided to ensure proper growth and survivality of fishes stocked.

Fertilization of Rice Fields

The plots utilized for rice-cum-fish culture is mainly based on organic fertilization with a varieties of animal excreta such as poultry dropping, pig excreta, cow dung and waste of plants such as rice husks, waste product of local beer and ashes from household brunt and remains of burnt straws after the harvest is over and compost fertilizer like decomposed straws, weeds and rice stalks etc. The Fertilization of the paddy field is mainly practiced and was done with dried cow dung at the rate of 20-30 kg /100 m^2 at least in a fortnight. Broadcasting the fertilizer evenly on the fields from various points of the dykes is the usual practice.

Stocking of fish seeds

Before releasing of fish seed to paddy field the paddy transplantation from seed beds to main paddy fields is done in the month of April, and there after paddy is left for two weeks for strengthening of paddy roots there after the fish seed @ 2500 nos./ha area is released. The fish rearing period varied from 3-6 months and the paddy rearing period is 5-7 months.

Stocking of Fish Fingerlings

The fish fingerlings initially having average weights between 50-100g respectively are stocked @ 500-1000nos. of fish / 100m^2 respectively after flooding.

Table 3: The most promising deep water rice varieties chosen for different states.

Sl.No.	States in India	Varieties
01.	Andhra Pradesh	PLA-2
02.	Assam	IB-1, IB-2 , AR-1, 353-146
03.	Punjab	BR-14, Jisurya
04.	Kerala	AR61-25B, PTB-16
05.	Tamil Nadu	TNR-1, TNR
06.	Uttar Pradesh	Jalamagan
07.	West Bengal	Jaladhi-1, Jaladhi-2, IET4786 popularly known as Gitanjali, Marisal, IET20235 popularly knownas Sujala, IET17713 popularly known as Rajdeep, IET19886 popularly known as Kanak, Swarnali, Amala, IET23403 popularly known as Gosaba 5, WGL 20471, Swarna sub1, Swatabdi, Ajit, Neeraja(IET-11865), Pankaj, Biraj(CNM-539), Suresh (CNM-540), Sabita(IET-8970), IET-10222(Panidhan), Rajshree(TCA-80-4)(IET-7970), Sudha(TCA-72)(IET-8977) Ratnagiri-2, Saraswathi (IET-11271), Bhagirathi (IET-11272), Mahananda (IET - 11910), Barsha.
08.	Manipur	Thoddabi

Water Quality Monitoring

Essential water quality parameters such as temperature, transparency of water, pH of water, dissolved oxygen, conductivity etc. are usually determined using standard methods. Plankton population densities in water are also determined in accordance with the standard methods.

Feeding of stocked fingerlings

Feeding of the stocked fish in the paddies starts immediately after stocking at pre-determined feeding spots. The fish were fed once in a day. The desired composition and quantity of the feed ingredients are given to the fingerlings at proper scheduled period. The feed is given @ 3% of the total fish stocked, in the system everyday preferably during morning hours.

Table 4: The ingredients and their percentage in fish feed during Paddy cum fish culture.

Ingredients generally used as fish feed	(%)
De-oiled rice bran (DORB)	60%
Mustered Oiled Cake (MOC)	29%
Ground nut oiled cake(GNOC)	10%
Common salt	0.5%
Vitamin and mineral mixture	0.5%

Growth Rate

Monitoring of growth rate of fish in the pond is generally carried out randomly netting the fish using a hand net. The timely weight and the measurements were taken and the fish returned to the trenches. Feeding rates were adjusted accordingly. Feed was used to attract the fish before each sub-sampling. The mean growth rates (MGR) are also calculated for each fish species.

Harvesting

Gears use for harvesting fishes is simple bamboo made basket called cane/ bamboo. The fishes remain under culture conditions for the period of 3-4 months in rice field, and a production of ± 500 Kgs/ha are achieved.

Methodology used for harvesting: The water is drained through outlet pipe, and thus allowing fishes and water accumulated in mid channel of paddy field, thereby the fishes are caught by hand picking, kept in large plastic bucket in live condition. After completion of fish harvesting the paddy harvesting is followed. Normally paddy harvesting is made last part of September and October. The paddy production range from 3500-4500 Kgs/ha from the same plot of land.

Advantages and disadvantages of fish culture in paddy fields

Advantages

The socio-economic importance of fish culture in rice fields evidently indicates that the deficit of animal protein in densely populated rice growing areas. In these areas usually all the available field space is used for rice leaving little space for cattle, sheep and poultry, which do not produce enough meat (Schuster, 1955). Thus fish grown in the paddy fields will be ideal use of land and would also be an easy source of cheap and fresh animal proteins. Thus fish culture can greatly contribute to the socio-economic welfare of rural populations of especially developing countries. An added advantage also is that unlike sea fish or other animal proteins, the fish from the local paddy fields would cause no transport problem and would be most fresh and healthy.

The production of a fish crop between the rice crops gives the farmer an off-season occupation (Hora and Pillay, 1962).Hickling (1962) points out that this "increases the income without increasing expenses". Apart from the additional income available from paddy-cum-fish culture, the combined culture leads to a reduction in labour in weeding and an increase in the yield of paddy by 5 to 15%. Schuster et al (1955) reported an increase of 6.2% in rice production in Indonesia, 10% in Malaysia, 10 – 15% in China and in Zimbabwe 6%. Using an improved strain (CR 1108), Sinha Babu *et al.* (1983) observed that rice growth with Indian carps in paddy-cum-fish culture trials showed an increase of 3.8 – 6.2%. The increase in paddy production is ascribed to various factors, namely:

1. Increase in organic fertilization by fish excreta and remains of artificial feed.

2. Better tilling of the rice seedlings due to the activity of the fish.

3. Reduction in the number of harmful insects, such as paddy stem borers, whose larvae are eaten by fish.

4. Reduction in rat population due to increase in the water level.

5. Increased mineralization of the organic matter and increased aeration of the soil resulting from the puddling of mud by benthic feeders.

6. Control of algae and weeds (by phytophagous fish) which compete with rice for light and nutrients.

Disadvantages

Coche (1967) lists the following disadvantages in fish culture in paddy fields.

1. A greater supply of irrigation water and a greater water depth required for fish culture. Especially in view of the shortage of water due to increased anthropogenic demands might cause serious difficulty in future.

2. Extra investment and labour involved in raising and strengthening fields bunds.

3. The need for rice variety tolerant to deep water and to low temperature.

4. Fish may damage the young seedlings by uprooting them (carps) or eating them (*Tilapia* sp.).

5. Certain parts of the field are lost for fish culture by the construction of trenches and refuges for fish.

6. Additional costs are involved in fertilizing and feeding the fish.

7. In certain types of soil continuous inundation may not be possible.

General considerations

The greatest difficulty in paddy-cum-fish culture may be attributed to the indiscriminate use of herbicides and pesticides in growing paddy seedlings - most of these are harmful to fish. Judicious application of these, to an extent some reduction in damage is possible, but most insecticides even in the lightest dose may kill the fishes. The organochlorines are in this way more harmful than organophosphorus insecticides. For example, among several insecticides viz. the organochlorine, 'Endosulfan' kills carp and *T. mossambica* 100 – 1000 times faster than the organophosphorus (Kutty et al, 1977). Albeit many insecticides are easily bio-degradable and their toxicity is reduced greatly within a few days' time. Hence choosing some less harmful and quickly bio-degradable biocides is necessary to protect both the paddy and the fish.

Perhaps one may control the time of application i.e. by applying the biocides soon after transplantation before the fish are introduced, by then the potency of the biocides would have become least. But in spite of such an approach the possibility of accumulation of pesticides which may not kill, cannot be ruled out. As discussed earlier it appears the whole ecosystem of each crop, is upset drastically by the use of chemicals and biocides - the whole biological complex is changed. And several pests have developed immunity to various chemicals applied to control them. Thus the ultimate cannot be an increase of chemical applications to protect the crop, but to evolve more harmonious methods in line with sound biological principles. This has already been proved in some cases in crop protection, by use of biological control methods. Let us hope that such methods, which are within the reach of man, will be evolved so that fish and paddy can be grown together in the vast expanse of water which is available on the global surface as paddy fields.

The paddy -cum-fish culture is an innovative farming system in which, rice is the main enterprise and raising of fish as integration is taken as additional means to secure extra income. Paddy-cum- fish culture reduces poverty of the farmers and also improves the yield, create employment opportunity, and increase nutrient intake which brings food security for them. Other classic examples of composite fish culture along with vegetable crops, (Fig. 21& 21a).

Fig. 21&21a: Pumpkin- Fish integration & Fish-cum-other vegetable farming

C. Integrated fish-cum-cattle farming

Fish-cum-Dairy Farming is considered as an excellent enterprise where the organic wastes are utilized in a perfect way. Use of cow/buffalo manure in fish farming is a commonly prevailing practice. On an average, one cow/buffalo excretes 12000 kg of dung and 8000 liter urine per year. The cattle feces and urine are beneficial to the filter-feeding and omnivorous fishes. On an average, 3-4 cows/ buffaloes can provide sufficient manure to fertilize one hectare pond. In this system, farmer gets milk, fish and calf as well, which increases revenue and reduces input costs. The system gives a more profit per year from one hectare land (Fig. 22).

Cow requires 7-8MT of grass/yr and Grass carp utilizes leftover grasses. Fish also utilizes fine feed wasted by cattle in the form of dung/slurry 20-30MT of dung/slurry which is recycled in 1ha water area to produce over 4.0MT of fish without feed or and fertilizer application. The system gives a net profit of Rs.1, 14, 000/- per year from 1ha land. Besides, cow-dung gas plant at the farm site (Fig. 23) gives electricity at the fish farm reducing fuel cost.

Fish farming by using cattle manure has long been practiced in our country. This promotes the fish-cum-cattle integration and is a common model of integration. Cattle farming can save more fertilizers, cut down fish feeds and increase the income from milk. The fish farmer not only benefitted financially butalso can supply fish, milk and beef to the market.

Fig. 22: A unique example of cattle-cum-fish farming at Gajdharpur, Tarakpur Mouza, P.S. Baharampur, West Bengal.

Fig. 23: Cow dung gas plant at the integrated fish farm.

Pond management practices

These practices are similar to poultry or pig or duck integration with fish. Cow dung is used as manure for fish rearing. About 5, 000 -10, 000 Kg/ha can be applied in fish pond in installments. After cleaning cow sheds, the waste water with cow dung, urine and unused feed, can be drained to the pond. The cow dung promotes the growth of plankton, which is used as food for fish.

Cattle husbandry practices

The cow sheds can be constructed on the embankments of the fish farm or near the fish farm. The floor of the cow shed preferably made with bricks, sand and cement. The outlet of the shed is connected to the pond so that the wastes can be drained into the pond. Cultivable varieties of cows are black and white (milk) and Shorthorn beef, Simmental (milk and beef), Hereford (beef), Charolai (beef), Jersey (milk and beef) and Qincuan draft (beef).

Necessity and feasibility

1. Cows can provide cow dung, which is used as manure in fish ponds. The left over dung can be used for earthworm culture.

2. Cow manure is nutritionally rich. It is experimentally seen that if 0.024 kg of fresh cow manure is applied per cu.m of the water body, the inorganic N and P are congenial for plankton multiplication.

3. The average amount of phytoplankton in a manured pond can reach 19.15 mg/lit and the zooplankton amounts to 5.61 mg/lit. Such is an indication of the eutrophic condition of a pond.

D. Integrated Fish cum Pig Farming

Benefits:

1. The fish utilize the food spilled by pigs and their excreta which is very rich in nutrients and acts as a substitute for pond fertilizer more so enhances the natural feed e.g., the plankton. Hence, the cost of fish production is greatly reduced.

2. No additional land is required for piggery operations, since the pig house may easily be constructed at the fish farm itself.

3. Cattle fodder required for pigs and grass may easily be grown on the pond embankments.

4. Pond provides water for washing the pig - sties and pigs.

5. It results in high production of animal protein per unit area.

6. It ensures high profit through less investment.

7. The pond muck which gets accumulated at the pond bottom due to constant application of pig dung can be used as fertilizer for growing vegetables and other crops and cattle fodder.

Stocking of Fish

- The stocking rates vary from 8, 000 – 8, 500 fingerlings / ha and a species ratio of 40 % surface feeders, 20 % of column feeders, 30 % bottom feeders and 10-20 % weedy feeders are preferred for high fish yields.

- Mixed culture of only Indian major carps can be taken up with a species ratio of 40 % surface, 30 % column and 30 % bottom feeders.

- In the northern and north - western states of India, the ponds should be stocked in the month of March and harvested in the month of October – November. Due to severe winter the growth of fishes get affected.

- In the south, coastal and north - eastern states of India, where the winter season is mild, the ponds should be stocked in June - September months and harvested after rearing the fish for 12 months.

Use of Pig Waste as Manure

- Pig-sty washings including pig dung, urine and spilled feed are channeled into the pond.

- Pig dung is applied to the pond every morning. Each pig voids between 500-600 Kg dung / year, which is equivalent to 250-300 Kg / pig / 6 months.

- The excreta voided by 30 – 40 pigs are adequate to fertilize one hectare pond.

- When the first lot of pigs is disposed off after 6 months, the quantity of excreta going to the pond decreases. This does not affect the fish growth as the organic load in the pond is sufficient to tide over for next 2 months when new piglets grow to give more excreta.

- If the pig dung is not sufficient, pig dung can be collected from other sources and applied to the pond. Pig dung consists 69 - 71 % moisture, 1.3 - 2 % nitrogen and 0.36 - 0.39 phosphours.

- The quality and quantity of excreta depends upon the feed provided and the age of the pigs.

- The application of pig dung is minimized or deferred on the days when algal blooms appear.

Pig- Husbandry Practices

The factors like breed, strain, and management influence the growth of pigs.

a. Construction of Pig House

- Pig houses with adequate accommodation and all the requirements are essential for the rearing of pigs.

- The pigs are raised under two systems the Open Air and Indoor Systems. A combination of the two is followed in fish cum pig farming system.

- A single row of pig pens facing the pond is constructed on the pond embankment.

- An enclosed run is attached to the pen towards the pond so that the pigs get enough air, sunlight, exercise and dunging space.

- The feeding and drinking troughs are also built to keep the pens dry and clean.

- The gates are provided to the open run only. The floor of the run is cemented and connected via the drainage canal to the pond.

- A shutter is provided in the drainage canal to stop the flow of wastes to the pond. The drainage canal is provided with a diversion channel to a pit, where, the wastes are stored when the pond is filled with algal bloom.

- The stored wastes are applied according to necessity. The height of the pig house should not exceed 1.5 m.

- The floor of the house must be cemented.

- The pig house can be constructed with locally available materials. It is advisable to provide 1 - 1.5 m² space for each pig.

Fig. 24: Pig-cum-fish culture

b. Selection of Pigs

- Four types of pigs are available in our country - wild pigs, domesticated pigs or indigenous pigs, exotic pigs and upgraded stock of exotic pigs.

- The Indian varieties are small sized with a slow growth rate and produce small litters. Its meat is of inferior quality.

- Two exotic upgraded stock of pigs such as large - White Yorkshire, Middle - White Yorkshire, Berkshire, Hampshire and Hand Race are most suitable for raising with fish culture. These are well known for their quick growth and prolific breeding.

- They attain slaughter maturity size of 60 - 70 Kg within six months. They give 6 - 12 piglets in every year.

- The age at first maturity ranges from 6 - 8 months. Thus, two crops of exotic and upgraded pigs of six months each are raised along with one crop of fish which are cultured for one year.

- 30 - 40 pigs are raised per hectare of water area. About two months old weaned piglets are brought to the pig-sties and fattened for 6 months and when they attain slaughter maturity are harvested.

c. Feeding

- The dietary requirements are similar to those of the ruminants.

- The pigs are not allowed to go out of the pig house where they are fed on balanced pig mash of 1.4 Kg / pig / day.

- Grasses and green cattle fodder are also provided as food to pigs.

- To minimize food spoilage and to facilitate proper feeding without scrambling and fighting, it is better to provide feeding troughs. Similar separate troughs are also provided for drinking water.

- The composition of pig mash is a mixture of 30 Kg rice bran, 15 Kg polished rice, 27 Kg wheat bran, 10 Kg broken rice, 10 Kg groundnut cake, 4 Kg fish meal, 3 Kg mineral mixture and 1 Kg common salt.

- To reduce quantity of ration and also to reduce the cost, spoiled vegetables, especially the rotten potatoes can be mixed with pig mash and fed to pigs after boiling.

- The pigs are hardy animals. They may suffer from diseases like swine fever, swine plague, swine pox and also infected with round worms, tapeworms, liver flukes, etc.

- Pig - sties should be washed daily and all the excreta drained and offal into the pond. The pigs are also washed.
- Disinfectants must be used every week while washing the pig - sites. Piglets and pigs should be vaccinated.

d. Harvesting of fish:

- Fish attain marketable size within a few months due to the availability of natural food in this integrated pond.
- According to the demand of fish in the local market, partial harvesting is done.
- After the partial harvest, same numbers of fingerlings are introduced into the pond. Final harvesting is done after 12 months of rearing.
- Fish yield ranging from 6, 000-7, 000 Kg / ha / yr is obtained.
- The pigs are sold out after rearing for six months when they attain slaughter maturity and get 4, 200 – 4, 500 Kg pig meat.

e. Fish-cum-Duck culture system

Integrated fish-cum-duck farming system is though a common practice in China, but in India, West Bengal, Assam, Tamilnadu, Andhra Pradesh, Kerala, Bihar in particular, such culture system is gradually developing.Ducks use both land and water as their habitat, the integration with the fish is mainly attributed to mutual benefits of a biological relationship. It is not only useful for fattening the ducks but also beneficial to fish farming by providing more organic manures to fish. It is apparent that fish cum duck integration could result in a good economic efficiency of fish farms (Fig. 25 and 26).

Benefits of fish cum duck farming

1. For an efficient utilization of pond surface waters, raising ducks is an excellent enterprise.
2. Fish ponds provide an excellent environment to ducks which prevent them from infection of parasites.
3. Ducks feed on aquatic predators and help the fingerlings to grow. A fish pond being semi-closed biological system with aquatic animals and plants provide disease free environment for ducks.
4. Ducks consume juvenile frogs, tadpoles, dragon fly making safe environment to fish.
5. Duck raising in fish ponds reduces the demand for protein to 2 – 3 % in duck feeds.

6. Duck droppings go directly into water providing essential nutrients to increase the biomass of natural food organisms i.e. plankton.

7. The daily waste of duck feed (about 20 - 30 gm/duck) serves as fish feed in ponds or as manure, resulting in higher fish yield.

8. Manuring is conducted by ducks and homogeneously distributed without any heaping of duck droppings.

9. By virtue of the digging action of ducks in search of benthos, the nutritional elements of soil get diffused in water and promote plankton production.

10. Ducks serve as bio aerators as they swim, play and chase in the pond. This disturbance to the surface of the pond facilitates aeration in pond water.

11. The feed efficiency and body weight of ducks increase and the spilt feeds could be utilized by fish.

12. Survival of ducks raised in fish ponds increases by 3.5 % due to   the clean environment of fish ponds.

13. Duck droppings and the left over feed of each duck can increase the output of fish to 37.5 Kg/ha.

14. Ducks keep aquatic plants in check.

15. No additional land is required for duckery activities.

16. It results in high production of fish, duck eggs and duck meat in unit time and water area.

17. It ensures high profit through less investment.

Stocking Density of fish

•   The pond is stocked after the pond water gets properly detoxified.

•   The stocking rates vary from 6000 fingerlings/ha and a species ratio of 40 % surface feeders, 20 % of column feeders, 30 % bottom feeders and 10-20 % weedy feeders are preferred for high fish yields or, as per the market demands of the Indian Major and Minor carps.

•   Mixed culture of only Indian major carps can be taken up with a species ratio of 40 % surface, 30 % column and 30 % bottom feeders or as per the market demand.

•   In the northern and north - western states of India, the ponds should be stocked in the month of March and harvested in the month of October - November, due to severe winter, which affect the growth of fishes.

- In the south, coastal and north - eastern states of India, where the winter season is mild, the ponds should be stocked in June - September months and harvested after rearing the fish for 12 months.

Fig. 25 and 26: Duck-cum-Fish culture

Use of duck dropping as manure

- The ducks are given a free range over the pond surface from 9a.m to 5 p.m, when they distribute their droppings in the whole pond, automatically manuring the pond.

- The droppings voided at night are collected from the duck house and applied to the pond every morning.

- Each duck voids between 125 - 150 gm of dropping per day.

- The stocking density of 200-300 ducks/ha gives 10, 000 - 15, 000 kg of droppings and is recycled in one hectare ponds every year.

- The droppings contain 81% moisture, 0.91% nitrogen and 0.38% phosphorus on dry matter basis.

Duck husbandry practices

The following three types of farming practice are adopted.

1. Raising large group of ducks in open water

- This is the grazing type of duck raising.

- The average number of a group of ducks in the grazing method is about 1000 ducks.

- The ducks are allowed to graze in large bodies of water like lakes and reservoirs during the day time, but are kept in pens at night.

- This method is advantageous in large water bodies for promoting fish production.

2. **Raising ducks in centralized enclosures near the fish pond**

   - A centralized duck shed is constructed in the vicinity of fish ponds with a cemented area of dry and wet runs out side.

   - The average stocking density of duck is about 4 - 6 ducks/sq.m. area.

   - The dry and wet runs are cleaned once a day. After cleaning the duck shed, the waste water is allowed to enter in to the pond.

3. **Raising ducks in fish pond**

   - This is the common method of practice.

   - The embankments of the ponds are partly fenced with net to form a wet run.

   - The fenced net is installed 40-50 cm above and below the water surface, so as to enable the fish to enter into the wet run while ducks cannot escape under the net.

4. **Selection of ducks and stocking**

   - The kind of duck to be raised must be chosen with care since all the domesticated races are not productive.

   - The important breeds of Indian ducks are Sylhet Mete and Nageswari.

   - The improved breed, Indian runner, being hardy has been found to be most suitable for this purpose, although they are not as good layers as exotic Khaki Campbell.

   - The number of ducks required for proper manuring of one hectare fish pond is also a matter of consideration.

   - It has been found that 200 – 300 ducks are sufficient to produce manure adequate enough to fertilize a hectare of water area under fish culture.

   - 2 - 4 months old ducklings are kept on the pond after providing them necessary prophylactic medicines as a safeguard against epidemics.

5. **Feeding**

   - Ducks in the open water are able to find natural food from the pond but that is not sufficient for their proper growth.

   - A mixture of any standard balanced poultry feed and rice bran in the ratio of 1:2 by weight can be fed to the ducks as supplementary feed at the rate of 100 gm/ bird/day.

   - The feed is given twice in a day, first in the morning and second in the evening.

- The feed is given either on the pond embankment or in the duck house and the spilled feed is then drained into the pond.

- Water must be provided in the containers deep enough for the ducks to submerge their bills, along with feed.

- The ducks are not able to eat without water. Ducks are quite susceptible to aflatoxin contamination; therefore, moulded feeds kept for a long time should be avoided.

- The ground nut oil cake and maize are more susceptible to *Aspergilus flavus* which causes aflatoxin contamination and may be eliminated from the feed.

6. **Egg laying**

- The ducks start laying the eggs after attaining the age of 24 weeks and continue to lay eggs for two years.

- The ducks lay eggs only at night. It is always better to keep some straw or hay in the corners of the duck house for egg laying.

- The eggs are collected every morning after the ducks are let out of the duck house.

7. **Health care**

- Ducks are subjected to relatively few diseases when compared to poultry.

- The local variety of ducks is more resistant to diseases than other varieties.

- Proper sanitation and health care are as important for ducks as for poultry.

- The transmissible diseases of ducks are duck virus, hepatitis, duck cholera, keel disease, etc.

- Ducks should be vaccinated for diseases like duck plague. Sick birds can be isolated by listening to the sounds of the birds and by observing any reduction in the daily feed consumption, watery discharges from the eyes and nostrils, sneezing and coughing.

- The sick birds should be immediately isolated, not allowed to go to the pond and treated with medicines.

8. **Harvesting**

- Keeping in view the demand of the fish in the local market, partial harvesting of the table size fish is done.

- After harvesting partially, the pond should be restocked with the same species and the same number of fingerlings.

- Final harvesting is done after 12 months of rearing.

- Fish yield ranging from 3500 - 4000 Kg/ha/yr and 2000 - 3000 Kg/ha/yr are generally obtained with 6 - species and 3 - species stocking respectively.

- The eggs are collected every morning. After two years, ducks can be sold out for flesh in the market. About 18, 000 - 18, 500 eggs and 500 - 600 Kg duck meat are obtained.

F.   Integrated Fish-cum-Goat culture

Commercial goat farming in India is becoming very popular day by day. As goat farming is a proven highly profitable business idea so, the popularity of this business is increasing rapidly in India. It is also one of the finest and established livestock management departments in the country (Fig.27 and 27a).Huge market demand and proper spread ensures fast profitability and sustainability of this business for long term. Goat is considered as poor man's cow and a goat's excreta is considered as a very good organic fertilizer.

Fig. 27 and 27a: Goats at the pond site

The goat excreta contains organic carbon-60%, N-2.7%, P-1.78%, K-2.88% and its urine is also equally rich in both N & P. 50-60 goats are essential to fertilize 1 ha pond. The goats should be provided with dry, safe, comfortable house protected from excessive heat The goat breeds are Jamanapari, Beetal, Barbari for milk and Bengal, Sirihi, Deccani are used for meat purpose. Goats are selective feeders and consume Berseem, Napier grass, Cowpea Soybean, Mulberry etc., This integration can provide 3500-4000 kg fish/ha/year without supplementary feeding and fertilizers.

G. Integration of fish and poultry (Chickens)

Fish culture practice followed in integration of poultry-cum-fish farming is the "Composite fish culture system" which perhaps attracted maximum attention to the fish farmers at least in India. Integration of poultry and fish farming together needs more than one species of fish in the same pond.Integration of chicken with fish farming might be an economically viable and productive system for both rural farmers and commercial entrepreneurs.

Supplemental feed and fertilizers – the high cost inputs in fish farming are not needed in such systems and the cost of inputs is therefore reduced. Poultry manure is a complete fertilizer, with the characteristics of both organic as well as inorganic fertilizers.

Considerations for selection of fish species

- The selected species should be compatible with each other
- The species and their combination ratio should be adjusted according to the amount of feed stuff and manure that are expected to be made available by the other sub-system.
- As far as possible the species should fast growing.
- Selected fish should be hardy and resistant to common diseases and parasites.
- The species should be able to tolerate low oxygen levels and high organic content in the water.

Stocking density of fishes

Stock 600-1000 fingerlings of Indian carps, Catla (*Catla catla*), Rohu (*Labeo rohita*), Mrigal (*Cirrhinus mrigala*) and Chinese carps, Silver carp (*Hypophthalmichthys molitrix*), Grass carp (*Ctenopharyngodon idella*) and Common carp (*Cyprinus carpio*). Species stocking rate could be 40 percent surface feeders (catla and silver carp), 20 percent rohu, 30 percent bottom feeders (mrigal and common carp) and 10 percent grass carp.

Pond preparation for stocking of fish

Side slope should be 1.5:1. Embankment should be 1ft more than the high flood level of the selected site. In case of clayey soil, pond dyke's external side should be 1:1.5 & internal side slope should be 1:2. Construction of inlet and outlet should be done, fencing is also required. Fill the pond with fresh water to a depth of 2-2.5m.

Liming

Depending upon the soil and water pH, requirement of quicklime in kg/ha (mentioned in water and soil chemistry of water and soil in fish ponds chapter). ($1/3^{rd}$ of the required liming material is applied initially to the pond and rest amount is divided into 11 installments and applied monthly).

Fertilization

Done after 7-10 days of liming. Fertilizers used are both organic and inorganic.Cow dung @ 5300-5500 kg/ha/year is applied as organic fertilizer.

Inorganic fertilizers used @Urea @ 100-110 kg/ha/year, Single super phosphate @ 22-26 kg/ha/year, Murate of potash @ 95-100 kg/ha/year are applied after 7 days of application of organic fertilizer.

$1/3^{rd}$ of the required amount is applied initially and rest amount is divided equally in 11 installments and applied monthly i.e. same as liming materials.

Harvesting

Indian major carps (IMC) and exotic Carps (Grass carp) attain 1 kg weight in 7-8 months.Harvesting may be done by removing only the table size fish or the complete stock.

Poultry husbandry practice

Deep litter system: In the deep litter system the poultry houses are constructed on the pond embankment or any convenient places adjacent to the fish farm. In this system the floor of the pen is covered with 10- 15 cm thick easily available dry organic matter like- dry leaves of tree, chopped straw, hay, saw dust, lime, etc. The dropping of the birds which fall gradually combined with the materials used and bacterial action started. When the depth of litter becomes less, more organic matter is added to maintain the sufficient depth of litter. The litter is regularly stirred and after 2 months it becomes dip litter and in about 10-12 months it becomes fully built up litter. If the litter turns damp, application of lime may be made to make it dry. Height of the house from floor to roof should be 3.6m and height of the walls should be 2.72m.

For 1 ha pond i.e. 7000-8000 fishes, 500-600 birds are required. In 1 year, 25-30 birds can produce 1 ton deep litter. Daily 50 kg / ha water spread area poultry dung is applied to fish pond. Space required for 1 bird is 0.3 to 0.4 m^2.

So, one (1) house can accommodate 250 birds. For 1 ha pond, two (2) poultry houses with housing capacity of 250 birds may be prepared.

Selection of poultry birds: In the integration of fish and poultry farming both egg type and meat type birds are farmed. In the cage and dip litter system both egg type and meat type are grown. But in storied house (changhar) egg type birds are farmed. In the dip litter and cage system any of the high yield (egg and meat) variety of poultry birds like- Rhode island, Leg horn, etc. are suitable, but in storied house Kisten golden breed is only preferred.

Housing of birds: Just one month prior to stocking of fish pond with fish seed after vaccination against viral diseases and after providing all prophylactic measures about 8 week aged poultry birds are brought to the house for farming. Before introduction of birds into the house, the house and the utensils to be used in the poultry raising practice should be disinfected with disinfectant like- potash. When the poultry birds becomes 18 months aged then their egg laying capacity will be reduced and the old stock should to be sale out and a new stock should be introduced to the house after cleaning the house with disinfectant.

Feeding of birds: Under litter system the poultry birds are fed according to their age. The normal feeding practices are followed.

1. **Starter mash**– 40- 45 gm/ day in 3- 4 times a day up to the age of 8 week of birds.

2. **Grower mash**– 50- 70 gm/ day in 5- 6 times a day from 8- 18 weeks age of birds.

3. **Layers mash**– 80- 120 gm/ day in 3- 4 times a day from 18 weeks age of birds.

Along with feed there should be sufficient supply of drinking water to the farmed poultry birds is required. For dust bath of birds an earthen pot of about 2 feet diameter filled with clean and dry earth are placed in the house. Roosting starts from 8 weeks age of the chicken and so perches are provided in the pen for roosting of birds at the rate of 8 inch/ perch/ bird.

Egg laying management: After the birds are 22 weeks of age, egg nests are kept in the house. Every 5-6 birds require 1 nest.

Production: In this poultry-cum- fish farming from a hectare of water spread area of the pond, 3500- 4000 kg fish, 650 kg chicken meat and 120000 nos. of chicken eggs can be produced per year.

Problems encountered in Poultry-Fish integrated farming: RCC work, work shop, repair and spare parts, cost of construction materials, like- brick, rubble, steel, cement, etc. need to be surveyed. The major item of construction expenses goes for earth moving and RCC work. Earth moving can be done either manually or mechanically depending on the cost of labour.

Marketing facilities

The farmed product can be sold either to internal market or to export market. In both cases the taste of consumer and the provision for supplying the product to the consumer should be observed.

Growth of unwanted aquatic weed in the pond

Unwanted aquatic weeds are needed to be removed from fish pond as it reduces the pond productivity. These unwanted aquatic weeds could be removed- manually, mechanically, chemically and biologically. If possible manual removal method is better. Grass carp, java puti, tilapia, etc. are good biological agent in removing aquatic weed from fish pond. Chemicals such as 2-4 D, 4 D Ester, Simazine. Paraquat, Urea etc. can also be used.

Weed fishes, insects, unwanted organisms etc. in the pond

These need to be removed. Removal can be done by repeated netting or by using chemicals. Soap-oil emulsion (soap: oil = 1:3) over the pond water surface is most commonly used technique to kill the insects in fishponds. Mohua oil cake@ 200- 250 ppm or tea seed cake@ 750 –975 Kg/ha give encouraging result in controlling weed fishes of a pond and it also helps in reducing the quantity of organic fertilizer required by 50%. Commercially available bleaching powder@ 97- 113 Kg/ha can also be used as fish toxicant.

Algal bloom

Sometime a thick layer of algal bloom of brown or green colour is seen over the water surface of pond. This can be removed from fish pond by using a piece of split bamboo followed by liming based on water P^H as mentioned earlier. Chemicals like, copper sulphate @ 0.1- 0.5 mg/lit.of water or Diuron@ 0.3- 0.5 mg/lit.of water also helps in controlling this bloom.

Partial replenishment of water

Since total dewatering is not possible, so depending on the feasibility some percentage of pond water may be pump out and the same is refilled with new water. But the water depth should be restricted to 2- 2.5 m for good production of fish.

Ammonia in pond water

Ammonia occurs in pond water in 2 forms i.e. ionized and unionized. Unionized ammonia is toxic to fish. Ammonium (NH_4) is lethal to fishes only at a level above 16 ppm while ammonia (NH_3) is lethal at a level above 0.02 ppm. If the unionized ammonia concentration is increased then the fish may die. Water temperature, pH, concentration of phytoplankton in the pond is important and renovated measures have to be taken immediately.

Phytoplankton bloom

The sudden increase of population of certain planktonic algal group as thick mass in water is called phytoplankton bloom. *This may be identified by the deep green or blue green or reddish green colour of the pond water.* During the day time phytoplankton produces excess oxygen and during night and cloudy days they absorb dissolved oxygen from water for their self-respiration resulting dissolved oxygen depletion and fish mortality. The death and decay of algae also cause dissolved oxygen depletion. The reason for this algal bloom in pond water is the presence of excess nutrients in water. Therefore, if this problem encountered in the fish culture pond then supply of poultry manure to the pond should immediately be cut off and the remedial measures should be taken as mentioned in the case of algal bloom cited in the renovation measures of a pond in case the same cannot be dried.

Changes in water temperature

It leads to loss of appetite of cultured fish. Fish may exhibit poor growth and they become susceptible to diseases. If the water temperature changes to a markable level then supplying feed and fertilizer to the pond should immediately be stopped. Replenishment of water from a nearby source, harvesting the table size fish, etc. are some of the corrective measures to be taken for it.

Note

Apart from the above cited water quality parameters the other water quality parameters to be noted are- total alkalinity, turbidity, micronutrients, chemical pollutants, insecticide, organic matter, presence of aquatic vegetation, etc. are to be checked regularly for good production from a fish pond.

Fish health management: Cultured fish should be checked regularly for their health. If any deviation in their normal behaviour is seen then they should be treated with the advice of an expert.

Health management in poultry

Poultry birds suffer from viral, bacterial, parasitic, fungal and nutritional diseases. Keeping the poultry house clean and dry, vaccinating the farmed stock of the poultry birds against all the viral diseases can help in maintaining a healthy stock of chicken. Before bringing the poultry birds into the house, the house and the utensils to be used in the farming practices should be disinfected. The farm chicken should be fed with balanced feed. The poultry birds should be dewormed at least once in a month. The veterinary expert should be consulted for any type of poultry disease and accordingly steps may be taken.

Fig. 28: Poultry-cum-fish integration

Advantages of poultry –cum- fish farming

1. Procurement of fish, poultry meat as well as chicken eggs from same farming system.

2. Water from fish pond can be used for poultry husbandry practice.

3. The transportation cost of the manure is not involved

4. The nutritive value of applied fresh manure is much higher than dry and mixed with bedding materials e.g. saw dust or rice husk.

5. Some parts of the manure is consumed directly by the fish.

6. No supplementary feed is needed for the fish.

7. No extra space is required for chicken farming. Chicken sheds can be constructed over the pond water or on the dyke.

8. More production of animal protein will be ensured from the same area of land.

9. The overall farm production and income will increase.

In integrated fish-layer farming systems, direct deposit of fresh chicken manure to fish ponds can produce enough natural fish feed organisms, thus maximizing profit and reduces production and feed cost. Higher production of animal protein can be achieved from the same area of minimum land with this system. Poultry manure is a complete fertilizer. The most valuable poultry production systems for fish production are those systems which produce nutrient-rich and collectable waste. Layers produce more calcium and phosphorus-rich excreta than broilers. The direct use of egg laying where the birds are of constant weight and produce fairly constant levels of waste, are easier to manage than broilers in which waste availability is cyclical.

Benefits of fish cum chicken integration

Following are some of the additional advantages when fish culture is integrated with raising of chicken on/or near the pond dykes:

* *The direct discharge of fresh chicken manure to the fish ponds produces enough natural fish feed organisms without the use of any additional manure/fertilizer. The transportation cost of the manure is not involved.*

● The nutritive value of applied fresh manure is much higher than dry and mixed with bedding materials e.g. saw dust or rice husk.

● Some parts of the manure is consumed directly by the fish.

● No supplementary feed is needed for the fish.

● No extra space is required for chicken farming. Chicken sheds can be constructed over the pond water or on the dyke.

● More production of animal protein will be ensured from the same area of minimum land.

● The overall farm production and income will increase.

● Nutrient contents (%) of various live stock fecal matters generally used during integrated fish farming systems.

H. Integrated Fish–cum-Rabbit farming

Rabbit meat is preferred by most of the health conscious consumers owing to its low fat in comparison to other meats. The important meat breeds are Soviet Chinchilla, Grey Giant, and White Giant etc. Rabbits are reared in cage, hutch and floor system (floor should be cemented). Rabbit excreta contain organic carbon-50%, N-2%, P-1.33%, and K-1.2%. Its excreta are high in nitrogen content and low in moisture, thus quality manure for sustained plankton production. It is estimated that excreta from 300 rabbits would be enough for 1 ha pond fertilization.

Miscellaneous Advantages

As far as fish production is concerned, the integrated farming serves the multiple purpose of providing cheap feedstuffs and organic manure for the fish ponds, thereby reducing the cost and need for providing compounded fish feeds and chemical fertilizers. By reducing the cost of fertilizers and feedstuffs the overall cost of fish production is reduced and profits increased. It is evident that the profit from fish culture is often increased 30-40 percent as a result of integration. The overall income is increased by adding pig and/or poultry raising, grain and vegetable farming, etc., which supplement the income from fish farming.

By producing grain, vegetables, fish and livestock products, the community becomes self-sufficient in regard to food and this contributes to a high degree of self-reliance. The silt from the ponds which is used to fertilize crops increases the yield of crops at a lower cost and the need to buy chemical fertilizer is greatly reduced. It is estimated that about one third of all the fertilizer required for farming in the country comes from fish ponds.

Estimated nutrient content (%) of excreta of various animals considered for integrateted fishculture systems:

Table 5: Nutritional status of livestock fecal matters.

Sl.nos.	Source	Organic content	Nitrogen	Phosphorus	Potash	Protein
01.	Cattle	30	0.7	0.3	0.65	4.38
02.	Pig	30	1.0	0.75	0.85	6.25
03.	Poultry	50	1.6	1.26	0.90	10.0
04.	Goat	60	2.7	1.78	2.68	17.31
05.	Rabbit	50	2.0	1.33	1.20	12.50

12. COMPOSITE FISH CULTURE

A. What is Composite Fish Culture?

In order to obtain high production per hectare of water body, fast growing compatible species of fish of different feeding habits are stocked together in the same pond so that all its ecological niches are occupied by fishes. This system of pond management is called mixed fish farming or **composite fish culture**. The name is synonymous with **Polyculture**.

Absolute utilization of the pond's productivity for obtaining high production per hectare of water body is achieved through intensive culture of fast growing compatible species of fish of different feeding habits are stocked together in the same pond, so that, all its ecological niches are occupied by fishes and full utilization of the pond nutrients/resources is achieved. When compatible fishes of different feeding habits are stocked together, they secure for themselves in the most efficient manner, all life requisites available in the pond for fish production without harming each other. This technique of fish culture is called Composite fish culture or Polyculture or Mixed farming. The main objective of this Composite fish culture is to select and grow compatible fish species of different feeding habits, in order to exploit all types of food available in different region of the pond for maximizing fish production.

B. Why Composite Fish Culture?

A single species of carp culture in a pond is known as **mono-culture** and previously this mono-culture was the normal practice in many of the countries. Hence there was a small production of 300-600 kg of fishes per hectare of water in a pond. Subsequently there is a great change in the process of carp culture. Instead of single species, more than one species of major carps have been selected for rearing and culture in the same pond. They are living in the different water levels as well as they have different feeding habits. This process of rearing of fishes is known as composite fish culture. Recently three exotic carps such as silver carp, grass carp and common carp along with the three Indian major carps are selected for composite carp culture in the Asian countries. Thus the composite fish culture or composite mixed culture of six species is now introduced in the same pond. They are to be reared scientifically so that there will be very good production of fishes. To achieve this, appropriate amount of supplementary food as well as manure are to be supplied timely.

C.Indigenous and Exotic Carps for Composite Fish Culture

Fish husbandry with its great potential gained importance in recent years as a source of protein for the under nourished and expanding human population.The Indian Major Carps *Catla catla* (Catla), *Labeo rohita (Rohu)* and *cirrihinus mrigala* (Mrigal) are the most important food fishes in the sub-continent (Bhowmik, 1990).

The economically important as well as cultivable herbivorous scaly fishes (having scales on body but not on head) without any teeth which are bony with air bladder as well as under the order cypriniformes are known as carps. Carps are of two types, such as indigenous and exotic carps.

Indigenous carps

Carps that are normally available in the inland waters of the Indian region are known as indigenous fish. Indigenous carps are of two kinds, such as major and minor carps. The carps that are big-sized cultivable fishes with rapid growth and high demand in the market are known as major carps. The carps that are comparatively small-sized with slow growth are known as minor carps, such as Bata (*Labeo bata*), Punti (*Barbus ticto*) etc.

Exotic carps

Carps which have been imported from other countries and acclimatized to native place climate are known as exotic fish. The examples are: Silver carp (*Hypophthalmichthys molitrix*) grass carp (*Ctenopharyngodon idella*) and Cyprinus carp (*Cyprinus carpio*).

D. Advantages of composite fish culture

Selected species of indigenous and exotic carps are all compatible; they have different feeding habits and habitats in the same pond; so they are not competitors to each other for food and space. The foods of the different layers in pond-water are fully utilized by the carps in the composite fish culture or mixed culture. Induced breeding can also be done in the exotic carps and they are fast growing and the total production of fishes is very high and cost of production is less; this is also a profitable one.

E. Fish species involved in composite fish culture

Depending on the compatibility and type of feeding habits of the fishes, the following types of fishes of Indian as well as Exotic varieties have been identified and recommended for culture in the composite fish culture technology (Six species culture) (Fig.29):

Table 6: Feeding habit /feeding zone of Indian Major and Exotic Carps.

Sl.no.	Type of fish	Feed types	Food niche
Indian Major Carps			
01.	*Catla catla*	Zooplankton	Surface feeder
02.	*Labeo rohita*	Omnivorous	Column feeder
03.	*Cirrhinus mrigala*	Detritivorous	Bottom feeder
Exotic carps			
04.	*Hypophthalmichthys molitrix*	Phytoplankton	Surface feeder
05.	*Ctenopharyngodon idella*	Herbivorous	Surface, column and marginal area
06.	*Cyprinus carpio*	Detritus/Omnivorous	Bottom feeder

Polyculture of fish is the latest and improved technology of fish farming. There are many advantages and disadvantages of intensive polyculture of fish which are described below:

Advantages

- Maximum fish production is possible in intensive polyculture.
- More profit from fish farming is possible.
- Creates many employment opportunities.
- Intensive polyculture of fish is fully controlled by the farmer.
- More fish can be cultivated and produced from short place.

Disadvantages

- Intensive polyculture of fish is very expensive and risky.
- In this system the probability of occurring diseases is most.
- This farming system gets obstructed due to lack of better facilitated artificial farm.
- It is not possible to make the fish big sized in this system.
- Intensive polyculture needs highly experienced employees.

13. PROXIMATE COMPOSITION OF FISH FLESH

Fish have been a key source of food for humans (Ayoola, 2010) and fish protein occupies an important position in human nutrition (Nargis, 2006). A portion of 150 g of fish can provide about 50-60% of an adult's daily protein requirement (FAO, 2014). Fish are also a good source of all the essential amino acids, fatty acids, vitamins and minerals and the consumption of fish and fish products helps in preventing cardiovascular and other diseases (Cahu *et al.,* 2004). Flesh texture, protein and fat composition are usually the main factors that determine consumer acceptance (Pal and Ghosh, 2013).

Human populations include fish as a main part of their daily diet, a fact that has become more relevant in developing countries, whose dietary pattern reveals a large dependency on staple foods, and fish is the main source, accounting in 2010, about 19.6% of animal protein consumption (FAO, 2015).

Vila Nova, et.al (2005) pointed out that fish is of high nutritionalvalue due to its high protein content, excellent quality of lipids (Omega-3 and 6 series) and low levels of total fat, saturated fats and cholesterol. It also incorporates high levels of polyunsaturated fatty acids, important for the promotion and maintenance of health and minerals like calcium, phosphorus, sodium, potassium and magnesium (Ismail *et al.,* 2004, Porto *et al.,* 2016).

According to the World Health Organization (WHO), the fish provides the healthiest animal protein and the Indians (According to the latest data, Indians on an average consume just 269 grams of fish per month in rural areas while in urban areas it's 238 gram (Approx.6.1kg/year). Noticeably, just about 282 of 1000 households in rural areas consume fish, while the number is 209 households for urban areas. This apparently indicates a vast majority of the Indians prefer consuming vegetarian dishes).The fish consumption again is far below the recommended minimum consumption of fish, which is 12 kg inhabitant / year or approximately 250 g of fish per week. Admittedly, the fish stands out for having in its composition high quality protein, retinol, vitamin D and E, iodine and selenium. Evidence increasingly associates their consumption to greater brain development and learning in children, also improving eye health, and protection against cardiovascular disease and some cancers. The fats and fatty acids of fish are highly beneficial and difficult to obtain from other food sources (FAO, 2013).

In this context a proper understanding about the biochemical constituents of fish has become a primary requirement for the nutritionists and dieticians. Fish and fishery products are used in animal feeds. In this case also, proper data on

Catla-the *Catla catla*

Silver carp-the *Hypophthalmichthys molitrix*

Rohu-the *Labeo rohita*

Grass carp-the *Ctenopharyngodon idella*

Mrigal-the *Cirrhinus mrigala*

Common carp-the *Cyprinus carpio*

Fig. 29: The most common fishes involved in composite fish culture in India

the biochemical composition is essential for formulating such products. Another vital area where accurate information on biochemical composition is a must is processing and preservation of fish and fishery products. Fish is an easily perishable commodity and deterioration in quality is due to the changes taking place to the various constituents like proteins, lipids etc.

Information on the biochemical constituents will help a processing technologist to define the optimum processing and storage conditions, so that the quality is preserved to the maximum extent.

14. CHEMICAL COMPOSITION OF FISH

Fish is the third largest commodity consumed globally after rice and vegetables (Hels *et al.*, 2002). Fish provides protein and other essential nutrients for the maintenance of a healthy body (Andrew 2001). Proper knowledge on the biochemical composition of fish finds application in several areas. Today there is an ever-increasing awareness about healthy food and fish is finding more acceptances because of its special nutritional qualities. In general, fish is considered to be an excellent source of protein and minerals.

The four major constituents in the edible portion of fish are water, protein, lipid (fat or oil) and ash (minerals). The analysis of these four basic constituents of fish muscle is often referred to as 'proximate analyses. Even though data on proximate composition are critical, reliable data on proximate composition of most of the species of fish are difficult to obtain. But this is not the only or basic reason for the absence of accurate and reliable data on biochemical composition of fish. The contents of protein, lipids and total ash were similar in all the major carp irrespective of weight groups. Calcium content was higher in Rohu and Mrigal in all the weight groups. Sodium, potassium and iron contents were more in >2000 gm group of Rohu when compared to others. The Zn content of Rohu of 51-500 g group was higher *vis-à-vis* other groups. Mrigal of size groups 1-50 g and >2000 g contained the highest content of vitamin A. The vitamin D level was higher in 1-50 g of Rohu and 51-2000 g groups of Mrigal Paul *et al.*, (2016).

Fishes are a very heterogeneous and highly specialized group evolved through biochemical adaptation and evolution, consisting approximately of 24, 000 species, showing extreme variations in size, shape, appearance etc. The habitat and food intake of these species are equally diverse. Some species are exclusively marine while some are confined to freshwater habitats. Some survive in marine as well as freshwater environments (Wide salinity

Table 7: Some freshwater edible fishes of Eastern and North-East India.

Sl.nos.	Local names	Scientific names
01.	Kalbasu/kalbaus	*Labeo calbasu* (Hamilton)
02.	Mahasol/Mahseer	*Tor putitora* (Hamilton)
03.	Bata	*Labeo bata* (Hamilton)
04.	Mourala, Mola, Moya	*Amblypharyngodon mola* (Hamilton)
05.	Sarpunti	*Puntius sarana* (Hamilton)
06.	Chela punti	*Puntius chola* (Hamilton)
07.	Vanti punti	*Puntius stigma* (Cuv. & val.)
08.	Moina punti	*Puntius conchonius* (Hamilton)
09.	Tit punti	*Puntius ticto* (Hamilton)
10.	Punti	*Puntius phutunio* (Hamilton)
11.	Jat punti	*Puntius sophore* (Hamilton)
12.	Khudi punti	*Puntius gelius* (Hamilton)
13.	Gili punti	*Puntius terio* (Hamilton)
14.	Kharsa/ Angrot	*Labeo angra* (Hamilton)
15.	Bhangon bata	*Labeo boga* (Hamilton)
16.	Kurchi	*Labeo gonius* (Hamilton)
17.	Raikhore	*Cirrihinus reba* (Hamilton)
18.	Kursa, Kurusia	*Labeo dero* (Hamilton)
19.	Beitka, Langu	*Labeo pangusia* (Hamilton)
20.	Kalobata	*Crossocheilus latius* (Hamilton)
21.	Boroli	*Barilius barila* (Hamilton)
22.	Folui	*Notopterus notopterus* (Pallas)
23.	Chital	*Notoptetus chitala* (Hamilton)
24.	Bhetki	*Lates calcarifer* (Bloch)
25.	Sal	*Channa marulius* (Hamilton)
26.	Shole	*Channa striatus* (Bloch)
27.	Lyata	*Channa punctatus* (Bloch)
28.	Chang	*Channa orientalis* (Bloch)
29.	Magur	*Clarias batrachus* (Linnaeus)
30.	Singhi	*Heteropneustes fossilis* (Bloch)
31.	Koi	*Anabas testudineus* (Bloch)
32.	Boal	*Wallogo attu* (Bloch & Sneider)
33.	Bele	*Apocryptes bato* (Hamilton)

[Table Contd.

Contd. Table]

Sl.nos.	Local names	Scientific names
34.	Chuno	*Trichogaster chuna* (Hamilton)
35.	Pabda	*Ompok pabo*(Hamilton)
36.	Pangas (Basa)	*Pangassius pangassius* (Hamilton)
37.	Tilapia	*Oriochromis mossambica* (Linnaeus) *O.nilotica* (Peters)
38.	Stripped Tyangra	*Mystus vittatus* (Bloch)
39.	Pacu	*Piaractus brachypomus* (Cuvier)
40.	Pengba	*Osteobrama belangeri* (Valenciennes) (Delicacy of Manipur)

tolerant species). Some marine species migrate to fresh water for spawning whereas many freshwater species enter the sea for spawning (Anadromous and catadromous fish). These widely different environmental conditions of temperature, salinity, pressure, availability of food etc. have profound influence on the biochemical composition.

There may be group-specific or even species-specific differences in the biochemical composition. Even within a species, variations occur for individual fish or lots of fish taken at different times or under different conditions. Another type of variation in proximate composition occurs between different parts of the same fish. There is generally an increase in the oil content of the muscle from the tail portion towards the head. Similarly the light and red muscle will vary in the biochemical composition. It is against this background that we have to view the data on the biochemical composition of fish. Data available in literature for proximate composition of individual species will only indicate the range or average and these are not usually taken as absolute values. (Courtesy: Vikaspedia).

The percentage composition of the four major constituents of fish viz. water, protein, lipid and ash (minerals) is referred to as proximate composition (it may be noted that the term does not indicate any degree of inaccuracy in the analysis). These four components account for about 96-98% of total tissue constituents in most cases. The range of values for these constituents in the edible portion of common fish species from Indian waters are given below (Table 10):

Carbohydrates, vitamins, nucleotides, other non-protein nitrogenous compounds etc. are also present in small quantities. Though quantitatively minor components, play vital roles in maintaining the system and thus are essential for growth and development of the organisms.

Table 8: Proximate composition of fish tissue of some Indian Fresh water fishes.

Sl.no	Species	Moisture%	Fat%	Protein%	Ash%	Calcium mg/100m	Phosphorus mg/100gm	Iron mg/100gm
1.	Mrigal	75.0	0.8	19.5	1.5	350	280	109
2.	Hilsa	53.7	19.4	21.8	2.2	180	280	213
3.	Singhi	68.0	0.6	22.8	1.7	670	650	226
4	Bele	79.7	0.6	14.5	2.3	370	330	104
5	Lata	74.0	0.6	19.4	2.6	610	530	130
6	Koi	70.0	8.8	14.8	2.0	410	390	135
7	Folui	73.0	1.0	19.8	2.5	590	450	169
8	Rohu	76.7	1.4	16.6	0.9	680	150	85
9	Catla	73.7	2.4	19.5	1.5	510	210	76
10	Boal	73.0	2.7	15.4	1.3	160	490	62
11	Kalbasu	81.0	1.0	14.7	1.3	320	380	83
12	Bhetki	82.0	1.1	13.7	1.2	530	400	102
13	Magur	78.5	1.0	15.0	1.3	210	290	74
14	Punti	70.2	9.5	16.5	1.53	220	120	0.54
15	Pangas	72.3	10.8	14.2	0.96	180	130	0.52
16	Bata	79.0	2.48	14.3	2.0	790	200	1.09
17	Chital	75.0	2.32	18.6	1.01	180	250	2.98
18	Mahseer	70.3	2.26	25.2	1.20	130	280	3.83
19	Tilapia*	75.3	5.7	20.3	2.3	47.3	276.5	0.2
20	Chela	77.5	4.3	14.6	2.1	590	340	1.96

Source: Wealth of India. Vol-10. 1962. *Tilapia—Jour. Nat. Sci. Sri Lanka, 1996, 24(1):21-26.

Table 9: Comparative analysis of micro-nutrient contents of small Indigenous with cultured fish species (modified afterDey, et al., 2017 and Thilstead et al., 1997).

Local name	Scientific name	Vit-A (µg/100gm)	Calcium (mg/100gm)	Iron (mg/100gm)	Fat (gm/100gm)	Zinc (mg/100gm)
Mourala	A.mola	1960±214	1071±41	7±4	4.4	
Dhela	O.cotiocotio	937	1260			
Darkina	E.dancricus	1457				
Chanda	Chanda sp.	341	1162			
Punti	Puntius sp.	37±16	1059±161		7.1	
Kachki	Corica soborna	93±8				
Mrigal	C.mrigala	<30	<10	2.5	2.9	2.5
Ilish	T.ilisha	69	126	3		
Silver carp	H.molitrix	<30	268	4.4	2.7	4.4
Rohu	L.rohita	<30	36			
Tilapia	T.nilotica	<30		5.0		

Fish have been a key source of food for humans (Ayoola, 2010) and fish protein occupies an important position in human nutrition (Nargis, 2006). A portion of 150 g of fish can provide about 50-60% of an adult's daily protein requirement (FAO, 2014). Fish are also a good source of all the essential amino acids, fatty acids, vitamins and minerals and the consumption of fish and fish products helps in preventing cardiovascular and other diseases (Cahu *et al.*, 2004). Flesh texture, protein and fat composition are usually the main factors that determine consumer acceptance (Pal and Ghosh, 2013). Pal, *et al.*, (2016) while working on the proximate analyses of Indian Major Carps, emphasized the importance of different minerals present in fish flesh (Table 10).

Table 10: Range of some important constituents in fish flesh.

Water	65-90%
Protein	10-22%
Fat	1-20%
Minerals	0.5-5%

*In most of the cases the range is found from 20-40mg/100gm.

** In most of the cases the approx. values are found to be 0.04mg/100gm.Ref: FAO Corporate document repository. Title: The composition of fish flesh.

The significance of various vitamins present in fish flesh (Vit. A, D, K, E and C) have been emphasized by various workers around the world (Liu, 2003 Ozyurt *et al.*, 2009, Hewson, 2011, Halver, 2002).

Table 11: Mineral composition in fish flesh.

Sl.no.	Minerals present	Approx. Value, mg/100gm	Range mg/100gm
01.	Sodium	72.0	10-134
02.	Potassium	278.0	19-502
03.	Calcium*	79.0	19-881
04.	Magnesium	38.0	4.5-452
05.	Phosphorus	190.0	68-550
06.	Sulfur	191.0	130-257
07.	Iron	1.55	1.0-5.6
08.	Silicon	4.00	—————
09.	Manganese**	0.82	0.0003-25.2

[Table Contd.

Contd. Table]

Sl.no.	Minerals present	Approx. Value, mg/100gm	Range mg/100gm
10.	Zinc	0.96	0.23-2.1
11.	Copper	0.20	0.001-3.7
12.	Arsenic	0.37	0.24-0.6
13.	Iodine	0.15	0.001-2.73

*In most of the cases the range is found from 20-40mg/100gm.

** In most of the cases the approx. values are found to be 0.04mg/100gm.Ref: FAO Corporate document repository. Title: The composition of fish flesh.

15. DIETARY VITAMIN REQUIREMENTS IN FISH

Vitamins are a heterogeneous group of organic compounds essential for the growth and maintenance of animal life. Approximately 15 vitamins have been isolated from biological materials; their essentiality depending on the animal species, the growth rate of the animal, feed composition, and the bacterial synthesizing capacity of the gastro-intestinal tract of the animal. In general, all animals display distinct morphological and physiological deficiency signs when individual vitamins are absent from the diet.

Vitamins may be classified into two broad groups; depending on their solubility; the water-soluble vitamins and the fat-soluble vitamins.Fish have specific requirements for quantitative and qualitative proteins, amino acids, fatty acids, vitamins and minerals, which are derived from their diet. Several chemical compounds, characterizing the quality of food, are also known to influence the survival and development of fish larvae.

Towers (2014) while emphasizing on the good nutrition in aquaculture systems is essential to economical production of healthy and high quality products postulated that Vitamins are essential nutrients found in foods. They perform specific and vital functions in a variety of body systems and these are crucial for maintaining optimal health. The majority of vitamins are not synthesized by the fish body or at a rate sufficient to meet the aquatic animal's needs.

Vitamins are distinct from the major food nutrients (proteins, lipids, and carbohydrates) in that they are not chemically related to one another. They are present in very small quantities within animal and plant food stuffs.They are required by the fish body in trace amounts.

Table12: Water and Fat soluble vitamins essential for growth of fishes.

Sl.Nos.	Water-soluble vitamins	Fat-soluble vitamins
01.	Thiamine (vitamin B1)	Retinol (vitamin A)
02.	Riboflavin (vitamin B2)	Cholecalciferol (vitamin D3)
03.	Pyridoxine (vitamin B6)	Tocopherol (vitamin E)
04.	Pantothenic acid (vitamin B5)	Phylloquinone (vitamin K)
05.	Biotin (vitamin BH)	
06.	Folic acid (Synthetic form of B9)	
07.	Cyanocobalamin (vitamin B12)	
08.	Inositol (vitamin B8)	
09.	Choline	
10.	Ascorbic acid (vitamin C)	
11.	Nicotinic acid (niacin)) (vitamin B3)	

In general, all fish display distinct morphological and physiological deficiency signs when individual vitamins are absent from the diet. The two different types of vitamins are evident, and these are: (i) Fat-soluble vitamins and (ii) Water-soluble vitamins.

Fat-soluble vitamins

Dissolve in fat before they are absorbed in the blood stream to carry out their functions. Excesses of these vitamins are stored in the liver, and are not needed every day in the diet. The fat-soluble vitamins are absorbed from the gastrointestinal tract in the presence of fat and can be stored within the fat reserves of the body when ever dietary intake exceeds metabolic demands.In this category following vitamins are grouped: Vitamin A1 (Retinol, retinal, retinoic acid), Vitamin A2 (Dehydroretinol), Vitamin D2 (Ergocalciferol), Vitamin D3 (Cholecalciferol), Vitamin E (Tocopherol, tocotrienols), Vitamin K1 (Phylloquinone), Vitamin K2 (Menaquinone), Vitamin K (Menadione)

Water-soluble vitamins

Dissolve in water and are not stored by the body. Since they are eliminated in urine and require a continuous daily supply in fish/shrimp diet. Water-soluble vitamins are easily destroyed or washed out during food storage or preparation.

Body stores being rapidly depleted in the absence of regular dietary water-soluble vitamin sources.In this category following vitamins are grouped: Thiamin (Vitamin B1), Riboflavin (Vitamin B2), Niacin (Vitamin B3) Vitamin B6 (Pyridoxol, pyridoxal, pyridoxamine), Pantothenic acid (Vitamin B5) Biotin (Vitamin H, vitamin B8), and Folic acid Vitamin B12 (Cobalamin), Choline (Gossypine), Vitamin C (Ascorbic acid).

Benefits of Multivitamins

- Aids normal bone, tooth and exoskeleton development
- Maintains the health of the skin and membranes
- Helps in energy metabolism and tissue formation
- Enhances calcium and phosphorus absorption and utilization
- Promotes immune stimulation
- Decreases the diseases incidence
- Acts as antioxidant
- Acts like 'Cement' for connective tissues
- Helps for wound healing
- Makes iron absorption
- Benefits for healthy eye & vision
- Involve in a large number of biological processes
- Creates collagen in the body
- Makes the skin, joints, exoskeleton and bones strong
- Regulates moult phase in shrimps

Biological Functions of Water Soluble Vitamins

Thiamine (B1): aids growth, digestion, fertility, nervous system, carbohydrate metabolism and oxidation of glucose

Riboflavin (B2): vision, enzyme functioning, energy metabolism, respiration of poorly vascularised tissues, metabolism of carbohydrates, fats and proteins.

Pyroxidine (B6): enzyme secretion, protein and carbohydrate metabolism.

Pantothenic Acid (B5): adrenal functioning, cholersterol production, normal physiology and metabolism,

Nicotinic Acid (niacin, B3): lipid, protein and amino acid metabolism

Biotin (H): enzyme secretion, purine and lipid synthesis, oxidation of lipids and carbohydrates,

Folic Acid (B9): blood cell formation, blood glucose regulation and various metabolism

Cyanocobalamin (B12): enzyme systems, cholesterol metabolism

Inositol: cell membrane permeability, structural component of skeletal, heart and brain tissue, growth of liver and bone marrow cells, liver lipid (cholesterol) transport, synthesis of RNA.

Choline: better growth and FCR, maintenance of cell structure and the transmission of nerve impulses, transport of lipid within the body.

Biological Functions of Fat-Soluble vitamins

Retinol (Vitamin A)

Highest biological activites, normal vision, form a visual pigments, maintenance of the mucous secreting epithelial tissues of the reproductive tract, skin, bone and gastro-intestinal tract, protecting mucous membranes and developing bone tissue, epithelial cell metabolism, release of proteolytic enzymes from lysosomes.

Cholecalciferol (Vitamin D)

Calcium and phosphorus metabolism, growth of bone tissue, synthesis of the calcium binding protein, the conversion of organic phosphorus to inorganic phosphorus in bone, maintenance of blood calcium level, and the deposition and oxidation of citrate in bone.

Tocopherol: act as lipid-soluble extracellular and intracellular antioxidants, protect the highly unsaturated fatty acids, important role in cellular respiration, biosynthesis of DNA and coenzyme Q.

Phylloquinone (Vitamin K)

Maintenance of normal blood coagulation, electron transport and oxidative phosphorylation in micro-organisms.

Vitamin C

Vitamin C probably is the most important because it is a powerful antioxidant and immune modulator for fishes/ shrimps. The fish and shrimp body needs vitamin C (ascorbic acid or ascorbate) to remain in proper health condition.

- Vitamin C makes various benefits to the fish/ shrimp body by holding cells together through collagen synthesis. Collagen is a connective tissue that holds muscles, bones, and other tissues together. Collagen is also needed for the healing of wounds.

- Vitamin C also requires in wound healing, bone and tooth formation, strengthening blood vessel walls, improving immune system function, increasing absorption and utilization of iron, and acting as an antioxidant.

- Vitamin C works with vitamin E as an antioxidant and plays a crucial role in neutralizing free radicals throughout the body.

- Vitamin C reduces the effects of toxic chemicals in water and prevent negative effects of water temperature fluctuations.

- Vitamin C increases intestinal absorption of iron from plant-based foods.

- Vitamin C enhances in immune response.

- Vitamin C helps synthesize carnitine, adrenaline, epinephrine, the neurotransmitter serotonin, the thyroid hormone thyroxine, bile acids, and steroid hormones.

- Vitamin C regulates the moult phase and helps in quick formation of exoskeleton in shrimps.

Major Functions of Vitamin C

Water temperature

Vitamin C diets helps to fishes during low temperature period for enhancing digestion efficiency and there after promote growth rate.

Stress

The ascorbic acid concentrations in tissues in fish change during stressful periods. Vitamin C prevents from various stress such as: environmental stress, handling stress, pathogenic stress, transportation stress, osmotic stress etc, of fishes and shrimps.

Toxic Gasses Eliminator

The ascorbic acid helps to eliminate the adverse effect of ammonia on growth of fishes and reduces the toxic effect of nitrite levels in intensive aquaculture systems.

Wound healing

Vitamin C makes collagen formation in body and helps in quick for optimal wound repair in fishes and shrimps.

Reproduction

Vitamin C helps in reproductive systems of fishes, prawns and shrimps such as growth of healthy brooders, proper egg and sperm formation, optimum fecundity, improve hatchability, healthy larval and fry conditions etc.

Larval Nutrition

Diet with vitamin C acts fast development of embryo /larvae and early fry stages with reduction of mortality in early stages of fishes /shrimps.

Immunity

Vitamin C is at the top of the list among natural immune boosters for fish /shrimp body. Vitamin C has the immune-stimulating properties and compensates the immune-depression. It offers protection to fish / shrimp immune system by encouraging the activity of immune system cells including neutrophils and phagocytes. The collagen prevents various diseases caused by bacteria and viruses. Mucus helps in preventing microorganisms from entering to the body through the skin, gills and gastrointestinal mucosa. Vitamin C prevents bacteria from adhering to epithelial cells. High concentration of vitamin C in white blood cells enables the immune system to function properly by providing protection against oxidative damage from free radicals generated during their action against bacterial, viral, or fungal infections.

Antioxidant

Free radicals, such as super oxide, hydroxyl ions and nitric oxide all contain an unpaired electron. These radicals can have a negative effect on cells causing oxidative damage that leads to cell death. Antioxidants, such as vitamin E & Vitamin C, prevent cell damage by binding to the free radical and neutralizing its unpaired electron. Ascorbic acid is well known for its antioxidant activities. Ascorbate acts as a reducing agent to reverse oxidation in aqueous solution.

Energy

Energy can be defined as the capacity of work or ability to perform any work. Every organism requires energy for:

(i) Mechanical work (Muscle activity for movement),

(ii) Chemical work (The chemical processes which takes place in the body),

(iii) Electrical work (For nerve activity),

(iv) Osmotic work (For maintaining the body fluids in equilibrium with each other and with the medium, whether fresh, brackish or marine where the fishes live.

(v) Free energy which is left available for biological activity and growth after the energy requirements for maintaining body temperature (not necessary for fish since, fishes are poikilotherms) is satisfied.

(vi) The excess energy is dissipated as heat (New, 1987).

Fish, prawn and shrimps require food to supply the energy that they need for movement and all the other activities in which they engage, and the 'building blocks' for growth. In this way they do not differ from any other farm animals even the human beings. The metabolic rates of the fishes however, largely depends on the surrounding water temperature and the optimum temperature on which they can survive, eat and grow best. Thus, in areas where there is wide range of water temperature seasonally, the fish will consume much more food in the summer than in winter (New, 1987).

Proteins and amino acids

Proteins are one of the large complex organic compounds that usually perform an essential role in the structure and functioning of every living organisms. Proteins are composed of several amino acids, considered as the building blocks of proteins. Out of the twenty three (23) of them which could have been isolated from natural proteins only ten (10) are considered indispensable for fish. The amino acid composition of proteins from different sources varies widely. Some proteins have none of certain amino acids. Where as some can be synthesized by animals and those cannot be synthesized are referred to as "Essential Amino Acids" or EAA's. For fishes and some crustaceans the EAA's are the following: Arginine, histidine, isoleucine, leucine, lysine, methionine, phenylalanine, threonine, tryptophan and valine.

Carbohydrate: Carbohydrates are the most abundant and relatively least expensive source of energy which are in use in aquaculture activities. This include starches, sugars, cellulose and gums containing only the elements like carbon, hydrogen and oxygen. Successful fish feeds contain a certain amount of carbohydrates about 20-30%. Besides providing energy, they have the physical function of texturizing the manufactured feed as a binder in the formulation of pellets.

Freshness Grading System of Fish

Freshness of the fishes mostly depends when carbohydrate, protein, vitamin contents of fish flesh are in order which may be judged by following the freshness grading system (Table-13).

16. ECOLOGICAL NICHE OF A POND

A pond can be divided into three distinct zones, according to the depth of the pond. i) Upper surface zone ii) Middle column zone and, iii) Bottom zone.In order to achieve productivity of the pond and depending upon the feeding habit of the fish, selection of species combination is a must. In case of mono-species culture, only one zone is being utilized or exploited while, the other zones remain unutilized. Therefore, the entire ecological area is not utilized, resulting in poor production. For example: if only catla is cultured in a pond, only the surface zone will be utilized and the other zones remain unexploited. If all the ecological niches of the pond are exploited, greater production is possible. In mixed culture the fish usually stocked are a mixture of plankton feeders and macrophyte (aquatic weed) feeders. The nutrients added to the water are taken up by both phytoplankton and the macrophytic water weeds, but as with land plants, they may not grow at the same pace, so that one group may use up most of the nutrients leaving little for the other. In ponds we try to maintain a balance by using both the phytoplankton to feeders and the water-weed eaters. If we use only fish that eat waterweeds, these may be heavily grazed and the nutrients released may then be taken up by the phytoplankton, which becomes denser and denser, shading out the submerged water-weeds and preventing them from growing again. If we only use the plankton feeders, the phytoplankton may become so heavily grazed, that the ungrazed submerged water-weeds grow very fast and use up all the nutrients. The plankton feeders will then starve. Sometimes not all the phytoplankton is grazed and we then have zooplankton developing (minute crustacea, rotifers, etc.). These also can be grazed down. We try therefore to achieve a balance where both phytoplankton and aquatic-weeds can grow, and to have different species of fish grazing down both.

Table13: Freshness grading system of fishes.

Grade	Highly acceptable	Acceptable	Fairly acceptable	Not acceptable
Skin	Bright, shining, iridescent or opalescent	Waxy, slight loss of bloom	Dull	Dull, gritty and shrinkage
Outer slime	Transparent or water white	Milky	Yellowish grey, some clotting	Yellow-brown, heavy clotted and thick
Eyes	Convex, black pupil, translucent cornea	Plane, slightly opaque pupil, slightly opalescent cornea.	Slightly concave, grey pupil, opaque cornea	Completely sunken, grey pupil, opaque discolored cornea
Gills	Bright red, mucus translucent	Pink, mucus slightly opaque	Grey, mucus opaque and thick	Brown, mucus yellowish grey and clotted
Peritoneum	Glossy, brilliant difficult to tear from flesh	Slightly dull, difficult to tear from flesh	Gritty, fairly easy to tear from flesh	Gritty, easily torn from flesh
Smell	Fresh, strong weedy odour	Slightly weedy, no odour, trace of musty mousy etc	Definitely musty mousy etc, bready and malty	Acetic, fruity amines sulphide faecal
Vent	Closed	Closed in many cases	Open in most cases	Open and gut content oozing

17. AQUATIC WEEDS

Aquatic weed plants are defined as a plant in an aquatic system (ponds, lakes, reservoirs, rivers and seas) that is not valued where it is growing and is usually of vigorous growth.Holm *et al.,* (1977) listed just ten aquatic weeds, including the three most notorious weeds viz.*Eichhornia cressipes* (water hyacinth), *Pistia stratious* (water lettuce) and *Salvinia auriculata* later identified as *S. molesta* (water fern, giant salvinia or kariba weed).Interestingly enough, In almost half a century since this book was published, the number of world's worst aquatic weeds has grown to about more than couple of dozens(Charudattan, 2001).Collectively, these weeds cause serious problems in nearly all countries, affecting nearly all uses of water bodies such as for aquaculture, commercial and subsistence fishing , drinking and household consumption, hydropower generation, irrigation, transport, navigation and recreation.The more invasive species among these weeds affect biodiversity in replacing native flora and fauna, often causing irreversible changes to habitat.Loss of aesthetic value of waterfront communities due to weed growth is also a serious concern.Dead biomass from large weed infestations increases the rates of sedimentation and eutrophication and reduces water depth. Excessive algal blooms can also render the water not only undrinkable but also impart an unpleasant odor to fish, deplete oxygen levels and cause fish kills (Pieterse, 1990).

Removal of Aquatic Weeds from Fish Rearing Ponds

Aquatic weeds share some common characteristics that contribute to their success as weeds such as their prolific growth rates, high seed output, multiple mode of propagation including clonal and sexual prpagules(by vegetative fragments, tubers, turions and rhizomes), and high vegetative and physiological plasticity that imparts intense competitiveness and environmental fitness(Langeland, 1996). These qualities contribute to the difficulty and complexity of aquatic weed management.Aquatic weeds are unwanted and undesirable vegetation that reproduce and grow in water. If left unchecked may choke the water body posing a serious menace to pisciculture.

- They provide breeding grounds and harbor predatory insects.
- Provide shelter to predatory and weed fishes and mollusks.
- They restrict free movement of fry.
- They cause obstruction during netting.
- Limit living space for fish.
- Limit plankton production.
- Reduce sunlight penetration and nutrients.
- Upsets the equilibrium of physico-chemical properties of water.

- Cause imbalance in dissolved oxygen budget.
- Promote accumulation of deposits leading to siltation.
- Reduce water movement, thereby limits oxygen circulation in water.
- Some weeds release toxic gases that cause fish death and add foul smell to water.
- **Classification of Weeds**

   1. **Floating weeds**: *Eichornia, Pistia, Azolla, Lemna minor* or *major*, etc.(Fig.30b) cause problems by partially or completely blanketing large and small aquatic bodies, interfering with the normal use of the water. They increase water loss through the dual actions of evaporation and transpiration.They also impart differential effects on submerged plants by promoting plant species that are tolerant of shades and changes in water chemistry, especially pH and Dissolved Oxygen might have detrimental effect on more sensitive species (Janes *et al.*, 1996).

   2. **Marginal weeds**: *Colecasia, Typha, Cyperus, Marsilia*, etc. Marginal or the shoreline weeds especially the cattails, grasses which are rooted emergent plants are common in lakes, canals and rivers.In the tropical countries where seasonal drought and the incidence of highly turbid waters generally are the rule, floating and rooted emergent plants like *Typha spp.* (the cuttails) and grasses are important weeds in irrigation and drainage canals. These perennial shallow water plants can grow in large clonal masses and obstruct the flow of water (Da Silva *et al.*, 2000).

   3. **Emergent weeds:** *Nymphae, Myriophyllum, Nelumbo*, etc.

   4. **Submerged weeds:***Hydrilla, Valisnaria, Chara, Ceratophyllum*, etc.Submerged aquatic macrophytes are particularly difficult and costly to manage more so pose serious problems for pond/lake management.The submerged aquatic weeds are typically influenced by the turbidity, age, sediment composition, bottom topography, water transparency and the nutrient concentrations of the water body(Clayton, 2000).

Fig. 30a: Floating euglenoids (*Euglena viridis*) Dinoflagellates

5. **Algal weeds:** *Spirogyra, Microcystis, Oscillatoria, Dinoflagellates,* etc.Very often a brick red scum appears on the surface of pond water during day time and gradually disappears as the sun sets. These are microscopic euglenoids (Fig.30a) having a distinct vertical migrations.

Removal of aquatic weeds

Manual removal, mechanical control by various types of mechanical harvesters; physical control by the use of shading devices such as dyes and shade films, burning and water level fluctuations , chemical control by using herbicides and biological control are the principal methods of weed control used.All these weeds have to be eradicated using one or more of the following methods:

1. **Manual method:** Manual removal of weeds involves physical removal of the weeds by hand. This may be practical provided the pond is small and labour is cheap.

2. **Mechanical method:** Some machines or implements are used for removing aquatic weeds. This method is normally applicable for larger water bodies. It is capital intensive and beyond the means of average fish farmer.In general, mechanical control methods are extremely inefficient, often requiring repeated cuttings to tackle weed populations that quickly rebound (Lindsay and Hirt, 1999).Plants fragmented by mechanical harvesting can regrow from the fragments, promote weed reestablishment and worsen the weed problem(Sidorkewici *et al.*, 2000).

   In smaller ponds raking is another way, not only to remove the weeds but also helps in nutrient recycling which helps the stocked fishes immensely.

3. **Chemical method**
   - Weeds are eradicated using chemicals.
   - Different weedicide are used for removal of different weeds present in aquaculture pond.

Fig. 30b: The common aquatic weeds (Water hyacinth, Pistia, Salvinia).

Table 14: Common indigenous weeds and the weedicides for their control.

Sl.NO	Type of weeds	Herbicide	Dosage	Method of application
1	Water hyacinth	2, 4-D	8 - 10 kg/ha	Foliar spraying
2	Ipomoea spp.	2, 4-D	2 - 4 kg/ha	Foliar spraying
3	Sedges and rushes	2, 4-D	5 -10 kg/ha	Foliar spraying/ root zone treatment
4	Lotuses and lilies	2, 4-D	5 - 10 kg/ha	Root zone treatment
5	Ottelia, Vallisneria	2, 4-D	10 - 20 kg/ha	Root zone treatment
6	Aquatic grasses(in young stages)	Dalaphon	5-10 kg/ha	Foliar spraying
7	Aquatic grasses	Paraquat	2 kg/ha	Foliar spraying
8	Aquatic grasses	Diuron	4 kg/ha	Root zone treatment
9	Microcystis, other planktonic and filamentous algae	Diuron	0.1-0.3 ppm	Root zone treatment. Dispersal in water column
10	All submerged weeds	Ammonia	10-15 ppm	Root zone treatment. dispersal in water column
11	Pistia	Ammonia	1% aqueous solution with 0.25% wetting agent	Foliar spraying
12	Pistia	Paraquat	0.2 kg/ha	Foliar spraying
13	Salvinia	Ammonia	2% aqueous solution with 0.25% wetting agent	Foliar spraying
14	Salvinia	Paraquat	0.4 kg/ha	Foliar spraying

- Dinoflagellate (algal weeds) especially the *Euglena viridis* scum may be eradicated by applying lime@ 30-40kgs /acre.

4. Biological method.

*The method is more advantageous since the undesirable weeds are converted into fish flesh.

- It is cheap as no labour is involved and most suitable from the social and environmental point of view.

- The method employs certain organisms which feed on the weeds.

- The grass carp which can eat up much more aquatic vegetation is itself an excellent example of biological weed control.

- The common carp helps in uprooting of certain plants.

*Tawes, *Puntius gonionotus* is also a good feeder of aquatic weeds.The Yamuna turtle consumes water hyacinth in the ponds.

13. Eradication of Weed and Predatory Fishes from Rearing Ponds

- Fishes which predate the spawn, fry and fingerling of cultured fishes.

- They get into cultured ponds through water or seeds.

- They breed easily in confined water little earlier than carps and therefore their size will be bigger than the size of the carps.

- They compete for food, space and DO and result in poor growth of carps.

- Their complete eradication using physical methods is difficult.

- They have the habit of burrowing in the mud bottom.

- Use of pesticides is inevitable.

Table 15: Common weed eating fish and the weeds of their preference.

Fishes	Names	Feed upon
Grass carp	*Ctenopharyngodon idella*	Submerged weeds e.g: *Hydrilla, Najas, Ceratophyllum, Potamogeton, Ottelia* and duck weeds
Common carp	*Cyprinus carpio*	Tender shoots
Gaurami	*Osphronemus goramy*	Tender shoots of submerged weeds and filamentous algae
Pearl spot	*Etroplus suratensis*	Filamentous algae
Silver carp	*Hypophthalmichthys molitrix*	Algal bloom

- **Common predatory fishes are:** *Channa* spp., *Clarias batrachas, Heteropneustes fossilis, Pangsius* sp., *Mystus* sp. *Ompok spp., Wallago attu* and *Glossogobius giuris* etc.

Weed Fishes

i. These are uneconomical, small sized, naturally occurring or introduced accidentally in ponds along with other fish seeds.

ii. They competes for food, space and dissolved oxygen.

iii. They have high fecundity and breed well before major carps breed.

iv. Many of them breed throughout the year.

v. Therefore, fish seeds from wild contain seeds of weed fishes.

Common weed fishes are: *Puntius sp., Oxygaster sp., Ambassis sp., Amblypharyngodon mola. Colisa sp., Rasbora sp., Aplocheilus sp., Laubuca sp., Esomus danricus,* etc.

Removal of predatory and weed fishes:

1. Repeated netting:

A. Repeated netting is suitable for only those ponds having no other fishes except carps.

B. It is not possible to catch predatory and weed fishes simply by netting.

C. Remaining fishes will breed and have sizable population in the pond.

2. Dewatering the pond is the best method:

- Dewater the pond, catch all the fish and allow the pond bottom to dry till the bottom soil cracks.

- Summer is the best time for this.

- Poisoning the pond (in case where dewatering and drying is not possible).

18. FISH TOXICANTS

Fish poison/toxicants is also known as piscicides or ichthyotoxins.A multiple of plant species are known to possess chemicals that are toxic to fish.Application of synthetic pesticides is one of the methods used to control of fish population but due to their long-term persistence, slow degradability in the water, toxicity to other aquatic organisms and accumulation in fish body tissues, synthetic piscicides adversely affect the aquatic environment. Piscicides of plant origin are safe for users and also do not adversely affect the environment.A number of compounds (saponins, tannins, alkaloids, alkyl phenols, di and tri-terpenoids etc.) present in several plants belonging to different families with piscicidal activities (family Euphorbiaceae), are used to control fish (Vimal and Das, 2015).Fish farmers should use piscicides of botanical origin that are not hazardous to the environment and have shorter residual affects rather that chemical pesticides that prove to be very dangerous to the environment and to humans. The latex of many plants have the potential to be used as piscicide which can be an alternative to harmful chemical piscicides that are widely used today to eradicate fishes in the ponds (Neuwinger, 2004).

Fish toxicants can be grouped into (a) Plant derivatives, (b) Chlorinated hydrocarbons and (c) Organophosphates. Chlorinated hydrocarbons are the most toxic and stable compounds, not metabolised and remain stored in fish tissues. Although organophosphates are less toxic but adversely affect other aquatic biota. These are relatively less persistent in water and rarely stored in fish body. Prior to use of fish toxicants, certain criteria should be followed.

(1) It should be effective at minimum dosage

(2) It should be less costly

(3) The fish killed by the toxicant should be suitable for consumption

(4) The toxicity should remain only for short duration

(5) The toxicity should be non-cumulative

(6) The toxicants should be easily available.

Since the use of chlorinated hydrocarbons and organophosphates is not normally recommended in fish culture ponds, use of plant derivative piscicides is encouraged.

Commonly used piscicides are:

a. Mahua oil cake

- It is the most commonly used fish toxicant in our country.

- It contains about 4 -6% saponin.

- It kills fish at 200-250 ppm in 6-10 hours.

- Fishes killed by this are fit for consumption.

- The toxicity lasts for 15 -20 days in water

- Mahua cake subsequently becomes the manure for the pond.

- Applied at the rate of 2000 -2500 kg/ha at one metre average depth.

- Powdered cake soaked in water and broadcast over water surface.

- After its application, the pond is netted repeatedly to enable uniform mixing and also to take out dead fishes.

b. Tea seed cake

- Crushed and powdered tea seeds are effective for the control of the unwanted fishes in the fish ponds.
- Kills fish @ 75 -100 ppm
- Applied @ 750 to1000 kg/ha for every one metre average depth.
- Toxicity lasts for 10 -12 days.
- Ultimately the cake acts as fertilizer in the pond.
- The treated fishes are suitable for human consumption.

c. Ammonia

* Anhydrous ammonia is quite costly

*Kills fish @ 20-25 ppm

*Ammonia gas comes in cylinders.

- For application, the cylinder is held in position at the bottom of the pond and moved from one side to the other releasing the gas in a controlled way so that there is neither excessive cooling nor gas loss during the absorption by the bottom soil.
- Sufficient protection should be provided to persons who apply the gas.

d. Bleaching powder

- Bleaching powder or calcium hypochlorite is an effective fish toxicant.
- It is used in eradicating the unwanted fishes from the-fish pond.
- It is easily available and less costly.
- Kills fish @ 25-30 ppm within 3 - 4 hours
- Dead fish start floating to the surface of the water.
- Applied @ 350 kg/ha, commercial bleaching powder (30% chlorine).
- Combination of urea @ 100 kg/ha + 175 kg/ha of bleaching powder can also be used.
- The toxicity lasts for about 7-8 days in the pond.
- The additional advantage of bleaching powder is its disinfecting and oxidizing effect on the decomposing matter at the bottom of the pond.

Table 16: Recommended doses of fish poisons.

Poison	Dose (kg/ha/m)
Bleaching powder	350 – 500
Mohua oil cake	2500
Anhydrous ammonia	20 – 30
Powdered seed of *Croton tiglium*	30 – 50
Root powder of *Milletiapachycarpa*	40 – 50
Seed powder of *Milletia piecidia*	40 – 50
Seed powder of *Barringtonia acutangula*	150
Seed meal of tamarind (*Tamarindusindica*)	1750 – 2000
Tea seed cake (*Camellia sinensis*)	750

The respective doses of the plant seeds pest powder are to be mixed thoroughly in a bucketful of water and splash over the entire pond surface.

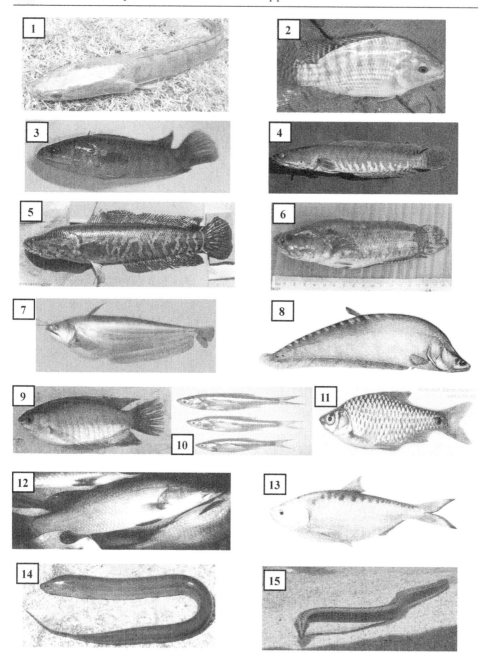

Fig. 31: Some piscivorous and weed fishes of fresh water ponds. (1. *Channa striata* (Shole) 2. *Oreochromis niloticus* (Tilapia) 3. *Anabas testudineus* (Koi) 4. *Channa striata* (Shole) 5. *Channa marulius*(Shal) 6. Channa punctatus (Lata) 7. *Wallago attu* (Boal) 8. *Notopterus chitala* (Chital) 9. *Colisa fasciata* (Kholse) 10. *Amblypharyngodon mola* (Mourala) 11. *Puntius ticto* (Punti) 12. *Lates calcarifer* (Bhetki) 13. *Gudusia chapra* 14. *Amphipnous cuchia* (Kuche) 15. *Mastacembelus* sp.)

SOIL AND WATER QUALITY FOR AQUACULTURE

1. INTRODUCTION

The living organisms and their non-living environment are inseparably connected and they constantly interact with each other. It is therefore, obvious that the entire dynamics as well as homeostasis of an ecosystem are largely regulated by the interplay of various physical and chemical parameters. Physico-chemical parameters of water viz. temperature, transparency, pH, alkalinity, dissolved gasses, available forms of bioactive nutrients etc. not only influence the occurrence and abundance of aquatic biota but also govern the productivity which form the cornerstone of community metabolism.

Soil is an integral and vital part of environment and may be defined as secrete bodies produced by interactions of climate, vegetation and surficial geologic materials on the earth surface. The bottom soil is the substratum on which water column stands. Obviously therefore, it plays a very significant role in influencing the productivity of ponds. In any type of aquacultural enterprise the farmer should have a basic idea of his pond soil. The knowledge of chemical characteristics and the nutritional status of pond soil (pH, available phosphorus, nitrogen, organic carbon and organic matter etc.) is a pre-requisite. Productivity of a pond is largely depended on the chemical properties of soil. In general the slothful nature of sandy-loam soil permits aeration and helps increasing the productivity of microbes. Presence of organic matter and their nutrients leech easily in the sandy soil. Mineralization process in presence of sunlight and congenial temperature generally do occur at the soil-water interphase. Such course of activities essentially makes available different types of minerals (both micro- and macro) in water. This process gets hampered when the quality of soil deteriorates and the overall pond productivity is badly affected.

An ideal pond bottom soil is that in which mineralization of organic matter is rapid and essential nutrients are released slowly in order to maintain an optimum trophic status.

2. POND SOIL CHARACTERISTICS

Soil composition and quality are extremely important factors in pond aquaculture. Soil is usually the main source of dyke building materials, and soil quality determines the productivity of a pond to a large extent. A pond soil, having the ability to absorb and release nutrients and as rightly been described by Hickling (1971) as the **chemical laboratory** of the pond that influence the natural productivity, water quality and, hence survival of fish and growth.

Macan *et al.*, (1942) classified pond bottom into three main types:

i) Inorganic bottom of gravel, sand or clay which are poor in nutrients, but can be improved by the application of manure or sewage/sludges;

ii) Peaty bottom formed by the accumulation of undecomposed vegetable debris which can be corrected by using heavy doses of lime to bring about decomposition and;

iii) Mud bottom which are the most productive one.

In order to utilize and maintain its productivity, biogeochemical cycles have to be set in involving replenishment of nutrients continuously consumed by aquatic life. The fertility of soil refers to its nutrient releasing properties since sediment is always a great source of minerals. As an absorber of toxic gases and as a filter of waste waters, soil is also important for hygiene with regard to waste water recycling.

Each and every soil type has its own history, all the conditions involved in its formation and development must be taken into account when studying a soil. Unfortunately, little is known about the direct or indirect relationship of soil with fish production (Jhingran, 2000).Hickling (1971) is of the opinion that soil is obviously the chemical laboratory where nutrients for aquatic inhabitants are produced by chemical process from the raw materials consisting of organic matter and mineral constituents of clay. Soil characteristics are related to biogeochemical processes including trophic status. An understanding of the chemistry of soil is of paramount importance in relation to fisheries (Banerjea, 1995).

It is needless to mention, sewage discharge into the water bodies carries a lot of organic and inorganic ingredients which will bring both qualitative and quantitative changes in the system. It is therefore, the study of important chemical parameters like pH, organic carbon, available nitrogen and phosphorus are of immense importance as the water quality parameters are dependant to a great extent on the quality of pond basin soil.

Banerjea (1967) from the study of large number of fish pond soil under different agro climatic conditions inferred that pond productivity depends on soil qualities such as its reaction status of available nutrients and OC. He also classified ponds into low, medium and high productive on the basis of nutrient status as given in the table below:

Table 17: Nutrient status of the pond soil.

Status	Av. N mg/100 gm	Av. P_2O_5 mg/100gm	OC%
Low	< 25	<3	<0.5
Medium	20 - 25	3-6	0.5-1.5
High	750	76	>1.5

Boyd *et al.*, 2002 studied organic matters and dynamics of soil organic matters and nutrients. He pointed out that pond soil containing more than 15 % organic carbon (OC) will be considered as organic soils. Organic soils again are divided into three different groups in connection with organic content, as low (25-25% OC) medium (26-40% OC) and high more than 40%. Pond sediments usually contain 0.5 to 5% OC and the optimum concentration is considered to be 1-3%. Thus the following categories have been suggested by Boyd *et.al*, 2002 for OC in pond sediment which are constructed on mineral soil as low < 1% OC, medium 1-3% OC and high more than 3% OC. Banerjee *et.al*, 2009, pointed out critical concentration of soil nutrients are: Av.N, 200 mg kg^{-1}, Av.P, 13mg kg^{-1} and K 80 mg kg^{-1}.

In any type of aqua cultural enterprise the farmer should have a basic idea of his pond soil. The knowledge of chemical characteristics and the nutritional status of pond soil (pH, available phosphorus, nitrogen, organic carbon and organic matter etc.) is a pre-requisite. Productivity of a pond is largely depended on the chemical properties of soil. In general the slothful nature of sandy-loam soil permits aeration and helps increasing the productivity of microbes. Presence oforganic matter and their nutrients leech easily in the sandy soil. Mineralization process in presence of sunlight and congenial temperature generally do occur in the soil-water interphase. Such course of activities essentially makes available different types of minerals (both micro- and macro) in water. This process gets hampered when the quality of soil deteriorates and the overall pond productivity is badly affected.

In acid-sulfate soils (Fig. 33) it is necessary to apply sufficient amount of agricultural lime (powdered calcium carbonate, $CaCO_3$) @ 90-100 kg/acre area which is the most common neutralizing material in such case. There are problems if even there is black soil (Fig. 33a). Black soil have low fertility and are poor in organic matter, nitrogen, available phosphorous and zinc. Cracking when dry and swelling when wet makes them difficult to manage, unless they are cultivated at appropriate soil moisture levels. This makes black soil difficult to manage hence should have to be carefully removed from the pond bottom. It is considered that the range of soil pH from 7.5 to 8.5 is perfect for aqua farming. Aquatic inhabitants (phyto- and zooplankton, fishes etc.) as a whole, are severely affected if the soil pH is acidic. Besides, soil pH makes phosphorus availability but pH more than 9.0 also restricts purveying phosphorus availability. For a meaningful aquaculture enterprise, available phosphorus in soil must be within 10 to 15mg/ 100gmof soil. In addition to this, the available nitrogen in soil should be between 8 to 10mg/100 gm of soil (Table-18a). The organic matter present in soil which helps in keeping the pond soil alive, effective and increase the water retention capacity. The excessive presence of organic substances in soil may bring down the pH of water and destroy the aquatic environment and may increase the autochthonous turbidity. The range of Organic Carbon (OC) should be between 1.5 to 2.5% and Organic Matter (OM) between 2.5 to 4.3% to enhance fertility and productivity of a pond. Generally it is believed that the available nitrogen in pond soil plays a pivotal role in productivity of pond water.

The carbon-nitrogen ratio (C: N) is also a decisive factor in fish production (Table-18b).This ratio is also playing a crucial role in the availability of various nutrients in water. In soil, not only the organic carbon but also the ratio of carbon and nitrogen (C: N), the decomposition of organic matter (OM) largely depends and the various bacteria have been found to play a crucial role in the process (Table.18c).

Table-18(a): Conditions of pond soil.

Pond Soil	Available Nitrogen (mg/100gm)	Productive nature of pond
Newly excavated soil	10.00	Non-productive
Old pond	50-75	Productive
General unproductive soil	25-40	Unproductive

Table-18(b): C: N ratio of soil vis-à-vis productivity of pond.

Carbon-Nitrogen ratio of soil	Productivity of pond
<10	Unproductive
10-15	Productive
>20	Unproductive

Table-18(c): Ideal range of chemical features of soil.

Chemical characteristics	Ideal range
pH of soil	6.5-7.5
Organic Carbon	1.5-2.5%
Organic matter	2.5-4.3
Available Nitrogen	59-75mg/100gm
Available Phosphorus	6-12mg/100gm

Primarily to understand the nature of the soil it is necessary to take out handful of the soil from the pond bottom and by the help of fingers, make it a ball and throw it upward. If the ball split and breakdown, it is to be understood that the soil is not ideal for aquaculture purpose. In such circumstances, to make it fit for aquaculture use, both time and money is a mere wastage. Water quality management has been considered as one of the most important aspects of pond aquaculture for many years, but less attention has been given to the management of pond bottom soil quality. There is increasing evidence that the condition of pond bottoms and the exchange of substances between soil and water strongly influence water quality (Boyd, 1995). It is needless to mention in this context that "soil beneath the pond water is the chemical laboratory of every pond". Hence, today more attention is being devoted to the study of pond soils.

3. POND PREPARATION TECHNIQUES

A. Preparation of soil

It is always advisable to excavate the soil at least once in two to three years of a perennial fish Pond (Fig. 32) either manually or by mechanically (Fig: 34, 34a and 35). The earth may well be utilized to renovate and streamlining the bunds around (Fig: 36& 36a). Due to anaerobic condition, presence of sulfur turns the sediment corrosive, otherwise called acid-sulfate soil or black soil (Fig. 33 and 33a) which is not desirable for an aquaculture pond.

Table 19: Ranges of bottom soil properties on Fish Production.

Variables	Range	Fish production
pH	<5.5	Low
	5.5-6.5	Average
	6.5-7.5	Optimum
	7.5-8.5	Average
Available Phosphorus	<3ppm	Low
	3-6ppm	Average
	>6ppm	Optimum
Organic Nitrogen	<25ppm	Low
	25-50ppm	Average
	>75ppm	Optimum
Organic Carbon	<0.5%	Low
	0.5-1.5%	Average
	1.5-2.5%	Optimum
	>2.5%	Declining

Fig. 32: A perennial fresh water carp culture pond

Fig. 33: Acid sulfate Soil indicating
anaerobic condition

Fig. 33a: Thick layers of black soil

Acidic bottom soil is a common problem in pond aquaculture, and fish farmers often as remedial measures apply agricultural limestone to such ponds. Aquaculture ponds are usually limed after draining for harvest and before refilling for the next crop (Fig.34a). However, sportfish ponds are usually not drained for liming, and agricultural limestone is spread over the water surface from a boat (Boyd, 1982; Boyd & Tucker, 1998).

The objective of liming is to neutralize acidity in the upper layer of bottom soil and to increase concentrations of total alkalinity and total hardness in the water (Thomaston & Zeller, 1961). Several studies have shown positive responses in phytoplankton productivity and fish production following liming of acidic ponds, and methods for determining the lime requirements of bottom soils have been developed (Boyd, 1995).

Nevertheless, liming often is applied to ponds indiscriminately, with no concern for bottom soil pH or total alkalinity and total hardness concentrations. It is doubtful that liming has a large influence where soil pH is above 7 or total alkalinity is above 50 mg L^{-1} (Boyd, 1995). Aquaculture ponds are ordinarily drained for harvest, bottoms are allowed to dry, and liming materials are applied. Lime is frequently blended with the bottom soil by tilling.

Figs.34 and 34a: Pond excavation in progress and after excavation

Fig. 35: Mechanized (Foklar) excavation of pond soil

Fig. 36&36a: Stream lined base & the bund

B. Water quality

Water quality is the totality of physical, biological and chemical parameters that affect the growth and welfare of cultured organisms. The success of a commercial aquaculture enterprise depends on providing the optimum environment for rapid growth at the minimum cost of resources and capital. Water quality affects the general condition of cultured organism as it determines the health and growth conditions of cultured organism. Quality of water is, therefore, an essential factor to be considered when planning for high aquaculture production (Mallya, 2007).

Although the environment of aquaculture, fish in particular, is a complex system, consisting of several water quality variables, only few of them play decisive role. The critical parameters are temperature, suspended solids and concentrations of dissolved oxygen, ammonia, nitrite, carbon dioxide and alkalinity. However, dissolved oxygen is the most important and critical parameter, requiring continuous monitoring in aquaculture production systems. This is due to fact that fish aerobic metabolism requires dissolved oxygen (Timmons *et al.,* 2001).

The living organisms and their non-living environment are inseparably connected and they constantly interact with each other. It is therefore obvious that the entire dynamics as well as homeostasis of an ecosystem are largely regulated by the interplay of various physical and chemical parameters. Physico-chemical parameters of water viz. temperature, transparency, pH, alkalinity, dissolved gasses, available forms of bioactive nutrients etc. not only influence the occurrence and abundance of aquatic biota but also govern the productivity which form the cornerstone of community metabolism.

Solar radiation, the arrival of which on the earth's surface marks the beginning of photosynthetic process, is the fundamental driving force behind the entire ecosystem dynamics because, it is the basic source of energy for all trophic levels. Detailed study of radiation incident upon water surface and its subsequent reflection by the attenuation in water was made by Anderson (1952). Water absorbs solar radiation very rapidly and as much as 53% of the radiation passing through the water is absorbed in the first meter in the water column including almost all the infra red energy (Krebs, 1985). The radiation transforms into heat and undergoes extinction. About 50% of the remaining light is attenuated with each additional meter depth (Reid and Wood, 1976).

Temperature variability is extremely important ecologically. Temperature in an aquatic ecosystem mainly (>75%) emanates from the absorbed radiation. Earlier, temperature was branded as a master limiting factor (Krebs, 1985) because it not only exerts powerful effect in the metabolism, reproduction and life cycles of organisms but is also a major determinant of the variability of several other important physico-chemical variables viz. salinity, turbidity, concentration of dissolved gases etc. Temperature also brings about thermal stratification of lakes but in ponds, such stratification cannot occur because of low depth.

Secchi disc transparency is essentially a function of the reflection of light in the water column. It is therefore, influenced by the absorption characteristics, both of water and of its dissolved and particulate matter. Transparency of water mainly depends on two factors:

i)   Autochthonous production of plankton, and

ii) Allochthonous input.

Both theoretical analyses as well as a large number of empirical observations have shown that the reduction in light transmission in relation to Secchi disc transparency is associated to a greater extent with increased scattering by particulate suspensoids. Reduction in transparency vis-à-vis increase in turbidity in water, caused by clay and silt particles, is often important as a limiting factor. However, if turbidity is the result of living organisms, measurement of transparency becomes an index of productivity (Odum, 1971).

The pH of natural waters is governed to a large extent by the interaction of H^+ ions arising from the dissociation of carbonic acid (H_2CO_3), and OH^- ions produced during the hydrolysis of bicarbonates. The value of pH provides an index of the general chemical environmental condition of any aquatic ecosystem, viz. the concentration of H^+ ions, amount of available gases, alkalinity, nutrients, dissolved salts, trace elements (Goldman and Horne, 1983). Although pH of natural waters ranges between < 2 to 12 (Wetzel, 1983), in most unpolluted water bodies, it is from 6 to 9 (Hutchinson, 1957). Santra and Deb (1996) while studying the pH of sewage fed aqua system in Kolkata, revealed that pH varied from 7.2 to 9.5.

Oxygen is the most fundamental chemical parameter of aquatic ecosystems and its measurement can help in feeling the pulse of the aquatic environment. DO is obviously essential to the metabolism of all aerobic aquatic organisms. Hence, the solubility and especially the dynamics of oxygen distribution in aquatic ecosystems are basic to the understanding of the distribution, behaviour and growth of aquatic organisms. The balance of DO content of water is determined by the interaction of a number of factors like, the amount of oxygen being dissolved, the activity of producers, hydromechanical distribution of oxygen, respiratory activity of the biota and the rate of decomposition of organic matter. The chemical characterization of sewages used for fish culture showed a distinct gradient, with a progressive rise in DO values (5.6-12.4mgl^{-1}) (Jana, 1998).

Free carbon-dioxide in water like oxygen, may be present in water in highly variable amounts but is difficult to make general statements about its role as a limiting factor for algae (Krebs, 1985). However, very high CO_2 concentration may be definitely limiting to animals because it is associated with low oxygen concentrations (Odum, 1971). The chief sources of CO_2 in water are respiration of biota, diffusion from atmosphere, microbial decomposition, soil and seepage from underground sources. Solubility of CO_2 in water is about 200 times greater than that of oxygen at same temperature and pressure (Wetzel, 1983). In water CO_2 may remain in any one or more of the three states:

i)   A free state (dissolved CO_2),

ii) Half bound state (HCO_3^-) and,

iii) Bound or fixed state ($CO_3^=$).

Autotrophs utilize both dissolved CO_2 and HCO_3^- for photosynthesis. CO_2 dissolves in water at about the same concentrations as in atmosphere. Dissolved CO_2 hydrates by a slow reaction to yield carbonic acid which, being a weak acid, dissociates into H^+ and HCO_3^- ions. At high pH, HCO_3^- ions further dissociates to H^+ and $CO_3^=$ ions. Therefore, the concentration of reactions is as follows:

CO_2 (air)	$\rightleftharpoons$	CO_2 (dissolved) + H_2O (Dissolved CO_2 predominates at pH 5 and below)
$CO_2 + H_2O$	$\rightleftharpoons$	H_2CO_3
H_2CO_3	$\rightleftharpoons$	$H^+ + HCO_3^-$ (HCO_3^- predominates between 7 and 9)
HCO_3^-	$\rightleftharpoons$	$H^+ + CO_3^=$ ($CO_3^=$ is prevalent above pH 9.5).

Thus, the above reactions establish a CO_2- HCO_3^- - $CO_3^=$ buffering system which prevents the pH from fluctuating widely and thus creates a relatively stable condition for the sustenance of aquatic life.

Alkalinity of water refers to the quantity and kinds of compounds present, which collectively shift the pH to the alkaline side. The property of alkalinity is mainly imparted by the presence of bicarbonates, carbonates and hydroxides. A mixture of bicarbonate and carbonate alkalinity is generally encountered in waters having a pH range of 8.2 to 10.5. Bicarbonates are sharply reduced at higher pH values while pH less than 8.2 but above 4.5 practically no carbonate is present (Jhingran, 1983).

PO_4-P is the least abundant but the most common limiting factor of biological productivity (Wetzel, 1983). Phosphorus occurs in a number of inorganic and organic compounds in both particulate and dissolved forms (Strickland and Parsons, 1968). The most significant form of inorganic phosphorus is orthophosphate ($PO_4^=$) which is assimilated rapidly by the biota in aquatic ecosystems (Hutchinson, 1957). A massive biological uptake of phosphorus also occurs during the growth of plankton and an increase in the amount of phosphorus leads to an increment of lake productivity. A principal source of autochthonous input of phosphorus is the result of death and decay of planktonic, benthic and nektonic organisms and this form a part of the phosphorus cycle. Exchange of phosphorus between sediments and the overlying water is also a major component of the phosphorus cycle in natural waters.

Nitrogen occurs in fresh waters in various forms viz. molecular nitrogen, a large number of organic compounds from amino acids, amines to proteins and humic compounds, dissolved ammonia (NH_4^+), Nitrites (NO_2^-) and nitrates (NO_3^-). Generally, concentration of nitrogen in unpolluted fresh waters ranges from undetectable levels to 19 mgl^{-1} but it is highly variable seasonally and spatially. The sources of nitrogen for aquatic flora are NO_3-N and NH_4-N and their utilization depends on their relative concentrations (Wetzel, 1983). Bacterial denitrification of NO_3-N removes nitrogen from the system in molecular N_2 form and it forms a major component of the nitrogen cycle. Nitrification and denitrification can occur simultaneously. Sedimentation of inorganic and organic nitrogen containing compounds to the soil also renders a large portion of nitrogen unavailable for biotic utilization.

SiO_4 is usually moderately abundant in fresh waters and it is relatively nonreactive. It is a major constituent of the frustule of diatoms and cysts of chrysomonadales (Yellow-brown algae). Both these groups of organism, the former in particular, remove large amounts of SiO_4 from their environment. Therefore, availability of SiO_4 can have a strong influence on the overall pattern of algal succession and productivity in aquatic ecosystem (Wetzel, 1983). Solubility of SiO_4 in fresh water is largely modified by the surface adsorption of silicic acid and formation of SiO_4 complexes (Ohle, 1964). Variations in silica content in eutrophic water bodies during summer and winter stratification and also at different depths have been discussed by Conway *et al.*, (1977). An inverse relationship between the concentration of silica and production of diatoms has been reported by Hutchinson (1957) and Kilham (1971).

The colloidal fraction of measured DOM is probably small in most water bodies. In fish ponds, organic matter is present as living plankton, suspended particles of decaying organic matter (detritus) and DOM. The organic matter present in natural waters is organic phosphorus, organic nitrogen, carbohydrate, vitamin etc. The organic matter in a water body may be either allochthonous or autochthonous in origin and may be in either dissolved or suspended states. The presence of organisms ensures a more or less continual addition of organic materials to the water body, some of which immediately or later go into solution.

Thus, all waters have certain content of DOM. Cole (1983) is of opinion that DOM in an ecosystem comes from four sources:

i)   Allochthonous input,

ii) Death and decay of aquatic organism,

iii) Excreted metabolites and / or photosynthates of phytoplankton,

iv) Littoral vegetation and excreted metabolites of fauna.

Since carbohydrates contribute predominantly to the major bulk of the total DOM, carbon appears to be of paramount importance as a limiting nutrient (King, 1970).

The BOD or biochemical oxygen demand (also known as biological oxygen demand) is an indication of the amount of biodegradable organic materials present in water. BOD is an essentially bacterial reaction by which organic matter present in any effluent are broken down to simpler substances and in that process oxygen is used up. For industrial effluents (which contain only chemical reducing agents that take up oxygen very rapidly by purely chemical action) and sterile effluents viz. phenols, wood pulp, saw dust etc. (which are very poor bacterial foods and are decomposed very slowly), measurement of BOD gives less satisfactory results. However, for sewage and other effluents containing pesticides, oxidisable organic matters, sugars, milk wastes, etc. BOD test is very useful and it yields highly satisfactory results (Hynes, 1960). Sewage is well inoculated with bacteria and is adequately supplied with wide range of compounds, so it is broken down easily. It has been found that, of all the various kinds of effluents wool scouring, pharmaceutical products, paper pulping, smokeless fuel carbonization and tanning industry effluents have extremely high BOD values.

Jana (1998) is of the opinion that a considerable reduction on BOD loads of sewage is a pre-requisite before its transfer to aquaculture facilities. Generally, the BOD load of raw domestic sewage ranges between 120 and 400mgl^{-1} and primary treatment by sedimentation is likely to reduce it by 33%. In oxidation ponds, most of the anaerobic bacteria and pathogenic forms are likely to be reduced considerably. Jana (1998) also opined that primarily treated sewage contains less organic loads than untreated sewage, but more nutrients than secondary and tertiary effluents. Because of the presence of certain obnoxious gases like un-ionized ammonia, CO_2 and H_2S in the treated effluents remain anaerobic and are not suitable for direct use. As a result, dilution with fresh water (1:1-1: 4) or proper loading of sewage provides congenial environment and aerobic conditions for fish culture. A significant contribution to the basic understanding on BOD is available from the works of Schreiber and Neumaier (1987). BOD test is however, purely arbitrary and takes no difference between the effluents of different kinds. Also it does not directly indicate the amount of other non-organic pollutants. Despite these short comings the BOD test, originally devised by Sir Edward Frankland at the beginning of the nineteenth century, is undoubtedly the most important contribution so far made to the study of water pollution.

C. Physico-chemical and Biological characteristics of pond water

Fishes are aquatic, exactly for the reason, whenever the word "aquaculture" being the question, the very first point will arise to that of the water and its quality. Depending on the quality, fish growth, production potential depends. In general, the water quality depends on the soil quality, source, geographical situation and natural surroundings.

1. Source of Water

Source of water is essentially important, if a pond is considered to be only a rain fed and if the pond is sandy in nature, there is every possibility of the same to be a temporary one. Quality of pond water seems to be a better one and productive, if the same is situated beside a canal with running water and by excavating a small drain or by passing the pond water may show a better productivity. The fish farms which are situated near a town or city, and the waste water when passes through a drain are pumped in also exhibit much better production. Since, the domestic waste water are enriched with essential nutrients. But a fish pond/farm situated near agriculture land may not be always suitable for farming since agricultural runoff might contain insecticides/pesticides. Aquafarming is not at all suitable for higher alkaline or acidic waters since the production of natural food for fishes is at a question.

2. Depth of Water

Depth of water plays an important role in fish culture /pond farming. In general both ponds having a stock of Indian and exotic carps should have a depth of 6 – 7 feet since the sunlight can reach upto that depth in fresh water. It is an absolute truth that so long the sunrays can penetrate in water is a photic zone and both phyto and zooplankton develops rapidly in this zone. The depth beyond this specified zone is aphotic zone and comparatively less productive.

In the following table (Table -20), the desired area and depth is given which shows better result:

Table 20: Types, area and depth of ponds.

Type of Ponds	Area	Depth	Remarks
Nursery	< 33 decimal	2- 4 ft	1 Bigha = 33 decimal
Rearing	1 Bigha -4 Bighas	3 – 5 ft	————
Stocking Pond	>4 Bighas	5 – 8 ft	Av depth – 6 feet

3. Pond water temperature

Water temperature is the single most important factor affecting the welfare of fish. Fishes are cold-blooded organisms or poikilotherms and assume approximately the same temperature as their surroundings. The temperature of the water affects the activity, behavior, feeding, growth, and reproduction of all fishes.

They don't regulate their body temperature, thus have variable metabolic rates dependent on the temperature of their body at the time. Basal metabolic rate (BMR) in mammals is the rate of oxygen consumption at a resting state. The concept of BMR is not appropriate for fish because the metabolic rate in a resting fish will vary depending on the ambient temperature. Standard Metabolic Rate (SMR) is the metabolic benchmark in fish. It is the metabolic rate of a resting fish at a specified temperature in the middle of its normal range. Therefore a trout's SMR15 would be the resting metabolic rate of a trout at 15°C. There is natural law called $Q_{10} = 2$: chemical reactions tend to double with every 10° C increase in temperature of the reactants. In poikilotherms, this is translated into biochemical events and in turn into metabolism. The metabolic rates of fish roughly double with every 10° C increase in temperature, except at the extreme ends of their temperature tolerance.

(a) Higher temperatures bring down the solubility of dissolved oxygen and thus decrease the availability of this essential gas.

(b) Elevated temperatures increase the metabolism, respiration and oxygen demand of fish and other aquatic life, approximately doubling the respiration for a 10° C. rise in temperature. Hence the demand for oxygen is increased under conditions where oxygen supply is lowered.

(c) The solubility of many toxic substances is increased as well as intensified as the temperature rises.

(d) Higher temperatures militate against desirable fish life by favoring the growth of sewage fungus and the purifications of sludge deposit, and

(e) Even with adequate dissolved oxygen, there is a maximum temperature that each species of fish or other organism can tolerate. Higher temperatures accelerates rate of death. The maximum temperatures that adult fish can tolerate vary with the species of fish, prior acclimatization, oxygen availability and the synergistic effects of other pollutants. Fish are generally categorized into warm water, cool water, and cold water species based on optimal growth temperatures (Table 21).

Table 21. General temperature ranges for coldwater, cool water and warm water species.

12.7-18.3⁰C	18.3-23.9°C	23.9-32.2⁰C
Cold water	Cool water	Warm water

Water temperature principally depends on climate, light and depth of the pond. Temperature has a direct impact on the organisms inhabit in soil and water as well as the growth of the fishes. The metabolic activity of the fishes greatly influenced by the surrounding water temperature. In general, appreciable growth of fishes is observed within $25\text{-}32^0$ C of water temperatures.

Ideally, species selection of fish farms should be based in part on the temperature of the water supply. Because any attempt to match a fish with less than their ideal temperatures will involve energy expenditures for heating or cooling. This added expense will subsequently increase production costs, so fish farms from different areas usually select different species of fish for the reason of temperature.

4. Transparency of water

Transparency of water or, conversely turbidity of water plays an important role in ponds and tanks. Turbidity of water is the resultant of higher growth of autochthonous planktonic production and surface run off from the catchment area i.e., allochthonous source. Due to turbid nature of natural waters, temperature also shows higher trend so also there is a reduction of penetration of sunlight. As a result photosynthetic activities of phytoplankton are greatly affected followed by the decrease of dissolved oxygen content of water.

Transparency of water plays an important role on the natural growth of fishes; Turbidity of pond water depends mainly on:

i) **Autochthonus matter:** Excessive growth of both phyto and Zooplankton.

ii) **Allocthonus matter:** Washing from the catchment area. The ponds where allochthonus turbidity is high, penetration of sunlight get obstructed affecting photosynthetic processes followed by depletion of dissolved oxygen. Besides smaller dust particles, after being thickened with planktonic biomass deposits on the gills of fish affecting normal respiration processes or even destroying the gill cells.

Transparency of water may be measured by a simple device called "Secchi Disc" (Fig. 37). It is a simple round shaped metallic plate having an attachment of a measured scale at the centre. If the disc is visible beyond 30-40cm, the water is moderately transparent, and if the visibility remains within 30-40cm, the pond water is suitable for aquaculture. It is presumed that much transparent waters do not harbor sufficient quantity of phytoplankton in water. Since planktonic communities as a whole constitute the natural food for the fishes it is necessary to take care of this parameter. Generally, 30-45cm.of transparency is permissible in ponds and tanks.

Fig. 37: Measurment of transparency of water of a pond by Secchi disc

Calculation of transparency of water

$$\text{Pond transparency (T)} = \frac{T_1 + T_2}{2}$$

Where, T_1 = Point of visibility of the disc, and

T_2 = Point of nonvisibility of the disc.

Much transparent pond water determines the absence of less plankton in water. Increased transparency of pond water evidently needs application of the organic manure @ 150 Kgs/Acre. Such application will boost up the natural food of fishes the planktonic communities. Excessive planktonic growth also has negative impact on fishes. In such circumstances it is necessary to apply Zeolite powder @15Kgs/ acre manure of the pond water having a depth of 4 – 5 ft. This will help the water lighter and congenial for pisciculture.

5. Suspended matters

Presence of debris, individual wastes, and domestic sewage sometimes may not be congenial for fish and fisheries. Ponds having an inlet connection from a waste fed canal should have to be drained in a separate sedimentation tank for settling the suspended matter for about 24 -48 hours the supernatant water is to be drained in culture ponds. The acceptability of suspended matter is < 300 Gms.

6. pH of pond water

The term "pH" is a mathematical transformation of the hydrogen ion (H+) concentration; it conveniently expresses the acidity or basicity of water. The lower case letter "p" refers to "power" or exponent, and pH is defined as the negative logarithm of the hydrogen ion concentration. Each change of one pH unit represents a ten-fold change in hydrogen ion concentration. The pH scale is usually represented as ranging from 0 to 14, but pH can extend beyond those values. At 25°C, pH 7.0 describes the neutral point of water at which the concentrations of hydrogen and hydroxyl ions (OH-) are equal (each at 10-7 moles/L). Conditions become more acidic as pH decreases and more basic as pH increases (Tucker and D'Abramo, 2008).The objective of liming of pond water is to neutralize acidity in the upper layer of bottom soil and to increase concentrations of total alkalinity and total hardness of the water (Thomaston & Zeller, 1961). Several studies have shown positive responses in phytoplankton productivity and fish production following liming of acidic ponds, and methods for determining the lime requirements of bottom soils have been developed (Boyd, 1995). Nevertheless, liming often is applied to ponds indiscriminately, with no concern for bottom soil pH or total alkalinity and total hardness concentrations. It is doubtful that liming has a large influence where soil pH is above 7 or total alkalinity is above 50 mg L^{-1} (Boyd, 1995).An optimal pH range is between 6.5 and 9 however this will alter slightly depending on the culture species. Aquaculture ponds are ordinarily drained for harvest, bottoms are allowed to dry, and liming materials are applied. Lime is frequently blended with the bottom soil by tilling. The pH is the measure of the hydrogen ion (H+) concentration in soil or water. The pH scale ranges from 0 to 14 with a pH of 7 being neutral. A pH below 7 is acidic and a pH of above 7 is basic. An optimal pH range is between 6.5 and 9 however, this will alter slightly depending on the culture species (Fig. 37).

pH will vary depending on a number of factors. Firstly pH levels of the pond water will change depending on the aquatic life within the pond. Carbon dioxide produced by aquatic organisms when they respire has an acidic reaction in the water. The pH in ponds will rise during the day as phytoplankton and other aquatic plants remove CO_2 from the water during photosynthesis. The pH decreases at night because of respiration and production of CO_2 by all organisms. The fluctuation of pH levels will depend on algae levels within the pond.

Sub-optimal pH has a number of adverse effects on culture animals. It can cause stress, increase susceptibility to disease, low production levels and poor growth. Signs of sub-optimal pH include increase mucus on the gill surfaces of fish, damage to the eye lens, abnormal swimming behaviour, fin fray, poor phytoplankton and zooplankton growth and can even cause death. In the case of freshwater crayfish low pH levels will cause the shell to become soft. This is due to the shell of the cray fish being composed of calcium carbonate which reacts with acid. Sub-optimal pH levels are usually caused by acidic water and soils, poorly buffered water (will be discussed further on) and increased CO_2 production. Treatment methods will depend on whether there is a high pH problem or a low pH problem. To treat a pond with low pH, a pond can be limed with agricultural limestone or fertilized to promote plant growth. To decrease a high pH, the pond can be flushed with fresh water, feeding rates can be reduced to decrease nutrient input into the pond, gypsum ($CaSO_4$) can be added to increase the calcium concentration, or alum ($AlSO_4$) can be added in extreme cases.

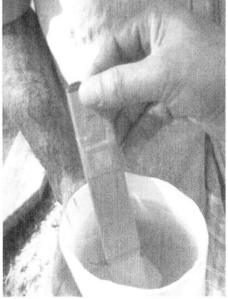

Fig. 38 &38a: Determination of pH by digital HANNA make pen.

Requirement of lime in aquaculture ponds

Liming is the first step in fertilization owing to the fact that it supplies calcium, one of the essential nutrients. The major functions of lime are the following:

(1) It corrects the acidity of soils and water.

(2) Speeds up the decomposition of organic matter, releasing carbon-di-oxide from bottom sediments.

(3) It raises the bicarbonate content and lack of CO_2 does not become a limiting factor.

(4) It counter acts the poisonous effects of excess magnesium, potassium and sodium ions and fixes inorganic acid like sulphuric acid.

(5) Keep the fishes, disease free to some extent.

If the soil of the stocking pond is found to be acidic, it has to be treated with quick lime to bring the soil pH to alkaline condition, which is ideal for fish production. Lime is applied when all the water is drained out and the bottom is well dried, exposed to sun for about 15 days.

During this period, the lime will help to kill all the pathogenic bacteria present in soil which may affect adversely on the fishes stocked in pond after filling the water. The doses of quick lime depend on the pH of the pond soil as given below. The quicklime is either sprinkled or spread on the pond water in the form of a paste or to act as an antiseptic and also to neutralize the toxic effect of old organic deposits at the bottom. It also stabilizes the pH of water at a slightly alkaline level, which enhances the growth of phytoplankton and fish, apart from increasing the calcium content of water. Further, it increases not only the bicarbonate content of the pond but also counteracts poisonous effects of ions like magnesium and sodium.

Table 22: Types of liming materials.

Sl.no	Types of lime	Chemical formulae
01.	Lime stone	$CaCO_3$
02.	Quick Lime	CaO
03.	Slaked Lime	$Ca(OH)_2$
04.	Gypsum	$CaSO_4, 2H_2O$
05.	Dolomite	$CaMg(CO_3)$

From the useful effects of liming given above, it is concluded that liming is an essential and successful tool for pond manuring. A pond containing lime is likely to be more fertile than the one without it (Fig. 41 and 42). The dose of lime depending on the characteristics of water is given below (Table-23 and 24):

Table 23: The doses of lime.

pH	Nature of soil	Quantity of Quicklime (kg/acre)	Quantity of quicklime (kg/Ha)
4.5-5.0	Strong acidity	668	2, 000
5.0-6.5	Medium acidity	334	1, 000
6.5-7.5	Neutral	167	500
7.5-8.5	Medium alkalinity	68	200
8.5-9.5	Strong alkalinity	No application	No application

Table 24: The doses of lime depending on acidic water (pH < 7.0).*

pH	Kg/acre/month	Kg/ha/month
4.0-4.5	1500	4500
4.5-5.0	1135	3400
5.0-5.5	818	2450
5.5-6.0	485	1450
6.0-6.5	318	950

*The quantity may however, be increased by 50%, if the bottom is clayey and reduced by 50% for a sandy bottom):

Fig. 39: Liming pond bottom

Fig. 40: Liming over pond surface

Figs. 41 and 42: The ideal colours of pond water

Calcium is the major ingredient in any liming materials. The effectiveness of different liming materials also differs significantly. The effectiveness of quicklime is double than that of lime stone (Table 23). During the pond preparation period, when the water is dried, quick lime or slaked lime is much more effective and safe (Figs: 39&40).

If the soil pH is below 7.0 limestone or dolomite are much more effective but when higher and turbid, use of gypsum is recommended.

It is always advised to determine water and soil pH before application of lime. At present digital pH pen is available at a cheaper rate in the market (Fig-38 and 38a).

7. Dissolved Oxygen (DO) in pond water

Oxygen is the most fundamental chemical parameter of aquatic ecosystems and its measurement can help in feeling the pulse of the aquatic environment. DO is obviously essential to the metabolism of all aerobic aquatic organisms. Hence, the solubility and especially the dynamics of oxygen distribution in aquatic ecosystems are basic to the understanding of the distribution, behaviour and growth of aquatic organisms. The balance of DO content of water is determined by the interaction of a number of factors like, the amount of oxygen being dissolved, the activity of producers, hydro mechanical distribution of oxygen, respiratory activity of the biota and the rate of decomposition of organic matter.

$$CO_2 + 2H_2O \xrightarrow[\text{green plants}]{\text{light}} (CH_2O) + O_2 + H_2O$$

The net process of photosynthesis (Eq. 1) is often described simply as the fixation of CO_2 (or HCOH3 in water; Eq. 2) catalyzed by several enzymes, including Rubisco*, driven by light and resulting in production of organic matter, O_2 and OH^-:

$$CO_2 + H_2O \rightarrow CH_2O + O_2 \qquad ... (1)$$

$$HCO_3H + H_2O \rightarrow CH2O + O_2 + OHH \qquad ... (2)$$

Rubisco*=Ribulose-1, 5-bisphosphate carboxylase/oxygenase, commonly known by the abbreviations RuBisC, RuBPCase, or RuBPco, is an enzyme involved in the first major step of carbon fixation, a process by which atmospheric carbon dioxide is converted by plants and other photosynthetic organisms to energy-rich molecules such as glucose. In chemical terms, it catalyzes the carboxylation of ribulose-1, 5-bisphosphate (also known as RuBP). It is probably the most abundant enzyme on Earth.

Sources of dissolved oxygen in pond water:

In general there are two major sources of oxygen in pond water. (i) Surface air mixing and (ii) Photosynthesis of algae.

Dissolved oxygen concentrations in pond water vary throughout the day. Dissolved oxygen in the water is obtained through diffusion from air into water, mechanical aeration by wind or aeration systems, and via photosynthesis by aquatic plants. Plants utilize carbon dioxide, water and sunlight to form simple sugars and oxygen – a process known as photosynthesis. Oxygen is also lost from the system via respiration where oxygen is consumed by aquatic organisms (both plants and animals), and by decaying organic matter on the pond floor. It may be inferred in this context that the photosynthesis is light dependent but respiration is a continuous process. Declining oxygen levels can be caused by a number of factors. This includes large blooms of phytoplankton and zooplankton, high stocking rates, excessive turbidity that will limit the amount of photosynthesis occurring and high water temperatures. Levels of dissolved oxygen will also decrease after a series of warm, cloudy, windless days.

In general, the oxygen produced during the day hours, being consumed by respiration throughout 24-hours by all the living organisms present in the aqua system.

In most of the aquaculture ponds, the supplementation of oxygen (Figs. 43a & b and 44a & b), at least where the intensification of composite culture is in practice, diffusion from the atmosphere and production from underwater photosynthesis exceeds the amount used in oxygen-consuming processes, and fish seldom have problems (surfacing of fish: Fig. 45) obtaining enough oxygen to meet normal metabolic demands. In aquaculture ponds, however, the biomass of plants, animals and microbes is much greater than in natural waters hence, oxygen is sometimes consumed faster than it is replenished. Depending on how low the dissolved oxygen concentration is and how long it remains low, fish may consume less feed, having slower growth followed by less efficient feed conversion, resulting more susceptibility to infectious diseases. As a result of which the fishes suffocate and die. In such a situation the intelligent aqua farmers in general, used to install aeration devices (Figs. 43a & b and 44a & b).

By the process of diffusion oxygen moves into or goes out of water. The rate of diffusion (as per Fick's Law of diffusion) depends on the difference in oxygen partial pressures between the water and gas phases- the greater the difference, the greater the driving force moving oxygen from one phase to the other. The maximum rate of oxygen transfer into water occurs when the dissolved oxygen concentration in water is 0 mg/L, the point at which the maximum difference in oxygen partial pressures between water and air occurs (Tucker, 2006).As dissolved oxygen concentrations increase from 0 mg/L, the oxygen partial pressure difference between air and water steadily decreases up to the point where the dissolved oxygen concentration equals the saturation concentration. At that point, there is no difference in oxygen partial pressure between water and air (super saturation of oxygen).Because there is no driving force compelling oxygen molecules to leave or enter water, no oxygen can be added to water no matter how much effort is made to increase turbulence or air water border area. When the dissolved oxygen concentration is greater than the saturation concentration (the water is supersaturated with oxygen), the oxygen partial pressure in water is greater than in air and, oxygen moves out from water to air. In other words, aeration causes the dissolved oxygen concentration to decrease. This process is called degassing.

When and how much to aerate the pond water

The need to aerate the pond varies seasonally because water temperature affects the rates of respiration and photosynthesis. Problems with low dissolved oxygen concentrations are rare when water temperature is consistently below 15^0 C. Problems are common when water temperature is above 27^0 C. Low dissolved oxygen concentration usually occur at night during the summer. Aeration continues until past dawn when measurements of dissolved oxygen indicate that phytoplankton photosynthesis is producing oxygen. Usually next day early morning when the sun starts rising.

When dissolved oxygen concentrations are low, fish congregate in that area and remain there until oxygen conditions improve throughout the pond. The practice of aerating only a portion of the pond and only when concentrations fall to critically low levels has the disadvantage of routinely exposing fish to suboptimal concentrations of dissolved oxygen. Despite this drawback this is the most common aeration practice in commercial fish ponds and is the only practice that has proved to be economically rational. Maintaining dissolved oxygen concentration above a critical threshold throughout the pond has not been shown to be economically justifiable using currently available aeration technology (Tucker, 2006). Putting all the aerators in one end of the pond also reduces the cost of the electrical supply and makes routine maintenance easier (Fig.44a and b).

Fig. 43a

Fig. 43b

Fig. 44a

Fig. 44b

Fig. 43a and b: Aeration of pond water through pumps.
Fig. 44a and b: Multi-wheel and four wheel aerator for oxygenation in ponds.

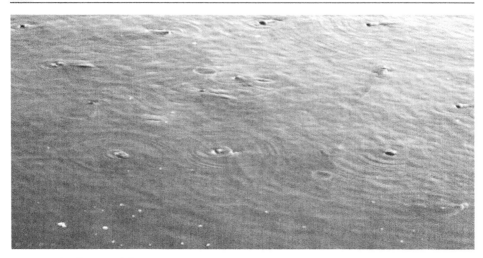

Fig. 45: Surfacing of fishes due to severe depletion of dissolved oxygen in water.

8. Ammonia

Ammonia is toxic to fish if allowed to accumulate in aquaculture production systems. When ammonia accumulates to toxic levels, fish cannot extract energy from feed efficiently. If the ammonia concentration gets high enough, the fish will become lethargic and eventually fall into a coma and die. Even in properly managed fish ponds, ammonia may accumulate to sub lethal concentrations resulting reduced growth, poor feed conversion, and reduced disease resistance— even might at lower than lethal concentrations (Hargreaves and Tucker, 2004).

Ammonia in ponds is produced from the decomposition of organic wastes resulting in the breakdown of decaying organic matter such as algae, plants, animals and uneaten food. Ammonia is also produced by fish and crayfish as an excretory product.

Ammonia is present in two forms in water: as a gas NH_3 or as the ammonium ion (NH_4^+). Ammonia is toxic to culture animals in the gaseous form and can cause gill irritation and respiratory problems.

Ammonia levels will depend on the temperature of the pond's water and its pH. For example at a higher temperature and pH, a greater number of ammonium ions are converted into ammonia gas thus causes an increase in toxic ammonia levels within the freshwater pond.

If high levels of ammonia is present within the pond's water, a number of measures can be taken. These include:

- Reduce or stop feeding,
- Flush the pond with fresh water,
- Reduce the stocking density,
- Aerate the pond,
- In emergencies – reduce the pH level.

Ammonia is the major waste product in the elimination of excess amino acids from the breakdown of proteins in fish feed (Durborow *et al.*, 1997). When fish is fed with high protein diet, they utilise the amino acids from protein digestion and excess amino acids are converted to ammonia which is excreted through the gills and in the faeces. The amount of ammonia excreted by fish varies with the amount of feed that is put into the pond or the species of fish being cultured and it increases with feeding rates (Durborow *et al.*, 1997, Sishula *et al.*, 2011). Ammonia is acutely toxic to fish and can cause loss of equilibrium; increased breathing, cardiac output, and oxygen uptake, and in extreme cases; convulsions, coma and death. Fish exposed to low levels of ammonia, overtime, are more susceptible to bacterial infections and will not tolerate routine handling (Svobodova *et al.*, 1993).

The first signs of ammonia toxicity include slight restlessness, increased respiration and the fish congregating close to the water surface. Affected fish lie on their side and spasmodically open their mouths and gill opercula wide, followed by a short period of apparent recovery. The fish return to normal swimming and appear slightly restless. The skin of ammonia poisoned fish is light in colour, and covered with a thick or excessive layer of mucus. Even at very low concentrations, ammonia still produces many effects in fish including; bacterial infections, reduction in egg hatching success, reduction in growth rate and morphological development, and pathological changes in the tissue of the gills, liver and kidney (Svobodova *et al.*, 1993).

Effects of pH and temperature on ammonia toxicity

Ammonia in water is either unionized ammonia (NH_3) or the ammonium ion (NH_4^+). The techniques used to measure ammonia provide a value that is the sum of both the forms. In the present communication the value as indicated is "total ammonia" or simply "ammonia." pH of pond water directly affects both unionized and the ammonium ion. Un-ionized ammonia is the toxic form and predominates when pH is high. Ammonium ion is relatively nontoxic and predominates when pH is low. In general, less than 10% of ammonia in pond water remains in the toxic form when pH is less than 8.0. However, this proportion increases dramatically as water pH increases. In general, the water pH increase with the increase of photosynthesis and reduction of pH is associated with respiration of the aquatic biota.Therefore, the toxic form of ammonia predominates during the late afternoon and early evening and ammonium predominates from earlier than sunrise through early morning. The equilibrium between NH_3 and NH_4^+ is also affected by temperature. At any given pH, more toxic ammonia is present in warmer water than in cooler water.

Ammonia sources

Fish excretion is principally accounted for the source of ammonia in fish ponds. The rate at which fish excrete ammonia is directly related to the feeding rate and the protein level in feed. As dietary protein is broken down in the body, some of the nitrogen is used to form protein (including muscle), some is used for energy, and some is excreted through the gills as ammonia. Thus, protein in feed is the ultimate source of most ammonia in ponds, where fish are fed.

Another main source of ammonia in fish ponds is diffusion from the sediment. Large quantities of organic matter are produced by algae or added to ponds as feed. Fecal solids excreted by fish and dead algae settle to the pond bottom, where they decompose. The decomposition of this organic matter produces ammonia, which diffuses from the sediment into the water column (Hargreaves and Tucker, 2004).

When is ammonia most likely to be a problem?

Ponds with high and dense algal blooms dominated by one or two species suddenly dies often called as 'algal crash' may lead to higher ammonia concentration because algal uptake is eliminated. Rapid decomposition of dead algae also reduces the dissolved oxygen concentration and pH and increases ammonia and carbon dioxide concentrations. After the crash of an algal bloom, ammonia concentration can increase to 6 to 8 mg/L and the pH can decline to 7.8 to 8.0.

Occasionally during the late afternoons in late summer or early fall

In the late summer or early fall, ammonia concentration begins to increase but daily changes in pH remain large. In these situations, fish may be exposed to ammonia concentrations that exceed the acute criterion for a few hours each day. If late afternoon pH is about 9.0, the acute criterion is about 1.5 to 2.0 mg/

L total ammonia-nitrogen. Total ammoniacal-nitrogen concentrations during summer are typically less than 0.5 mg/L, so fish are unlikely to be stressed if the late afternoon pH is less than 9.0(Hargreaves and Tucker, 2004).

Ammonia Management

Even though practical ammonia management actions may be limited in a large pond aquaculture setting there may be some ways to reduce ammonia levels but others may exacerbate the situation - no method is a complete long-term solution in and of itself.

Reduce feeding rate

Since excess feed and fish excretion are the main culprits of ammonia build up it seems reasonable to feed only what the fish need. This is not a short term fix but better managed over time to help keep ammonia levels reasonable.

Increase aeration

Aeration can be ineffective at reducing overall pond ammonia concentrations due to the relatively small area of the pond being aerated. However, it does increase DO levels causing fish to be less stressed. Avoid vigorous aeration to prevent stirring bottom sediment which can actually increase ammonia concentrations.

Lime

Using lime agents such as hydrated lime or quick lime could actually make a potentially bad situation much worse by causing an abrupt and large increase in pH. Increasing pH will shift ammonia toward the form that is toxic to fish. In addition, the calcium in lime can react with soluble phosphorus, removing it from water and making it unavailable to algae.

Fertilize with phosphorus

Under normal pond conditions, algal blooms are very dense and the rate of algae growth is limited by the availability of light, not nutrients such as phosphorus or nitrogen. Adding phosphorus does little to reduce ammonia concentrations because algae are already growing as fast as possible under the natural conditions in the pond.

Adding bacterial amendments

Common aquatic bacteria are an essential part of the constant cycling of ammonia in pond ecosystems. Typical pond management creates very favorable conditions for bacterial growth. This growth and activity is limited more by the availability of oxygen and by temperature than by the number of bacterial cells. In most amendments, abundant bacteria are responsible for the decomposition of organic matter. Therefore, if bacterial amendments accelerate the decomposition of organic matter, the opposite deleterious effect could occur and ammonia levels could actually increase.

9. Hydrogen sulphide (H_2S)

Fresh water fish pond should be free from H_2S because even at concentration of 0.01 ppm fish lose their equilibrium. Frequent exchange and increase of pH through liming can reduce its toxicity.

10. BOD, COD and DOM in aquaculture ponds

The productivity of freshwater community that determines the fish growth is regulated by the dynamics of its physico-chemical and biotic environment (Wetzel, 1983). For successful fish production from a water body, it is important to study the physico-chemical factors, which influence the biological productivity of the same. The property of water regulate the fish food organism and other bio-community in an aquatic ecosystem and thereby helps in the process of fish productivity, because the biological productivity of any water body is mostly influenced by the quality of water column of aquaculture system. The pollution caused on the other hand is also an important criterion for successful fish production. Biological Oxygen Demand (BOD), Chemical Oxygen Demand (COD) and Dissolved Organic Matter (DOM) are used to know the pollution status of an aquaculture system (Deka, 2015).

BOD is the measurement of total dissolved oxygen consumed by microorganisms for biodegradation of organic matter such as food particles or sewage etc. The excess entry of cattle and domestic sewage from the non-point sources and similarly increase in phosphate in the village ponds may be attributed to high organic load in these ponds thus causing higher level of BOD (Bhatnagar and Devi, 2013).

Desirable limits of BOD

According to Clerk (1986) the BOD range of 2- 4 mg L^{-1} does not show pollution while levels beyond 5 mg L^{-1} are indicative of serious pollution. However, Bhatnagar *et al.*, (2004) opined that the BOD level between 3.0-6.0 mgL^{-1} is optimum for normal activities of fishes; 6.0-12.0 mgL^{-1} is sub lethal to fishes and >12.0 gmL^{-1} can usually cause fish kill due to suffocation. Santhosh and Singh (2007) recommended optimum BOD level in aquaculture ponds should be less than 10 mg L^{-1} but the water with BOD less than 10-15 mg L^{-1} can be considered for fish culture. Bhatnagar and Singh (2010) suggested the BOD <1.6mg L^{-1} level is suitable for pond fish culture and according to Ekubo and Abowei (2011) aquatic system with BOD levels between 1.0 and 2.0 mg L^{-1} considered clean; 3.0 mg L^{-1} fairly clean; 5.0 mg L^{-1} doubtful and 10.0mg L^{-1} definitely bad and polluted.

Remedies

1. Liming at higher doses, suspending use of fertilizers, removal of non-biodegradable / floating organic matter from the pond surface, aeration, screening or skimming to reduce BOD level.

2. Before stocking, pondwater may be allowed to stabilize for few days (5-15 days).

3. Add safe quantities of manure accordingly local conditions of pond in terms of differences in type of manure, water temperature and normal dissolved oxygen.

Chemical Oxygen Demand

COD represents to the oxygen requirements to oxidize all of the organic matter in a water sample to CO_2 and water. COD is highly correlated with chlorophyll 'a' concentration of phytoplankton. Since plankton is the major causative factor for changes in COD and plankton density. Boyd (1982) also showed that the hourly rate O_2 consumption in ponds is closely correlated with turbidity or Secchi disc visibility.

Dissolved Organic Matter

Accumulation of organic matter in pond soils during the grow-out cycle causes severe oxygen depletion at the sediment-water interface. A small amount of organic matter in pond soils is beneficial. However, too much organic matter in pond soils can be detrimental because microbial decomposition can lead to the development of anaerobic conditions at the sediment-water interface, under which organic compounds are often decomposed to reduce substances such as NO_2, H_2S, NH_3, and CH_4, which are toxic to fish at relatively low concentrations (Boyd and Bowman, 1997). It is of primary importance to prevent such situations in fish ponds. Two methods commonly practiced by fish farmers are: 1) polyculture with detritivorous fish and 2) pond drying between cycles of production. Detritivores consume organic matter, but also disturb bottom sediment while feeding, which may increase turbidity and reduce water quality (Pillay, 1992). The drying process enhances oxidation of organic material as well as nutrient regeneration in pond soils, and also allows photo-oxidation and microbial decomposition of organic matter (Fast, 1986).

The colloidal fraction of measured DOM is probably small in most water bodies. In fish ponds, organic matter is present as living plankton, suspended particles of decaying organic matter (detritus) and DOM. The organic matter present in natural waters is organic phosphorus, organic nitrogen, carbohydrate, vitamin etc. The organic matter in a water body may be either allochthonous or autochthonous in origin and may be in either dissolved or suspended states. The presence of organisms ensures a more or less continual addition of organic materials to the water body, some of which immediately or later go into solution. Thus, all waters have certain content of DOM. Cole (1983) is of opinion that DOM in an ecosystem comes from four sources:

i)   Allochthonous input,

ii) Death and decay of aquatic organism,

iii) Excreted metabolites and / or photosynthates of phytoplankton,

iv) Littoral vegetation and excreted metabolites of fauna.

Since carbohydrates contribute predominantly to the major bulk of the total DOM, carbon appears to be of paramount importance as a limiting nutrient.

11. Specific Conductivity of pond water

Conductivity is an index of the total ionic content of water and therefore, indicates freshness of the water (Ogbeibu and Victor, 1995). Conductivity can be used as indicator of primary production (chemical richness) and thus fish production. Conductivity of water depends on its ionic concentration (Ca_2+, Mg_2+, HCO_3^- ,

$CO3^-$, NO_3^- and PO_4^-), temperature and on the variations of dissolved solids. Distilled water has a conductivity of about 1 μ mhos/cm and natural waters have conductivity of 20-1500 μ mhos/cm (Abowei, 2010). Conductivity of freshwater varies between 50 - 1500 hs/cm (Boyd, 1979), but in some polluted waters it may reach 10, 000 hs/cm and seawater has conductivity around 35, 000 hs/cm and above (Bhatnagar and Devi, 2013).

Desirable limits

As fish differ in their ability to maintain osmotic pressure therefore, the optimum conductivity for fish production differs from one species to another. Sikoki and Veen (2004) described a conductivity range of 3.8 -10 hs/cm as extremely poor in chemicals. Stone and Thomforde (2004) recommended the desirable range 100-2, 000 mSiemens/cm and acceptable range 30-5, 000 mSiemens/cm for pond fish culture.

Specific Conductivity of pond water

Conductivity is an index of the total ionic content of water and therefore, indicates freshness of the water (Ogbeibu and Victor, 1995). Conductivity can be used as indicator of primary production (chemical richness) and thus fish production. Conductivity of water depends on its ionic concentration (Ca_2+, Mg_2+, HCO_3^-, $CO3^-$, NO_3^- and PO_4^-), temperature and on variations of dissolved solids. Distilled water has a conductivity of about 1 μ mhos/cm and natural waters have conductivity of 20-1500 μ mhos/cm (Abowei, 2010). Conductivity of freshwater varies between 50 - 1500 hs/cm (Boyd, 1979), but in some polluted waters it may reach 10, 000 hs/cm and seawater has conductivity around 35, 000 hs/cm and above (Bhatnagar and Devi, 2013).

12. Primary productivity of pond water

The basis of all biosphere function is primary productivity which evidently means the formation of organic matter by photosynthetic plants incorporating solar energy. More explicitly this is referred to in the test of plant physiology as phtosynthesis or the conversion of carbon dioxide, water and the kinetic energy or sunlight into oxygen and sugars which is obviously an energy binding reduction process.

The autotrophic component of any ecosystem, functionally known as the producers, forms the first and fundamental step of a food chain. Its importance on terrestrial ecosystem is well known since the produce of primary productivity is being largely used as the basic food of man and fodder for the primary consumers. Virtually, the entire biotic world depends on primary production directly or indirectly. Of course, the primary production of aquatic ecosystem is no less important although the produce is not widely used presently as basic food by man. The future potentialities of aquatic production are not bleak, if not promising. Idyll (1978), in his book entitled, **'Sea against hunger'** explored the possibilities of utilizing aquatic produce as food for man. Again it may be emphasized that the aquatic primary productivity is the keystone of aquaculture.

The rate of energy transformation from solar radiations to chemical energy of producers by chlorophyll bearing organisms gives a dependable parameter for assessing the productivity of aquatic ecosystems. The process of photosynthetic oxygenation is represented by the basic equation:

$$6CO_2 + 6H_2O + 709 \text{ cal solar energy} = C_6H_{12}O_6 + 6O_2$$

From the above equation it is evident that the energy required to liberate 1 mg of oxygen through algal photosynthesis is approximately 3.6 calories and hence the amount of oxygen produced during photosynthesis gives a measure of solar energy trapped during photosynthesis.

One of the aspects of ecosystem dynamics that has gained considerable importance in modern ecological research is the primary productivity. Primary productivity of an ecological system, community or any part there of is defined as the rate at which the radiant energy is stored by photosynthetic and chemosynthetic activity of producer organism (chiefly green plants) in the form of organic substance which can be used as food materials (Odum, 1971). Primary productivity thus denotes the rate of primary production, i.e., primary production per unit of time and area. Primary production which refers to the quantity of new organic matter produced by photosynthesis (or its energy equivalent fixed by the autotrophs), is the initial process in the ecosystem functioning. It is the main spring of life and cornerstone of community metabolism in any ecosystem.

Photosynthetic production of carbon in the inland aquatic ecosystems occurs in various plant communities such as phytoplankton, periphytic algae, benthic algae and macrophytes. Production by the phytoplankton, the prime synthesist, is the most important phenomenon and this reflects the nature and degree of productivity in aquatic ecosystems. This has received much attention in limnological studies during the first few decades and it has been measured by several workers in various aquatic ecosystems of the world. The historical survey of primary productivity research was traced by Leith (1975) and according to him there are at least three major periods in the history of primary productivity study. These are:

(1) Before Liebig

(2) From Liebig to International Biological Programme (IBP)

(3) IBP to consequence.

Primary productivity in an aquatic ecosystem depends on a number of physico-chemical, biological, climatological and geographical factors. In an extensive study programme of productivity under the auspices of IBP (summarized by Brylinsky, 1980) almost all possible climatological, geographical, physico-chemical and biological variables were taken into consideration. All these variables were classified under three major groups:

(1) Driving variables which include those factors having an effect on a system but which are not in turn affected by the system. Examples include incident radiation, precipitation, wind and nutrient input from the drainage area.

(2) Site parameters (or site constants) which include those variables which are specific for a particular site, i.e., they do not vary within a particular site. Their magnitude however, may vary from site to site. Examples include latitude and longitude, altitude, morphometic characteristics of the water body, viz. mean depth, surface area, volume etc.

(3) The site variables are those which we wish to make predictions about. Theoretically their variations are the result of the interactions between driving variables and site constants. Examples of this kind include all the chemical and biological variables and physical variables such as water temperature, Secchi disc transparency, duration of summer stratification etc.

The subject of under water attenuation of light has been thoroughly studied by a number of authors and the effects of all possible interfering elements have been examined (Roberts and Zohary, 1992; Mukhopadhyay *et al.*, 1997). Vallentyne (1965) reported that in terms of the visible radiation incident upon the earth surface an overall efficiency of the net primary productivity are approximately 0.4% for land, 0.2% for the ocean and 0.2-0.3% for the earth as a whole. In some cases, all the energy available for heterotrophs in the aquatic system does not come through the fixation by autotrophs but a large amount of energy also comes from allochthonous sources (Gaedke *et al.*, 2002, Silva, 2007 and Nath, 2010).

Table 25: Basic guideline of various parameters for aquaculture practice for the fish farmers in obtaining high fish yield in low input following water, soil and plankton quality maintenance of their ponds. (Bhatnagar and Devi, 2013. and Banerjea, S.M.1967.)

Sl.no	Parameters	Desirable limits	Remedial measures
01.	Water temperature(^{0}C)	i) 30-35 is tolerable to fish. ii) 28-32 good for tropical major carps. iii) <10for cold water species. iv) 25-30 ideal for shrimps. v) < 20 sub lethal for growth and survival for fishes and > 35 lethal to maximum number of fish species.	1. By water exchange, planting shady trees or making artificial shades during summer's thermal stratification can be prevented. 2. Mechanical aeration can prevent formation of ice build-up in large areas of the pond
02.	Transparency (cm)	30 and 40 cm	Addition of more water or lime (CaO, alum Al_2 (SO4)$_3$14H2O @ 20 mg/l and gypsum on the entire pond water @ 200 Kg/ 1000m3 of pond increase transparency.
03.	Water color	Pale color, light greenish or greenish waters suitable for fish culture.	Application of organic and inorganic fertilizers in clear water ponds may increase productivity.
04.	Dissolved oxygen (mg/l)	>5ppm is essential to support good fish production	(i) Avoid over application of fertilizers and organic manure to manage DO level (ii) Physical control aquatic plants and also management of phytoplankton biomass (iii) Recycling of water and use of aerators. (iv) Artificially or manually beating of water. (v) Avoid over stocking of fishes. (vi) Introduction of the hot water gradually with pipes to reduce if DO level is high.
05.	Biochemical oxygen demand (BOD)	2 to 4 mg/l but 3.0-6.0 mg/l is optimum for normal activities of fishes;	1. Add lime suspending use of fertilizers, and removal of nonbiodegradable / floating organic matter from the pond surface, aeration, screening or skimming to reduce BOD level. 2. Before stocking, pond water may be allowed to stabilize for few days (5-15 days). 3. Add safe quantities of manure accordingly local conditions of pond in terms of differences in type of manure, water temperature and normal dissolved oxygen.

[Table Contd.

Contd. Table]

Sl.no	Parameters	Desirable limits	Remedial measures
06.	Carbon-dioxide (CO_2)	Ideal level of CO_2 in fishponds is less than 10 mg/l.	1. Proper aeration can "blow" off the excess gas 2. Check organic load and reduce the same by adding more water (no fish) and add Muriatic acid (swimming pool acid) to adjust the pH to about 5 or if possible remove the matter by repeated nettings. 3. Use of lime ($CaCO_3$) or sodium bicarbonate ($NaHCO_3$) (iv) Application of potassium permanganate at the rate 250 g for 0.1 hectare.
07.	pH	7.5 to 8.5 is ideal	1. Add gypsum ($CaSO4$) or organic matter (cowdung, poultry droppings etc.) and initial pre-treatment or curing of a new concrete pond to reduce pH levels. 2. Use of quicklime (CaO) to rectify low pH of aquatic body.
08.	Alkalinity	50–150 mg /l	1. Fertilize the ponds to check nutrient status of pond water 2. Alkalinity can be increased by calcium carbonate, concrete blocks. oyster shells, limestone, or even egg shells depending upon soil pH and buffering capacity.
09.	Hardness	30–180 mg/l	1. Add quicklime/alum/both and add zeolite to reduce hardness. 2. During heavy rainfall avoid the runoff water to bring lot of silt into the fish pond.
10.	Calcium	Free calcium in culture waters is 25 to 100 mg L–1 (63 to 250 mg/l $CaCO_3$ hardness).	Calcium is generally present in soil as carbonate and most important environmental, divalent salt in fish culture water. Fish can absorb calcium either from the water or from food.
11.	Conductivity	Varies between 50 to 1500 hs/cm in fresh waters. Desirable range 100–2.000 mSiemens/cm and acceptable range 30–5.000 m Siemens/cm for pond fish culture.	Conductivity is an index of the total ionic content of water and indicator of primary production (chemical richness) and thus fish production

[Table Contd.

Contd. Table]

Sl.no	Parameters	Desirable limits	Remedial measures
12.	Salinity	Desirable range0- 2 ppt for carps, 10-28ppt for shrimps and brackish water species.	1. Salinity is increased or diluted by replenishment of water. 2. Aeration is essential to equalise the water salinity all over the water column.
13.	Chloride	Chloride (in the form of salt) is required at a minimum concentration of 60 mg /l and a ratio of chloride to nitrite of 10:1 reduces nitrite poisoning of fish.	Control harmful bacteria and is useful to fish in maintaining their osmotic balance.
14.	Ammonia (NH_3)	Between 0.6 and 2.0 mg/l for pond fish, 0.01-0.5 ppm is desirable for shrimp; >0.4 ppm is lethal to many fishes & prawn species.	1. Increase pond aeration. 2. Addition of liming agents such as hydrated lime or quick lime decreases ammonia and this technique is effective only in ponds with low alkalinity. 3. Regular water exchange.
15.	Nitrite-N (NO_2 -N)	Nitrite is an intermediate product of the aerobic nitrification bacterial process, produced by the autotrophic Nitrosomonas bacteria combining oxygen and ammonia. It oxidizes haemoglobin to methemoglobin in the blood, turning the blood and gills brown and hindering respiration also damage for nervous system, liver, spleen and kidneys of the fish. Desirable range 0-1 mg/l. It should not exceed 0.2 mg/l in freshwater and 0.125 mg/l in seawater.	1. Reduction of stocking densities, Improvement of feeding, biological filtration and general husbandry procedures. Increase aeration to maximum, Stop feeding. 2. Addition of small amounts of certain chloride salts, regular water change out. 3. Use of biofertilizers to accelerate nitrification.

[Table Contd.

Contd. Table]

Sl.no	Parameters	Desirable limits	Remedial measures
16.	Nitrate-N(NO_3-N)	Favourable range of 0.1 to 4.0 mgl in fish culture water, It is harmless and non-toxic.	Dilution by water change.
17.	Phosphate-Phosphorus (PO_4-P) in surface water mainly present as bound to living or dead particulate matter and in the soil is found as insoluble $Ca_3(PO_4)_2$ and adsorbed phosphates on colloids	SoilPO_4-P:<3: Productive, 3-6 is average production and >6 is unproductive. According to some school of thought 0.05-0.07ppm is optimum and productive, 1.0ppmis good for plankton and shrimp production.	Use inorganic fertilizers to increase phosphorus level (N: P=15:30)
18.	Primary productivity	This is the rate at which photosynthesis takes place in per unit of time. 1.60-9.14 mgC/l/d (GPP)-as optimum status and 20.3 mgC/l/d (GPP)-as poor productivity of a pond culture. According to some school of thought ideal value of primary productivity is 1000-2500 mgC/ M3/d (=1.0-2.0 mgC/ L/ d). If the productivity value is less than 0.5mg /m^3 (<1.5mg'Ch-a'/m^3) it is considered as oligotrophic. 1-5gm/m^3 (1.5-10mg ' ch-a'/m^3) it is mesotrophic. 5-10gm/m^3 (10-25mg'ch-a'/ m^3) it is eutrophic and >10 gm/m3 (>25mg'ch-a'/m^3) as highly eutrophic i.e.. polluted one.	1. Productivity can be improved by use of organic/ inorganic fertilizers in ponds. 2. In case of plankton bloom / swarm; feed/manure application can be suspended for some time.

Table 26: At a glance: The optimum ranges of various water quality parameters.

Sl.Nos.	Parameter	Acceptable range	Desirable range	Stress
01.	Temperature (°C)	15-35	20-30	<12 >35
02.	Transparency(Cm)	—	30-80	<12 > 80
03.	Water colour	Pale to light green	Light green to light brown	Clear water, Dark green &Brown
04.	Dissolved oxygen (mg/l)	3-5	5	5- 8
05.	BOD (mg/l)	3-6	1-2	>10
06.	CO2 (mg/l)	0-10	<5, 5.8	>12
07.	pH	7.0-9.0	7.5-8.5	<7.0 >9.0
08.	Alkalinity (mg/l)	50-200	25-100	>20 <300
09.	Hardness (mg/l)	>20	75-150	>20 <300
10.	Calcium (mg/l)	4-160	25-100	>10 <250
11.	Ammonia (mg/l)	0-0.05	0-0.025	>0.03
12.	Nitrite (mg/l)	0.02-2.0	<0.02	>0.2
13.	Nitrate (mg/l)	0-100	0.1-4.5	>5 <0.01
14.	Primary productivity(mgC/l/d)	1-15	1.6-9.14	<1.6 >20.3
15.	Phosphorus (mg/l)	0.03-2.0	0.01-3	>3
16.	H2S(mg/l)	0-0.02	0.002	Any detectable level
17.	Plankton (No./l)	2000-6000	3000-4500	<3000 >7000

4. BIOTIC COMPONENTS OF NATURAL WATERS

a. Plankton

Plankton is a general term for the "floaters", the small or microscopic organisms in a body of water that drift with the movement of water. This includes zooplankton (animal origin), phytoplankton (plankton that is capable of photosynthesis or of plant origin), and bacterioplankton (bacteria). Plankton is an important food source for fish and other larger organisms.

Plankton Size Groups

Although most people think of plankton as microscopic animals, there are larger plankton. With their limited swimming capability, jellyfish are often referred to as the largest type of plankton. In addition to being categorized by life stages, plankton can be categorized into different groups based on size.

These groups include:

- Femtoplankton - Organisms under 0.2 micrometers in size, e.g., viruses

- Picoplankton - Organisms 0.2 micrometer to 2 micrometers, e.g., bacteria

- Nanoplankton - Organisms 2-20 micrometers, e.g. phytoplankton and small zooplankton

- Microplankton - Organisms 20-200 micrometers, e.g., phytoplankton and small zooplankton

- Mesoplankton - Organisms 200 micrometers to 2 centimeters, e.g., phytoplankton and zooplankton such as copepods. At this size, the plankton are visible to the naked eye.

- Macroplankton - Organisms 2 centimeters to 20 centimeters, e.g., like ctenophores, salps, and amphipods.

- Megaplankton - Organisms over 20 centimeters, like jellyfish, ctenophores, and amphipods.

The categories for the smallest plankton sizes were needed more recently than some others. It wasn't until the late 1970's those scientists had the equipment available to help them see the great number of planktonic bacteria and viruses in the ocean.

Development of phytoplankton (algae) is a common phenomenon in any natural water systems (Primary producers). Depending on the presence or absence of the phytoplankton the fish farmers usually can determine the extent of productivity potential of the pond waters.

Table. 27: Composition of phytoplanktonic associations.

01.	Blue-green algae	Myxophyceae or Cyanophyta
02.	Green algae	Chlorophyceae
03.	Yellow-green algae	Xanthophyceae
04.	Golden-brown algae	Chrysophyceae
05.	Diatoms	Bacillariophyceae
06.	Cryptomonads	Cryptomonadinae
07.	Dinoflagellates	Dinophyceae
08.	Euglenoids	Euglenophyceae
09.	Brown algae	Pheophyceae
10.	Red algae	Rhodophyceae

b. Phytoplankton

Blue-green algae or, Myxophyceae or Cyanophyceae

These occur in unicellular, filamentous, and colonial forms, and most are enclosed in mucilaginous sheaths either individually or in colonies. A majority of the planktonic blue-greens consists of members of the coccoid family Chroococcaceae (e.g., *Anacystis* = Microcystis, Gomphosphaeri= *Coelosphaerium*, and *Coccochloris*) and filamentous families belonging to Oscillatoriaceae, Nostocaceae , and Rivulariaceae (e.g., *Oscillatoria*, Lyngbya, *Aphanizomenon*, *Anabaena*).

Blue-green algae, also known as Cyanobacteria, are a group of photosynthetic bacteria that many people refer to as "pond scum." Blue-green algae are most often blue-green in color, but can also be blue, green, reddish-purple, or brown. Blue-green algae generally grow in ponds, and slow-moving streams when the water is warm and enriched with nutrients like phosphorus or nitrogen. When a blue-green algae bloom dies off, the blue-green algae cells sink and are broken down by microbes. This breakdown process requires oxygen and can create an increase in biological oxygen demand resulting in decrease in oxygen concentration in the water, and this can adversely affect fish and other aquatic life, and can even result in fish kills.

Blue-green algal toxins are naturally produced chemical compounds that sometimes are produced inside the cells of certain species of blue-green algae. These chemicals are not produced all of the time and there is no easy way to tell when blue-green algae are producing them and when they are not. When the cells are broken, the toxins may be released. Sometimes this occurs when the cells die off naturally and they break open as they sink and decay in a pond. Cells may also be broken when the water is treated with chemicals meant to kill algae, and when cells are swallowed and mixed with digestive acids in the stomachs of animals.

Many different species of blue-green algae occur in waters, but the most commonly detected include *Anabaena* sp., *Aphanizomenon* sp., *Microcystis* sp., and *Planktothrix* sp. It is not always the same species that blooms in a given waterbody, and the dominant species present can change over the course of the season.This group of algae is harmful to shrimp. Some of the algae cause scum on the water surface, like Microcytis sp., it makes shrimp smell fishy and foul; discharges gel through cell membrane, and cause blockage in shrimp's gills.

Green algae or, Chlorophyceae

These are an extremely large and morphologically diverse group of algae that is almost totally freshwater in distribution. A majority of the planktonic green algae belong to the orders Volvocales (e.g., *Chlamydomonas, Sphaerocystis, Eudorina,* and *Volvox*) and Chlorococcales (e.g., *Scenedesmus, Ankistrodesmus, Selenastrum,* and *Pediastrum*). Many members are flagellated (2 or 4, rarely more) at least in the gamete stages; in the desmids (Conjugales or Desmidiales).There are more than 4, 000 species of green algae. Green algae may be found in marine or freshwater habitats, and some even thrive in moist soils.

The most important algal species are Chlorella, which are used as nutrient source in aquaculture production worldwide. Chlorella belongs to the green algae group and is being granted GRAS (Generally Recognized as Safe) status.

Chlorella minutissima is a unicellular microalga without flagella recognized as an oil-rich green alga that exhibits many attractive features, such as easy cultivation, fast growth, and high levels of amino acids and polyunsaturated fatty acids (PUFA's). These attractive characteristics make *Chlorella minutissima* as a potential microalga in pharmaceuticals and health foods .Falaise *et al.* (2016) have reviewed that Chlorella are versatile as they play important roles as antimicrobial agent's like anti-bacterial, antifungal and antiviral activities against related diseases in aquaculture. For this end the "green water" technique is a useful tool against bacterial disease in aquaculture.

Antibacterial compounds from microalgae can be lipids or fatty acids. An anti-marine bacterial activity was demonstrated in vitro with the polyunsaturated fatty acid (PUFA) such as the PU free FA in *P. tricornutum*, identified as eicosapentaenoic acid.

PUFAs, such as Arachidonic Acid (AA) (20:4 omega-6), eicosapentaenoic acid (EPA) (20:5 omega-3) and docosahexaenoic acid (DHA) (22:6 omega-3) are essential for the growth and survival of fish larvae. In fact, fish larvae have a very limit ability to synthesize PUFAs, which must be derived from zooplankton such as rotifers that consume algae. In rotifer production, microalgae could increase the DHA and EPA contents of the rotifers even with a short-term enrichment period. However, to observe more positive effects on growth and survival of fish larvae using rotifers with short term enrichment in microalgae, microalgae need also to be added as "green water".

This group contains no toxicity, usually has small size and doesn't make shrimp smell and this types of algae in the water can greatly affect color and turbidity of aquaculture water.

Yellow-green algae or Xanthophyceae

These are unicellular, colonial, or filamentous algae that are characterized by conspicuous amounts of carotenoids in comparison to chlorophylls. Nearly all of the motile cells possess two flagella. A majority of them are associated with substrata, and many are epiphytic on larger aquatic plants. A few members are planktonic and include common genera such as *Chlorobotrys*, *Gloeobotrys*, and *Gloeochloris*.

Golden-brown algae or Chrysophyceae

These have a dominance of carotene and specific xanthophyll carotenoids, in addition to chlorophyll. Most of them are unicellular, a few are colonial; they are rarely filamentous. A number of them from important components of the phytoplankton. The unicellular species with a single flagellum (e.g., *Chromulina*, *Chrysococcus, and Mallomonas*) are usually very small algal constituents of the nannoplankton. Larger colonial forms such as *Synura*, *Chrysosphaerella*, *Uroglena* and particularly Dinobryonare widely distributed. Certain species of *Dinobryon* and Uroglena develop in lakes of very low phosphorus concentrations, while other species of Dinobryon and *Synura* have high phosphorus requirements.

Diatoms or Bacillariophyceae

They are a most important group of phytoplankton even though most species are sessile and associated with littoral substrata. Both unicellular and colonial forms are common.The group is commonly divided into the centric diatoms (Centrales), which have radial symmetry, and the pennate diatoms (Pennales) which exhibit essentially bilateral symmetry.The four major groups of pennate diatoms are differentiated on the basis of cell thickenings and dilations:

a. the **Araphidineae** (e.g., Asterionella, Diatoma, Fragilaria, Synedra) possess a pseudoraphe (i.e., a depression in the axial areas of the cell wall).

b. the **Raphidioidineae** (e.g., Actinella, Eunotia) in which a rudimentary raphe (i.e., a slit traversing all or part of the cell wall) occurs at the cell ends.

c. Monoraphidineae (e.g., Achnanthes [10 μm], Cocconeis [10 μm]) which have a raphe on one valve and pseudoraphe on the other, and

d. The **Biraphidineae** (e.g., Amphora, Cymbella, Gomphonema, Navicula, Nitzschia, Pinnularai, Surirella) in which the raphe occurs on both valves.

Diatoms are a major group of algae, specifically microalgae, found in the oceans, waterways and soils. Living diatoms number in the trillions: *they generate about 20 percent of the oxygen produced on the planet each year; take in over 6.7 billion metric tons of silicon each year from the waters in which they live; and contribute nearly half of the organic material found in the oceans. The shells of dead diatoms can reach as much as a half mile deep on the ocean floor.*

Movement in diatoms primarily occurs passively as a result of both water currents and wind-induced water turbulance however, male gametes of centric diatoms have flagella, permitting active movement for seeking female gametes. Similarly to plants, diatoms convert light energy to chemical energy by photosynthesis, although this shared autotrophy evolved independently in both lineages. Unusually for autotrophic organisms, diatoms possess a urea cycle, a feature that they share with animals, although this cycle is used to different metabolic ends in diatoms.

Aquaculture ponds with significant levels of diatoms are desirable for shrimp production because of the diatoms' high nutritional value, particularly for younger shrimp.

Silicon is second only to oxygen in abundance in the earth's crust. Much sand consists of silica (silicon dioxide or SiO_2), and clay minerals are hydrous aluminum silicates. Natural waters contain silicon because of the dissolution of silicate minerals with which they come in contact. For example, silicon dioxide reacts in water to form silicic acid, a weak acid that is largely unionized within the pH range of most natural waters. When calcium silicate reacts with carbon dioxide in water, the resulting dissolved substances are calcium ions, bicarbonate ions (alkalinity) and silicic acid.

Silicon concentrations in natural waters typically are reported in terms of SiO_2 and usually range from 5 to 25 mg/L in freshwater bodies. The global average for silica in river water is 13.1 mg/L. Normal seawater contains 6.4 mg/L silica. A silica concentration can be converted to silicon concentration by multiplying by the factor 0.467, the proportion of silicon in SiO_2.

Plants take up silicic acid from water. Silicon in higher plants is incorporated into cell walls, making stems and leaves more rigid and strong. Among the phytoplankton, diatoms have a particular need for silicon, because their frustules – the hard but porous cell walls – are composed almost entirely of silica. Depending upon the species, plants contain from less than 0.1% to as much as 10.0% silicon on a dry-weight basis. Diatoms contain the greatest amount of silicon.

The ratio of carbon: nitrogen: silicon: phosphorus in diatom cells averages 106:15:16:1. Thus, diatoms require about the same amounts of nitrogen and silicon for growth. There is evidence that nitrogen: silicon ratios above 3:1 lessen the growth rate of diatoms.

Diatoms have very short life cycles in the ponds; algae crashes and low-oxygen are major threats and brown gill problem. They are responsible for changing water colour.

Cryptomonads or Cryptomonadinae

Most of these algae are naked, unicellular, and motile. This class is very small and most of the planktonic members belong to the Cryptomonadineae (e.g., Cryptomonas, Rhodomonas, and Chroomonas). Dense populations of these algae often develop during cold periods of the year under relatively low light conditions.

Dinoflagellates

They are unicellular flagellated algae, many of which are motile. A few species of the order Gymnodiniales (e.g., Gymnodinium) are naked or without a cell wall; but most develop a conspicuous cell wall, the Peridiniales (e.g., Ceratium, Glenodinium, Peridinium). Although many of this group may change in size and pattern seasonally, only a few undergo seasonal polymorphism or cyclomorphosis e.g. dinoflagellate ceratium. Presumably, these changes are of adaptive significance in that they reduce the rate of sinking out of the photic zone.

The dinoflagellates (Greek δίνος dinos "whirling" and Latin flagellum "whip, scourge") are a large group of flagellate eukaryotes that constitute the phylum Dinoflagellata. Most are marine plankton, but they also are common in freshwater habitats. Their populations are distributed depending on sea surface temperature, salinity, or depth. Many dinoflagellates are known to be photosynthetic, but a large fraction of these are in fact mixotrophic, combining photosynthesis with ingestion of prey (phagotrophy). In terms of number of species, dinoflagellates are one of the largest groups of marine eukaryotes, although this group is substantially smaller than diatoms. About 1,555 species of free-living marine dinoflagellates are currently described. Another estimate suggests about 2, 000 living species, of which more than 1, 700 are marine (free-living, as well as benthic) and about 220 are from fresh water. The latest estimates suggest a total of 2, 294 living dinoflagellate species, which includes marine, freshwater, and parasitic dinoflagellates.

A bloom of certain dinoflagellates can result in a visible coloration of the water colloquially known as red tide, which can cause shellfish poisoning if humans eat contaminated shellfish. Some dinoflagellates also exhibit bioluminescence—primarily emitting blue-green light.

Unfortunately, a small number of species also produce potent neurotoxins that can be transferred through the food web where they affect and even kill the higher forms of life such as zooplankton, shellfish, fish, birds, marine mammals.

Some species, such as the dinoflagellate *Alexandrium tamarense* produce potent toxins, which are liberated when the algae are eaten and many types of algae in this group conveying toxins that can make fishes and shrimp die.

Euglenoids or Euglenophyceae

Few species are truly planktonic. Almost all are unicellular, lack a distinct cell wall, and possess one, two, or three flagella. Their development in the phytoplankton occurs most often in seasons, strata, or lake/pond systems in which concentrations of ammonia and especially dissolved organic matter are high. However, these algae are found most often in shallow water rich in organic matter such as farm ponds.

Brown algae or Phaeophyta

They are mostly filamentous or thalloid, and are almost exclusively marine. The genera that are found in fresh water are attached to substrata, such as rocks. No species is planktonic.

Red algae or Rhodophyta

None is planktonic and sparsely represented in fresh water. The thalloid or filamentous species (e.g., Batrachospermum) are nearly all restricted to fast-flowing streams of well-oxygenated, cool waters.

Table 28: Characteristics of common major algal associations of the phytoplankton in relation to increasing pond water fertility (Wetzel, 1983).

General lake trophy	Water characteristics	Dominant algae	Other commonly occurring algae
Oligotrophic	Slightly acidic	Desmids *Staurodesmus, Staurastrum*	*Sphaerocystis, Gloeocystis, Rhizosolenia, Tabellaria*
Oligotrophic	Neutral to slightly alkaline; nutrient-poor lakes	Diatoms, especially *Cyclotella* and *Tabellaria*	Some *Asterionella* spp., some *Melosira* spp., *Dinobryon*
Oligotrophic	Neutral to slightly alkaline; generally nutrient poor.	Dinoflagellates, especially some *Peridinium* and *Ceratium*spp.	Small chrysophytes, cryptophytes, and diatoms
Eutrophic	Usually alkaline; nutrient enriched; common in warmer periods of temperate lakes or perennially in enriched tropical ponds and lakes.	Diatoms much of year, especially *Asterionella* spp., *Fragilaria crotonensis, Synedra, Stephanodiscus*, and *Melosira granulate.* Blue-green algae, especially *Anacystis* (= *Microcystis*), *Aphanizomenon, Anabaena*	Many other algae, especially greens and blue-greens during warmer periods of year; desmids if dissolved organic matter is fairly high.Other blue-green algae; euglenophytes if organically enriched or polluted.

Phytoplankton, better known as algae, is the basis of all biosphere function as a primary producer passes the stored energy to its next trophic level i.e the primary consumers (Zooplankton), thus forming a food chain in a water body. However, excessive growth of this phytoplankton in a waterbody (Bloom) some times becomes a nuisance. These nasty, smelly filamentous mats on the surface of the water can lead to fishing frustration and reduced aesthetics, and can negatively impact recreational activities like swimming and boating. Nuisance phytoplankton can also be suspended in the water column, often called unicellular blooms. These blooms typically occur later in the season, reducing the water clarity and often giving the water a "pea soup" green appearance. They are usually associated with cyanobacteria (blue-green algae) and can negatively impact water clarity and water chemistry. Cyanobacteria can also be toxic and cause serious health problems not only to the stocked fishes in the water body but also to the other animals those are used to drink, bath or otherwise make use of it and so also the humans (Fig. 46).

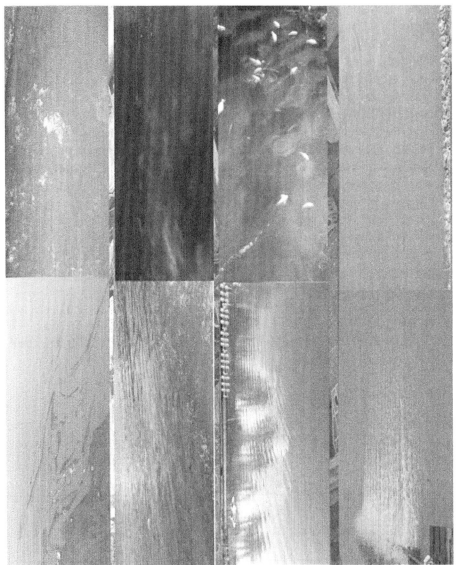

Fig. 46: Bloom of phytoplankton in natural waters

c. Zooplankton

Fresh water zooplankton is extremely diverse, and includes representatives suspended in water with limited powers of locomotion. Like phytoplankton, they are usually denser than water, and constantly sink by gravity to lower depths. The distinction between suspended zooplankton having limited powers of locomotion, and animals capable of swimming independently of turbulence-the latter referred to as nekton- is often diffuse. Freshwater zooplankton is dominated by four major groups of animals: protozoa, rotifers, and two subclasses of the crustacea, the cladocerans and copepods. The planktonic protozoa have limited locomotion, but the rotifers, cladoceran and copepod microcrustaceans, and certain immature insect larvae often move extensively in quiescent water. Many pelagial protozoa (5-300 μm) are meroplanktonic, in that only a portion, usually in the summer, of their life cycle is planktonic. These forms spend the rest of their life cycle in the sediments, often encysted throughout the winter period. Many protozoans feed on bacteria-sized particles (most cells <2μm), and thereby utilize a size class of bacteria and detritus generally not utilized by large zooplankton. Although most rotifers (150μm-1mm) are sessile and are associated with the littoral zone, some are completely planktonic; these species can form major components of the zooplankton. Most rotifers are nonpredatory, and omnivorously feed on bacteria, small algae, and detrital particulate organic matter. Most food particles eaten are small (<12μm in diameter). Most cladoceran zooplankton are small (0.2 to 3.0 mm) and have a distinct head; the body is covered by a bivalve carapace. Locomotion is accomplished mainly by means of the large second antennae. Planktonic copepods (2-4 mm) consist of two major groups, the calanoids and the cyclopoids. These two groups are separated on the basis of body structure, length of antennae, and legs.

Types of Zooplankton

Zooplankton may be classified according to their size or by the length of time they are planktonic (largely immobile). Some terms that are used to refer to plankton include:

- **Microplankton**: organisms that are 2-20 μm in size — this includes some copepods and other zooplankton.

- **Mesoplankton**: organisms that are 200 μm-2 mm in size, which includes larval crustaceans.

- **Macroplankton**: organisms 2-20 mm in size, which includes euphasiids (e.g., krill) - an important food source for many organisms, including baleen whales.

- **Micronekton**: organisms 20-200 mm in size. Examples include some euphasiids and cephalopods.

- **Megaloplankton**: planktonic organisms greater than 200 mm in size, which includes jellyfish and salps.

- **Holoplankton**: organisms that are planktonic throughout their entire live, such as copepods.

- **Meroplankton**: organisms that have a planktonic stage, but grow out of it at some point, like fish and crustaceans.

Zooplankton and the Food Web

Zooplankton are basically the second step of the aquatic food web. The food web starts with the phytoplankton, which are primary producers. They convert inorganic substances (e.g., energy from the sun, nutrients such as nitrate and phosphate) into organic substances. The phytoplankton, in turn, is eaten by zooplankton, which are eaten by smaller fish and even gigantic whales present in marine waters.

Composition of zooplanktonic associations

Fresh water Zooplankton is an important component in aquatic ecosystems, whose main function is to act as primary and secondary links in the food chain. Primarily the physical and chemical environment shapes their community structure. However, these communities are also influenced by biological interactions, predation and their specific competition for food resources (Nevas *et al.*, 2003). The zooplankton community composition and structure is affected by eutrophication, these communities can also be used as the indicator of changing trophic status of an aquatic ecosystem. Zooplankton has long been used as indictors of the eutrophication (Weber Mona *et al.*, 2005).

The fresh water zooplankton can broadly be devided intofour major groups:

Rotifera

Rotifers also called as rotatoria or wheel animalcules are a group of small, usually microscopic, pseudocoelomate animals which have been variously regarded as a separate phylum.The species of rotifers are very common in summer and monsoon.The importat species are: *Brachionus calyciflorus, Brachionus angularis, Brachionus caudatus, Kertella tropica, Kertella cochlearis, Kertella Spp., Filina Spp., Filina longiseta, Rotaria, Monostyla bulla, Trichocerca Spp.*

Cladocera

Cladocerans popularly called as "water flea" prefer to live in deep water and constitute a major item of food for fish. Thus they hold key position in food chain and energy transformation (Uttangi, 2001).*Alona rectangula, Chydorus reticulata, Daphina carinata, Daphnia magna, Moina Spp., Bosminopsis deitersi.*

Copepoda

Fresh water copepods constitute one of the major zooplankton communities occurring in all types of water bodies and ranging from free living to parasitic forms. They serve as food to several fishes and play a major role in ecological pyramids. Water temperature and availability of food organisms affect the copepoda population. *Naupli, Cyclops viridis, Paracyclops Spp., Mesocyclops* Spp., *Diaptomus* Spp.

Ostracoda

Ostracodes are small crustaceans having the bivalve carapace enclosing the laterally compressed body. The freshwater ostracodes occur in lakes, tanks, swamps, streams, and even polluted waters. The higher population of ostracodes during monsoon may be due to the abundance of fine detritus to which omnivorous organisms switch over form their natural benthic habitat and bacteria, mould and algae as food (Allen, 1955*). Cypris, Metacypris, Cyprinotus, Stenocypris* etc. are the common examples.

Zooplankton of various types

Benthic zone: The **benthic zone** is the ecological region at the lowest level of a body of water such as ponds, lake, rivers and seas etc. including the sediment surface and some sub-surface layers. Organisms living in this zone are called benthos, e.g., the benthic invertebrate community, including crustaceans , annelids, mollusks etc. The organisms generally live in close relationship with the substrate bottom and many are permanently attached to the bottom. The superficial layer of the soil lining of a given body of water, the benthic boundary layer, is an integral part of the benthic zone, as it greatly influences the biological activity that takes place there.

Some abundant species of rotifera

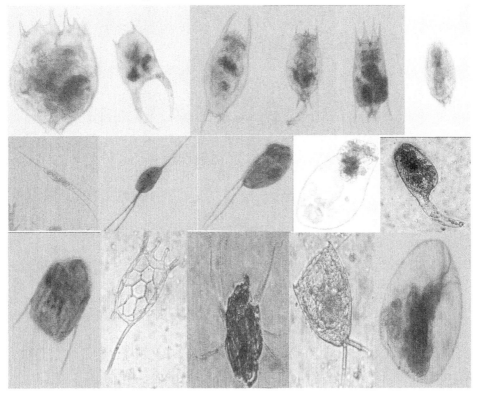

Fig. 47: Top row : *(Brachionus rubens, B. forficula, B. diversicornis, B. caudatus, B. calyciflorus and B. angularis* Second row: *Filinia opoliensis, F. terminalis, F. longiseta, Asplanchna brightwelli and Epiphanes* sp. Third row:*Polyarthra multiappendiculata, K. tropica, Hexarthra mira , Lecane* sp *and Testudinella patina.*

Benthic fauna: Benthos is the organisms that live in the benthic zone, constitute an integral part of an aquatic ecosystem and are different from those in the water column. Many have adapted to live on the substrate (bottom). In their habitats they can be considered as dominant creatures. Many organisms adapted to deep-water pressure cannot survive in the upper parts of the water column. The pressure difference can be very significant.

Benthic macro fauna are those organisms that live on or inside the bottom deposits of a water body. Large benthic animals, those readily visible through naked eye are collectively referred to as macrozoobenthos or macro-invertebrates. They are important and integral component of all aquatic ecosystems. These organisms also play a vital role in the circulation and recirculation of nutrients in aquatic environments. They constitute the linkage between the unavailable nutrients

Some abundant species of Copepoda

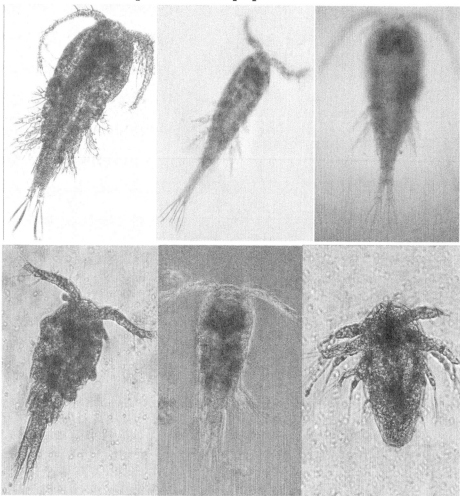

Fig. 48: Top row:*Mesocyclops leuckarti (female), M. leuckarti (male)* and *Cyclops* sp. Second row: *Mesocyclops hyalinus , Ectocyclops* sp. and*Nauplius* larva.

in detritus and useful protein materials in fish (Covich *et al.,* 1999). Most benthic organisms feed on debris that settle on the bottom of the water and in turn serve as food for a wide range of fishes (Chakma *et al.,* 2015)

Because light can penetrate only upto 2meters in fresh water and does not penetrate very deep, the energy source for the benthic ecosystem is often constitute of organic matters from higher up in the water column that drifts down to the depths. This dead and decaying matter sustains the benthic food chain; most organisms in the benthic zone are scavengers or detritivores. Some micro-organisms use chemosynthesis to produce biomass.

Some species of cladocera and ostracoda.

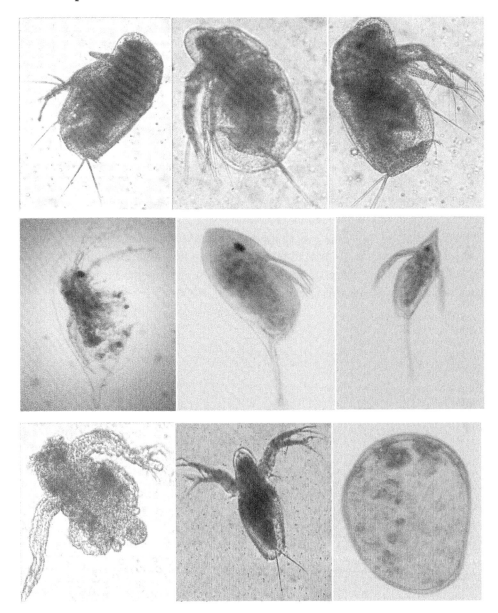

Fig. 49: Top row: *Moina flagellata*, *M.brachiata* and *M.micrura*, Middle row: *Daphnia magna*, *D.dubia* and *D.lumholtzi*, third row: Juvenile cladocera, *Ceriodaphnia regaudii* and *Cyprinotus* sp.

Benthic organisms can be divided into two categories based on where they make their home: floor or an inch or two into the floor. Those living on the surface of the bottom floor are known as epifauna. Those who live burrowed into the mud floor are known as infauna. A number of soil constituents such as colloidal clay, organic matter, nitrogen, phosphorus, potassium etc.and also trace elements get distributed in different proportions at various parts of fish ponds which ultimately determines the physico-chemical and biological characteristics of the water of fish pond (Banerjee and Lal, 1990).

Micro-organisms play an important role in determining the physico-chemical properties of soil beneath the water.Soil microbes are highly sensitive to the soil reaction. Soil pH at and around neutral is favourable for healthy growth of soil micro organisms and microbial activity. Atmosphere is a rich source of nitrogen (>79%) but can not be utilized by terrestrial crops, until and unless it is somhow trapped by the microbial activity in the soil.Several forms of microorganisms are present in soil and are of great importance as they help in the decomposition of organic wastes and release of nutrients. Nitrogen cycle is the best example for bacterial activity in the soil (Banerjee and Lal, 1990).

3

DIVERSIFICATION AND ALTERNATIVES OF CARP CULTURE

1. INTRODUCTION

Understanding the fact of saturations in the fresh water Indian major carp farming, it is now become imperative to diversify from Indian Major Carp culture to various other major and minor species of importance in fresh water culture since these have high nutrient value in terms of protein, micronutrients, vitamins and minerals. The fishes like:

i) *Pangassius hypopthalmus the Pangus,* ii) *Chanos chanos,* the milk fish, iii) *Clarius batrachus*, the Magur, iv) *Heterpneustes fossilis* the Singhi, v) *Puntius sarana,* the Olive barb, the Sar Punti, vi) *Oreochromis niloticus,* the nile tilapia monosex culture, vii) Climbing perch (*Anabas testudineus*), the Koi (Vietnam strain), viii) *Notopterus notopterus*, the Bronz Feather Back (Folui), ix) *Amblypharyngodon mola,* the *Mourala* x) *Mystus vittatus* xi) *Nandus nandus* the *Gangetic leaffish.* Among these fishes *N .notopterus, A.mola, M. vittatus* and *N. nandus* may well be cultured with Indian Major Carps in fresh water ponds and last but not the least, xii) the culture possibilities of Pacu – the *Piaractus brachypomus, xiii)* the fresh water giant prawn *Macrobrachium rosenbergii*and xiv) the slim bellied *cyprinus carpio haematopterus* and last but not the least xv) the delicacy of Loktak lake of Manipur – the Pengba, *Osteobrama belangeri.* The culture possibilities of these fishes are given here for the interest of farmers.

Apropos the need of diversification of Indian Major Carp farming, it may be mentioned here that an enormous possibility do exist in the farming of exotic Amur-China type of wild carp, *Cyprinus carpio haematopterus.* There is an increasing trend of interest among some sector of farmers who are able to procure seed of this particular species unlike the traditional common carp *Cyprinus carpio.* The speciality of this species is that it is a slim bellied one, unlike traditional common carp *Cyprinus carpio* as well as, has a very fast growth rate. An attempt has also been made to discuss the breeding biology, food and feeding and growth performance of this species since this has been newly introduced in India.

2. SMALL INDIGENOUS FRESHWATER FISH SPECIES OF INDIA

India, one of the 17 global mega biodiversity hotspots, is inhabitant to 2246 fin fish species out of which 838 are from fresh water (Mohanta *et al.,* 2008). Among them, total 450 fish species could be categorized as small indigenous freshwater fish species (SIFFs) out of 765 native freshwater species, and 239 of them are from West Bengal. In West Bengal the total production of inland fish was 15.30 lakh ton and marine fish was 2 lakh metric ton (Dey, *et al.,* 2017).

Small indigenous freshwater fish species (SIFFS) are defined as fishes which grow to the size of 25-30 cm in mature or adult stage of their life cycle. They inhabit in rivers and tributaries, foodplains, ponds and tanks, lakes, beels, streams, lowland areas, wetlands and paddy fields. Although rural population depend highly on indigenous species of fish for nutrition in many parts of India, very little attention has been paid on their role in aquaculture enhancement, nutrition, processing, biology, captive breeding, livelihood security and conservation needs. Consequently, many small indigenous fishes have become threatened and endangered due to pollution, over exploitation coupled with habitat destruction, water abstraction, siltation, channel fragmentation, diseases and introduction of exotic varieties. In order to achieve sustainable utilization, appropriate planning for conservation and management strategies are of utmost importance. This article addresses the untapped potential of the small indigenous fishes of India and challenging issues for sustaining biodiversity, management, aquaculture, nutrition and livelihood security and highlights the future priorities.

A large diversity of small indigenous fish species are found in freshwater systems in Indian sub-continent. These small indigenous freshwater fish species (SIFFS) form a major component of food consumed by families, especially those living closer to freshwater resources. Among traditional communities indigenous knowledge about the health benefits of such species exists, for example, mola (*Amblypharyngodon mola*), commonly found in eastern and northeast India, is often included in the diet of pregnant and lactating mothers, for their nutritive values.

i) Culture possibilities of *Pangasianodon hypopthalmus*

Culture possibilities of *Pangasianodon hypopthalmus* as a candidate species in fresh water ponds

Pangasianodon hypopthalmus, earlier popularly known as *Pangassius sutchi* (Fig.50&51) is a silver stripped catfish, and is basically a riverine species with migratory habits. *Pangasius* emerged as a major culture species in the 1990. It is cultured due to its good market demand, faster growth; few countries dominate the culture production, and have attained most important freshwater fish within aquaculture sector.It grows to 1kg in four months and a minimum harvest of 12, 000kg/ha/year is achievable from culture ponds.In India, farming of this fish has increased since 2004 due to the commercial importance and there is a growing interest among fish farming community in both Krishna and Godavari districts of Andhra Pradesh to take up its culture in a large extent, thus paving way for demand for its seed. About 300-400 million seeds of *P. hypopthalmus* is produced every year in West Bengal. This fish exhibits remarkable growth rate even in semi-clean grow-out ponds. It has been observed that there is a shift of culture practice from major carps to Pangas catfish in considerable areas of the country especially in the southern states of India.The production of the fish is reported to be 7 metric tonnes/hectare/year to 20 metric tones/hectare/year and the average production are found to be higher than IMC production in some of the areas. *Pangassius* is an air-breathing fish that can tolerate low Dissolved Oxygen (DO) content in the water and can be cultured in ponds, concrete tanks, fish cages or pens.

Fig. 50: The fry stage of *P. hypopthalmus*

Fig. 51: Adult *P. hypophthalmus*

The systematic position and other necessary details of the species is given here under.

Phylum- Chordatata.

Class- Actinopterygii [which is a characteristic feature of all Ray-finned fishes]

Order- Siluriformes [this is a characteristic feature of all catfishes]

Family- Pangassidae [this character goes to all shark cat fishes]

Genus- Pangassius.

Species- Pangassius.

Binomial name: *Pangassius pangassius* (Hamilton, 1822).

Morhology

The body is elongated and laterally compressed.The upper surface of the head remains unpolished and the snout obtusely rounded.The upper jaw is lower than the lower jaw and the gape of the mouth is moderate.Two pairs of barbells are present. The dorsal spine is serrated anteteriorly. The pectoral spine is comparatively strong than dorsal spine and serrated internally. Caudal fin is deeply forked. The lateral line is complete.

The colour on the abdominal side of the fish is silvery, side of the head contains golden tinge while the above portion of the lateral line is whitish grey and on the flanks it is silvery purple. The back porion of the fish is however, dark or yellosish green.

Habitats

The fish is found to inhabit both in fresh and brackish waters. Some of the common habitats are big rivers, flood plains, estuaries and canals etc. Usually inhibits at the lower portion of the large rivers and estuaries.

Some features of pond management

The cherry plum fruit of chebulic myrobalan *Terminalia chebula* (**Haritaki**) extract is used in *P.Hypopthalmus* ponds. It destroys the dense mass of green algae in fish ponds and also reduces the turbidity of ponds caused due to mud particles of allochthonus origin. On application of chebulic myrobalan, mud particles coagulate and settle down at the bottom of pond.When required, it is applied @50-60kg/Acre area. A mixurc of soap-oil emulsion (preferably Life Buoy and Mustered oil) is used in the nursery ponds. Besides the killing of aquatic insects, since the mustered oil has fat content, this also boosts up zooplankton population in the ponds. Soap is first boiled over the furnace, there after oil is added. A mixture of 15kg soap and 45kg of oil is applied in every acre of pond area preferably 48-hours prior to release of spawn (*P.hypopthalmus*). The depth of the nursery ponds should be ± one meter.

After 2-3 crops, the nursery ponds should be sun dried and 75-100 kg/acre of Mahua Oil Cake (MOC) should have to be broadcasted on the top soil. On the same day water soaked MOC (preferably 24 hours before) has also to be sprayed over the top soil. Next day morning liming@ 50-60kg/acre area has to be made and kept as such for a day, while in the evening water is allowed to fill the pond area.Utmost care is to be taken while allowing water from outside reservoir/ponds. The inlet pipe is to be opened in a hapa made up of fine meshed bolting cloth in order to trap larger zooplankton viz. *Cyclops* sp, *Diaptomus* sp. etc. inside the hapa (Fig.52). This larger copepod zooplankton has a tendency to predate the fish spawn stocked in the nursery ponds. As many as twenty lakhs spawn of *P. hypopthalmus* may be stocked in a nursery pond covering an area of 1.0 acre meter size.

Fig. 52: Ideal technique of water filling in the nursery pond.

Food and feeding habit

This fish is not only carnivorous in nature but also voracious feeder. It also feeds on decaying animal and vegetative matters. When found in rivers, it shows predatory on snails and other molluscan populations.

Diseases and remedial measures (Table 29): Large-scale disease outbreaks and mortality of *Pangasianodon hypophthalmus* seldom occur. The following diseases, however, have been encountered.

Parasite infections including *Ichthyophthirius* and *Trichodina* can occur at the fingerling stage and are usually treated with 15-20 ppt of formalin for 30 minutes with aeration before water exchange.

Table 29: Diseases and remedial measures of *Pangasianodon hypophthalmus*.

Disease	Agent	Type	Syndrome	Measures
Bacillary Necrosis of Pangasius (BNP)	*Edwardsiella ictaluri*	Bacterium	Haemorrhages on eyes and fin bases; white spots in the kidney, spleen and liver; some cellular necrosis.	Antibacterial and antifungal agents @300gms/acre area(6ft.depth)
Motile Aeromonad Septicaemia (MAS)	*Aeromonas* spp. (mainly *A. hydrophila, A. sobria* and *A. caviae*)	Bacterium	Haemorrhages on eyes, body and fins; bloody peritoneum, leading to swollen belly.	Improve water quality and nutrition; Iodin 20%@300ml/ acre

ii) Culture possibilities of *Chanos chanos*

Culture possibilities of *Chanos chanos,* the milk fish as a candidate species in fresh water ponds.

The euryhaline nature of **milkfish (*Chanos chanos* Forskal)** attracted attention of the fresh water aquaculture professionals both from Chennai and West Bengal. The fish however, is also widely cultured in brackish water ponds largely in South East Asian countries.

Fig. 53: The milk fish (*Chanos chanos*)

Milkfish (*Chanos chanos*) is euryhaline in nature and can tolerate the salinity of water ranging from 0 to 80 ppt. The larvae are to be slowly acclimatized from brackish water to fresh waters where the salinity is 0.0 ppt. After acclimatization the fry can well be reared in fresh waters.

The systematic position of the fish

Phylum-Chordata.

Class- Actinopterygii.

Order- Gonorynchiformes.

Family- Channidae.

Genus- Chanos

Species-Chanos.

Binomial name: *Chanos chanos* (Forskal, 1775)

The milkfish can grow to 1.80 m (5 ft 11 in), but are most often no more than 1 m (39 in) in length. They can reach a weight of about 14.0 kg and an age of 15 years. They have an elongated and almost compressed body, with a generally symmetrical and streamlined appearance, one dorsal fin, falcate pectoral fins and a sizable forked caudal fin. Mouth is small and toothless. Body color is olive green, with silvery flanks and dark bordered fins. They have 13-17 dorsal soft rays, 8-10 anal soft rays and 31 caudal fin rays.

Food and feeding habit

Milk fish in its natural habitat, apparently feeds on planktonic microorganisms and is most frequently designated as microphagus carnivore. Milkfish has fine, almost membranous gill rakers that suggest filter feeding habit of the fish. It also has a specialized epi branchial organ above and behind the gills that may help to concentrate microplankton. Milkfish aquaculture sounds well in the ponds where predominantly of unicellular algae, diatoms, and other organisms associated with the algal community and locally known as *lablab*, and have fibrous filamentous green algae. To obtain better *lablab* in the milkfish ponds the farmers are advised to apply 6kg/acre/meter Multi-micronutrient mineral fertilizer and humic acid together @5kg/acre/meter each which greatly helps formation of the *lablab* in the cultured ponds. Since much work related to culture of Milk fish in fresh water ponds has yet to be done, occurrence of disease outbreak during culture have not been noticed till now. However, interested farmers are advised to contact CENTRAL INSTITUE OF BRACKISH WATER AQUACULTURE (CIBA) at Chennai (HQ) or Kakdwip Regional Centre, Dist-24-Parganas(S), West Bengal, where the rearing technology of this species in fresh waters have been developed successfully.

Since the species has newly been introduced in fresh waters from marine sectors, detail information on the biology, diseases etc. needs further investigations.

iii) Culture possibilities of *Clarias batrachus*

Indian Sub-continent is endowed with vast numbers of villages where there are significant numbers of derelict water bodies. Air breathing fishes especially Magur (*Clarias batrachus*) and Singhi (*Heteropneustes fossilis*) deserve special mention due to their easy digestibility and protein and iron content (Tab-8). The other air breathing fishes are murrels of various species. The specialty of these fishes, they remain alive even when are out of water for a couple of hours. This is possible, due to the presence of accessory respiratory organs.

Fig. 54: *Clarias batrachus* (Magur)

Systematic Position:

Phylum-Chordata

Class- Actinopterygii.

Order- Siluriformes.

Family- Clariidae.

Genus- Clarias.

Species- Batrachus.

Binomial name: *Clarius batrachus* (Linnaeus, 1758)

Clarias batrachus is a black, slippery fish with mustache to aid it in swimming. The availability of wild-caught magur seed is insufficient to meet demand due to a combination of over-exploitation, aquatic pollution, spread of disease, uncontrolled introduction of exotic fishes and habitat modification (Mahapatra B.K 2004:Conservation of the Asiatic catfish, *Clarias batrachus* through artificial propagation and larval rearing technique in India . Sustainable Aquaculture **Vol. IX No. 4:** *Clarias batrachus* (Linnaeus, 1758).

Habitat

Catfish are usually found in marshes, rice fields, swamps, streams, rivers, lakes irrigation canals, or in any body or fresh water. Size of the ponds depends. The minimum size is 50 square meter (sq. m) and should be located in low and flat areas. Land where pesticides have been regularly used should be avoided. A shady area where the fish will have shade when the sun is intense and lumut or moss will grow easily. The area should also have a good supply of water either from wells, spring or run-off ponds.

Characteristics of air breathing fishes

1) The hardy nature and tolerance to adverse environmental conditions particularly low oxygen level.

2) Intensive culture with high production rate.

Behaviour and Compatibility

C. batrachus is highly predatory and will eat any fish it can fit into its large mouth. It can be kept with other similarly-sized, robust species, such as large barbs.

Stocking rate

Catfish farming requires extremely heavy stocking (75 to 100 fingerlings per square meter).During a study conducted in Maharashtra on the stocking density of *Clarius batrachus,* it is evident that stocking @200 individual causes nomortality. However, the extremely heavy stocking method to the tune of 300-400 individual/sq. meters produces mortality and minimal production of fish per unit area. Catfish farming produces 2 crops a year at an average rearing period of 5 to 6 months.

It is advisable to stock the seeds of magur in the late afternoon or early morning when the climate remains cool. The stocking rate depends on size of fish and depth of water (Table 30).

Table 30: Stocking of *C.batrachus* depending on size and depth of water.

Size of fish	3-4cm	5-6cm	7-10cm
Number of fish	60pcs/kg	50pcs/kg	30pcs/kg
Stock/sq.metre	80-100	60-80	40-60

Different Culture types

1) Monoculture, 2) Culture with other fishes, 3) Culture in cages, 4) Culture in cement tanks, and 5) Culture in paddy fields.

Fig. 55: High density stocking of *C. batrachus*

In monoculture, the pond size should be less than a hectare, preferably an acre in size having one meter depth of water. After eradication of all unwanted fishes by Mahua Oil Cake (MOC) (Required quantity of MOC is about 1000 kg/Acre / meter). By observing the pH of water application of lime @ 120 to150 kg/Acre/ meter to be made to make the pond water alkaline. Fingerlings of air breathing fishes especially Magur or Singhi preferably (60 – 100 nos /per Kg) to be taken in a hundi. Raw formalin @ 0.5 ml/litre to be added to the water of the container (Hundi) as disinfectant. After 15 – 20 minutes of emersion in formalin water of the fishes, they may be released in the pond water. In monoculture, the pond size should be less than a hectare, preferably an acre in size having one meter depth of water.

Stocking density of the fishes in monoculture may be 15, 000 to 22, 000 / Acre/meter. From the next day supplementary formulated feed (30% Protein: 5 %Fat) pelletized sinking feed @ 5 – 7 % of the total body weight may be given to the stocked fishes (Table-31). The total quantity of feed should be divided into two parts. The first part (30% of the total) should have to be broadcasted during the morning hours and the rest 70% in the evening. This is because the air breathing fishes especially both the magur and singhi are nocturnal feeders.The protein percentage of the feed should be minimum 30%. Almost all the air breathing fishes prefer animal Protein. The fishes grow well and faster with such feed. The following table depicts the percentage of feed to be given in monoculture.

Table 31: Quantity of feed per 100 kgs of fish in Monoculture.

Weight of fish (gms)	Kg/day
<25	8-10
<50	7-8
<100	5-7
<200	3-4
200 and above	2-3

Problems in rearing air breathing fishes and their remedial measures:

(1) Water temperature: During summer months when water water temperature is high and to avoid evaporation infestation of duck weeds or water hyacinth is encouraged. It is always advisable to maintain the depth of water maximum to the tune of 1.2 meters.

(2) Autochthonous matters may decrease the overall transparency of water. In such case water exchange or pumping water from other source is advocated.

(3) The pH of water should remain within 8.0.

(4) In case of higher algal growth stop supplementary feeding.

The fishes are omnivorous in nature and eat almost anything which is offered. A varied mixture of dried pellets, meaty frozen foods and vegetable matter is recommended. Adult specimens do not need to be fed every day. Take care when feeding, as this is one fish that just doesn't know when to stop eating. Adult magur are carnivorous so their feed should contain higher amount of animal protein. These can be from ground fresh trash fish, worms, insects, slaughter house buy-products, and chicken entrails, and dried or fresh water shrimp, fish and by-products of canning factories.Table-32 indicates the food source at the different life stages of magur. The remaining 10 % may be composed of boiled broken rice mixed with vegetables or rice bran. To augment food supply, install strong light over pond. Feed the catfish twice a day. To avoid waste, give the feeds slowly, by handful, until the fish stop eating. Daily feed ration is 6-7 per cent of the fish stock's body weight. Never over feed since the excess would only pollute the water causing death or stunted growth.

Sexual Dimorphism

In the wild form, male fish show spotting in the dorsal fin. This distinction cannot be made for the aquarium varieties but sexing can be determined by examining the genital papillae of the fish. This is elongate and pointed in the male, and shorter and blunted in the female. Female fish are also much more rounded in the belly than males (Fig. 56a&b).

Table 32: Food preferences and natural food source of *Clarias batrachus*.

Food Source	Life Stage
Benthic algae	Adult/Fry/Larval
Chironomid larvae	Adult/Broodstock/Fry/Larval
Detritus	Adult/Fry/Larval
Dragonfly nymphs	Adult/Broodstock/Fry/Larval
Finfish	Adult/Broodstock
Insect eggs and pupae	Adult/Broodstock/Fry/Larval
Molluscs	Adult/Broodstock
Ostracods	Adult/Broodstock/Fry/Larval
Other zoobenthos	Adult/Broodstock/Fry/Larval
Weeds	Adult/Fry/Larval
Worms	Adult/Broodstock

Spawning season

Between May and October. In natural condition, the female makes a small round hollow nest with grassy bottom abouy 30cm deep in shallow waters. The eggs are deposited in the nest and attached to the roots of aquatic vegetation in the nest. The males take charge of these eggs until they are hatched out within twenty (20) hours at temperature between the ranges of $25-30^0$C. Female weighing300-800gms can produce between 5000-10, 000 eggs.

Mating and spawning

Female catfish ready to spawn or to produce offspring when it build its nest of debris or roots of aquatic plants like water hyacinths or filamentous algae. It spawns in shallow water, 30 cm to 60 cm deep. Male and female catfish starts courting by chasing each other, darting sideways, pressing their abdominal regions together. This constantly movement is repeated several times until the females releases the eggs and male milt or sperms is simultaneously ejaculated. Fertilization takes place at 27^0 to 30^oC water temperature. The number of eggs laid, range from a few hundred to several thousands. Never scrimp on feeds especially during this period, remember that catfish are cannibalistic and quarrelsome. If the parent fish are very hungry, they may gobble up the young as quickly as any other food that comes their way.

One drawback that discourages people from catfish farming is non-availability of fingerlings. The following process will assure you a continuous supply of fingerlings.

The two methods effecting spawning is by hormone injection. Both methods use the most gravid or pregnant females and healthy mature males, each weighing at least 200 grams (g).

The body of the pregnant is distended prominently, the genital part pinkies, and the blood vessels on its belly prominent. Breeders should be conditioned first in the concrete or semi-concrete tanks/vats 2 to 5 months before they are injected with hormone.

The natural method entails injecting hormones (Gonadotropin or Synahorin @200-250IU and 50IU) to gravid females and males respectively through their bases located on the posterior side of the pelvic fin. Wrap fish in a small net so they will not struggle during injection. After injection, put the male and female together in an oxygenated tank. The stripping method requires several male catfish milters to be killed. Use forceps to remove their testes which are pinkish yellow and soak in Ringer's solution. Extract sperm by macerating the testes in the distilled water. Use sperm to fertilize breeder's eggs. Inject hormone (mentioned earlier) into the gravid breeder's body. Inject at the side of the fish's body, a little above lateral line, with following dosage:

Gonadotropin = 750 to 1, 500 IU; Synahorin = 1, 000 to 1, 500 IU. After 12 hours, squeeze the breeder's abdomen to force the eggs out. A 250-gram breeder produces 8, 000 to 15, 000 eggs. Mix eggs with sperm and stir for a minute.

Technology of induced breeding with special reference to Magur (*Clarias batrachus*).

The major constraint for wide spread culture of this species is attributed to the non-availability of seed both from hatcheries and natural resources due to depletion of natural stocks. Several attempts have been made in past to breed magur under captivity through induced breeding method. Synthetic hormones viz .Ovaprim, Ovatide, Pond Pro etc. were tried and found to be effective. However, differences of opinion do exist and information's are also available that fish pituitary extract as inducing agent gives better result.

The latency period of 16-18 hours after the injection of the inducing hormonal agent (dose: 1-2 ml/kg body weight in case of female and 0.5-1.0 ml/kg body weight of male) found to be suitable for the ovulation of magur.

Table 33: The range of the abiotic factors of water suitable for ovulation of *C.batrachus*.

Sl.nos.	Water quality	Range
01.	Water temperature	28+/-2^{0}C
02.	pH	6.9-7.6
03.	Total Alkalinity	128-136mg/liter
O4.	Dissolved oxygen	6.8-7.4mg/liter

It is obligatory to select the mature males and female brooders of magur (*Clarias batrachus*) at least 2-3 months before the induced spawning operations. It is also a pre-requisite to segregate the males and females and rear them in separate enclosures or ponds. To be a good brooder they must be a year old weighing about 150gms in weight. The sex can be distinguished by the shape of the genital papilla where the males have pointed reddish papilla while the females have an oval reddish shape with bulging abdomen and soft to touch (Fig. 56 a-c). In general the female brooders weighing between 120-148gms weight are selected for induced breeding. The cloacal opening is reddish and prominent (Courtesy: Prof. S.K. Das of Assam Agricultural University College of Fisheries, Raha, Nagaon, Assam, India.). The sperm from the males are generally collected by stripping method (Fig. 56c).

Selection of spawners of *Clarias batrachus* and artificial insemination

Artificial insemination: Ovulation generally takes place about ten (10) hours of second injection. The milt from the males is made to drip on the eggs pouring the water through the fine meshed cloth. The eggs and sperms are to be mixed and stirred gently with a feather. Next to that a little clean water is added and gently mixed the same again. After a couple of minutes, light flow of water is added two to three times to clean the fertilized eggs. The fertilized eggs are then transferred to the hatching hapa (Figs. 57. a-e).Most of the fertilized eggs hatch out within 24-30 hours.

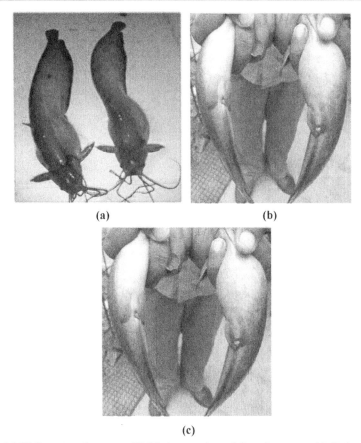

(a)   (b)

(c)

Fig. 56 :(a) Well matured magur, (b) Mature male and female magur (c) Stripping method

Nursing of the early fry of Magur

It takes about two days to absorb the yolk when the larvae are transferred from hapa to nursery ponds. Immediately after absorbing the yolk, the magur fry changes the feeding behavior and up to 21 days they prefer to consume the zooplankton, especially daphnids. By this time they attain 2-3cm.in length when they are distributed to other growers. A suitable feed is the basic requirement at this time although majority of them prefer to consume planktonic fauna in the early stages but nutritionally balanced formulated feed in bulk quantities at the later stage of their life is a pre-requisite. Generally, Magur (*Clarius batrachus*) inhabit at the soil-water interface of the pond and grow well where the aquatic weeds predominate. Logically, like all other aquatic vertebrates like fishes the magur fish are also prone to various diseases. And these are: Parasitic, Bacterial, and Fungal diseases of different types. Depending on the severity, these are categorized under three categories.

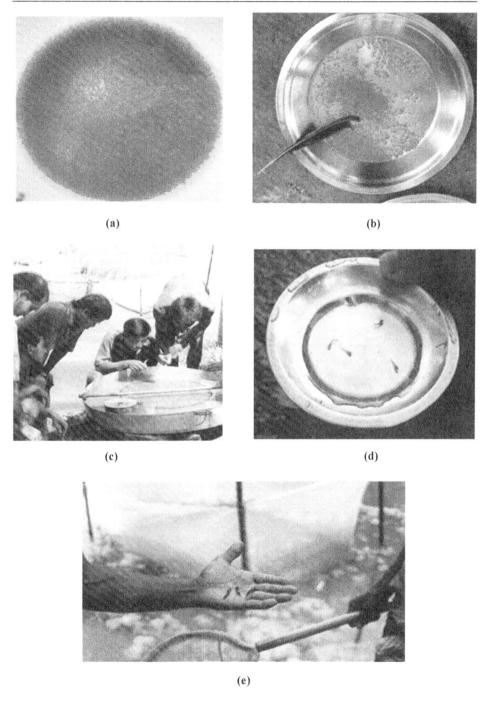

Figs. 57: (a) Healthy fish eggs (b) Mixing of eggs & sperm (c) Production of hatchling
(d) Early fry (e) Advanced fry.

Table34: Diseases and remedial measures of *Clarius batrachus.*

VARIUS DISEASES OF MAGUR (*Clarius batrachus*)

Sl.nos	Type of disease	Caused by	Symptoms	Treatment
01.	Parasitic (monogenians)	*Trichodina and Trichodinella*	Mass mortality of fry	Water and soil probiotic.
02.	Digenians	*Dactylogyrus*	Ill growth	Iodin 20%@300ml/acre.
03.	Myxosporida.	*Myxosoma/Myxobolus* cysts	Ill growth	Iodin 20%@300ml/acre
04.	Bacterial Septicaemia.	*Aeromonas hydrophila, Pseudomonas fluorescens, Edwardsiella tarda.*	Ill growth followed by mortality	Antibacterial antifungal products@300gm/acre each
05.	Bacterial Ulcerative Disease. (Category A.)	Number of different aetiologies.	Traumatic damage. The lesions are deep dermal ulcers with severe bacterial necrosis of skin and muscle.	Antibacterial and antifungal agents@300gm/acre each.Apply 750gms of
06.	Viral diseases	Not reported.		
07.	Fungal infections	*Saprolegnia* sp.		Antifungal agent@300mg/acre
08.	Nutritional Diseases	i) Fatty Liver Syndrome ii) Vitamin C Deficiency Syndrome	i) Crack head ii) Failure of fibrous wound healing.	Multivitamin@ 250300gms/ 100kg of feed

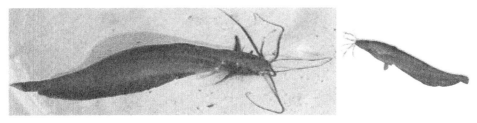

Fig. 58

(iv) Culture possibilities of Singhi *Heteropneustes fossilis* (Bloch, 1794) (Fig. 58)

Systematic position:

Phylum-Chordarta.

Class- Actinoptrygii.

Order- Siluriformes.

Family- Heteropneustidae.

Genus- Heteropneustes.

Species- Fossilis

Binomial name: *Heteropneustes fossilis* (Bloch, 1794).

Hereropneustes fossilis, commonly known as "Singhi" (Fig.58), comes under the category of air-breathing fishes and contribute significant role in aquaculture. They differ from other teleosts by possessing a tubular accessory respiratory organ run backwards on either side of the vertebral colmn, almost upto caudal peduncle from gill chamber through the myotome muscles in the back, for direct aerial respiration by periodically gulping air at the air-water interface and may survive for about 16hours when out of water.. The air-breathing fish can therefore, may well be cultured in waters having low or deficient in dissolved oxygen content such as swamps, wetlands, oxidation ponds and sometimes in muddy rivers. The fish is epibenthic and also compatible with the carps in mixed culture.

Morhology

The body is elongated and compressed. The head is depressed and covered with osseous plate at the top and the sides of the head. Four pairs of barbels in which the the maxillary pairs extended to end of the pectorals or to the commencement to anal and the mandibular pairs extended upto the base of pelvis but the nasal pair is comparatively shorter than the mandibular pairs. The presence of air sac or the accessory respiratory organs which extends backward from the gill chamber on either side of the vertebral column.The caudal is rounded.The body colour is reddish brown or purplish brown but during mature conditions it appears to be black in colour.

It may be mentioned here that *Heteropneustes fossilis*is protein enriched fish species. Each 100 gm of fish contains 23.0 gm of protein, the fat is comparatively low (0.6gm) while the Calcium content is about 670 mg and the Phosphorus content is 650 mg and the water content is about 79.3%.

These fishes breed in confined waters during monsoon months. Spawning has been observed in swampy patches falling within the flood prone regions of rivers. It is capable of breeding in ponds, derelict water-bodies and ditches when sufficient rainwater accumulates. Sexual maturity is attained when both male and female are one year old and are about 5.5 cm and 12 cm in length, respectively. Induced spawning is successful using homoplastic pituitary glands. Fertilized eggs are adhesive, demersal and spherical in form. The fish, which attains a length of 30 cm, is in great demand because of its medicinal value. Fishermen generally stock singhi in ponds during the rainy season. The fish is much dreaded because of its aggressive behavior and can inflict painful wounds with its potentially dangerous pectoral spines (Nayak *et al.,* 2000).The fish is omnivorous in feeding habit and also have a distinct sexual dimorphism like *Clarias batrachus*. The female *H. fossilis* has well-rounded abdomen filled with matured ova. The males, on the other hand, look lean. In a mature female, the genital papilla, refrains in the form of a raised prominent structure, looking round and blunt with a slit-like opening in the middle while in males. it remains in the form of a pointed structure.Fecundity varies from fish to fish depending upon size and growth. Generally, one gram of ovary contains 1500 - 2000 eggs. *H. fossilis* can be successfully induced to breed under controlled conditions through administration of hormones or inducing agents.

Dry stripping method and nursing of larvae

Very often it is observed that the females of *H.fossilis* refuse to release the eggs due to the plugging of genital pore.During such condition fresh water taken from a tubewell is boiled. The fishes are kept in mild hot water for a period of 20-30 minutes facilitating the loosening of plugged point. As much as 75% impregnated ovum will come out are kept in clean dry rectangular enamel trays which happens to look like semolina pudding or concentrated sweet porridge.Brood fishes are ready to be stripped on 4th. Hour of 2nd.injection to females.Pair of testes from male fish is dissected out and homogenized or squeezed with the help of cotton cloth using 2% normal saline solution (NSS). This milt is further diluted with with 1 part water added to every 2parts of NSS-associated milt; it is spread over the stripped eggs. Sperm and eggs are mixed well with afeather for 3-4 minutes, fertilized eggs are washed with fresh water, which are greenish-blue/greenish-brown in colour and demersal in nature.Those are incubated in individual mini rectangular trays of dimension 25cmx20cmx5cm.

Hardness of water plays an important part in the process. Hardness ranging from 250-400 ppm.favors the incubation of *H. fossilis*, whereas in case of *Clarias batrachus* the ideal range is 90-110 ppm.

As observed, after an incubation period of 9.5-10 hours at $29\text{-}30^{0}C$ water temperature, *H.fossilis* larvae begins to hatch out and on 3^{rd}day they attain 7-8 mm. At this stage these are stocked in a rectangular concrete chamber (3mx1.8m with water depth 30cm). About 30, 000 nos larvae are obtained from every four females. Early fry/advanced larvae are stocked @45, 000-60, 000 nos/chamber. In each chamber, four bubble diffuser-type oxygenation machines (aerators)are introduced.Growing larvae are fed two times a day, it comprise mixture of 2-3 boiled chicken eggs and 100gm of Amul milk powder mixed with 3 liter water in each such chamber. In summer (April-June) during day hours 30-40 gm glucose powder is mixed in 2 liters of water which is allowed to sprinkle over water on each chamber. In addition to this, oxygen powder/tablets are applied as per the recommendation of the aquaculture consultant which helps increasing dissolved oxygen content in the chamber water. *H.fossilis* larvae attain 15-18mm on 6th day of stocking, and are harvested from the chambers and stocked in pre-prepared productive nursery ponds@2, 00, 000/1335m^3. After three weeks these attain 3.8-4.8 mm in length when they are either sold out or transferred in rearing ponds.

Utmost care has to be taken during larval development. The advanced fry stocking density may be raised to the tune of 3000 to 5000 fry/m^2. Stocking of ponds may be done 1 - 2 lakhs per ha ensuring prophylactic treatments at the time of stocking. They should be fed 3 - 4 times per day with artificial feed and unconsumed food to be removed immediately. It is important to check the health of fry and fingerlings every week and diseased fry/fingerlings should be segregated, treated and kept separately preventing further spread of disease. The fingerlings may be harvested after about 2 months of stocking (Nayak *et al.,* 2000).

v) Culture possibilities of *Puntius sarana*

The olive barb *P.sarana* is a (Fig. 59 and 60) high priced but critically endangered species which is mainly attributed to overfishing, destruction of natural spawning grounds etc.

Fig. 59: Brooders of *Puntius sarana*

Fig. 60: Fingerling of *Puntius sarana*

Systematic position:

Phylum- Chordata.

Class- Actinopterygii.

Order- Cypriniformes.

Superfamily- Cyprinoides.

Family- Cyprinidae.

Sub family- Barbinae.

Genus- Puntius.

Species- Sarana

Binomial name: *Puntius sarana* (Hamilton, 1822).

This barb is very widely distributed all over India in rivers and tanks. It attains a length of 31 cm. It breeds during monsoon in running waters amongst submerged boulders and vegetation. Spawning occurs in two stages once between May to Mid-September but prominent in June and the second spawning time in the months of August and September.

Morphology

The body of this minor carp is elongated and laterally compressed with a small head.The head is 4-4.7 times shorter than the standard length of the body. The mouth and eyes are moderate. There exist two pairs of barbells: the maxillary and the rostral. The maxillary barbells are longer than the rostrals.The scales are medium and the lateral line is complete.

The body colour is olive on the back and flanks silvery with golden reflections. Barbels are reddish –brown. Fins are dusky brown to orange. A dull blotch appears to be on the lateral line before the caudal base.

Feeding habit

It normally forms groups of four or five to several dozen. The species is omnivorous and feeds on aquatic insects, fish, algae and small prawns. Food of *P. sarana* is 27% algae, 45% higher plants, 20% protozoan, 8% mud and sand. The food groups Chlorophyceae, Bacillariophyceae and Cyanophyceae were recorded highly dominant in the gut content of all the size groups of fish by both average indexes of fullness and average points per fish.

Among the phytoplankton groups Chlorophyceae was the most dominant food group (22.39%). Next to the Bacillariophyceae (21.85%), Cyanophyceae (19.28%), Euglenophyceae (16.17%), Rotifera (10.03%), Crustacea (9.38%) occupied the successive position. These findings indicate that the phytoplankton is the most important food item of fish.

Economic importance

Flesh of olive barb contains 17.5% crude protein, 2% fat and 74% water. The digestibility and biological value of flesh of this species is very high. Though it is with inter-muscular bones yet it is highly esteemed as food both in eastern and north eastern part of India including Bangladesh. It can grow up to 400-500 gm in a year and fetches around Rs. 100-150 per kg. This species can also be used as an ornamental species due to its attractive silver-coloured body and hardy nature. This species is considered as the "biological control" in aquacultural practices, since it can be used for eradication of aquatic weeds (*Lemna* species) from the water bodies like ponds and tanks.

Cultural importance

The olive barb, *P.sarana* that attain 100-200 gm. is comparatively smaller species when compared to Indian Major Carps. Due to high consumer preference, makes the species a suitable candidate for diversifying the carp culture and also for short-term culture in seasonal water bodies. Even the species were available in plenty in the natural waters in entire South East Asian countries but the poor seed survival and over-exploitation over the years have reduced its natural population to the extent of placing it under vulnerable group. Since the species possesses culture potential, its introduction into the carp polyculture system would not only help in diversification of culture practices, but also can serve for its conservation. Study reveals the compatibility of this species is high with other major carps and the production to the tune of 4200—4819 kg ha^{-1} from polyculture using olive barb at 30—35% of the stocked density of 9980 fingerling ha^{-1} with four other major carps.

Breeding behavior

P. sarana is a prolific breeder and breeds during monsoon in running waters amongst submerged boulders and vegetation. Spawning of this barb occurs in two stages, once between May to Mid-September but prominent in June and the second spawning time in the months of August and September.

Fecundity

It has been reported that the average fecundity of *P. sarana* is 3, 20, 438 Nos. /Kg body weight. As the eggs of *P. sarana* are adhesive in nature, massive aeration is needed in the hatching unit followed by suitable egg collectors.

Estimation of spawning fecundity and fertilization rate

The spawning of fish has to be confirmed the next day by examining the eggs and after spawning, the spent brooders were removed from tank.

Then the spawning fecundity and fertilization rate are estimated using random sampling method. The attached with eggs if any, are randomly collected from different places from the hatching tank after 8 hours of spawning. The fertilized and unfertilized eggs from the samples are counted separately. The spawning fecundity and percentage of fertilization are estimated by using the following formulae:

The spawning fecundity and percentage of fertilization may be estimatedbyusing the following formulae:

$$\text{Fecundity} = \frac{\text{No.of eggs per gm}}{\text{Wt. of fish in gm}} \times 100$$

(Note: Fecundity per Kg. body weight.)

$$\text{Fertilization rate (\%)} = \frac{\text{Number of fertilized eggs}}{\text{Total no of eggs in sample}} \times 100$$

Inducing agent and calculation spawning fecundity and % of fertilization

Pituitary gland extract, Ovaprim, Ovatide or Spawn pro etc. any of these four may act suitably as inducing agents.It is revealed that 'Ovatide' at 0.3 ml/240g female and 0.02 ml/180g maleis sufficient for induced spawning in *P.sarana*. As far as pituitary extract is concerned, a single dose of 5-8 mg pituitary gland/kg body weight is recommended for the female, while a single dose of 4.0 mg pituitary gland/kg body weight was recommended for the male.

Harvesting of spawn

In general, the yolk sac starts absorbing after 60 to 65 hours of hatching. After the absorption of yolk, the spawn are harvested and randomly counted from different locations of the pond. However, continuous aeration in the pond is suggested for the uniform distribution of spawn. The quantity of spawn is calculated by using the following formula:

No. of spawn harvested = No. of spawn per litre X volume of water in spawning tank.

Hatching rate is calculated after harvesting of spawn on third day after hatching. The hatching rate is calculated by using the following formulae.

$$\text{No. of spawn hatching percentage} = \frac{\text{Total no of eggs in sample}}{\text{No. of fertilized eggs}} \times 100$$

Preparation of egg custard

To feed the collected spawn it is necessary to feed them egg custard where chicken eggs being the major component. To prepare the same, the ingredients are mixed thoroughly in blender and cooked in a boiling water bath for about 15 to 20 minutes to make it semi-solid form.

The egg custard may be stored in a refrigerator until further use. The protein content of the egg custard is found to be 45.2%. The egg custard is passed through 0.2 to 1.0 mm size sieves to feed various larval stages.

Brood stock development

During maturation it is observed that males are smaller in size than females. Male and female brooders are identified by following morphological indicators. The females, with slightly smaller and brighter pectoral fins, bulging abdomen and discharge ova on applying slight pressure on abdomen where as males, with slightly larger and dull color pectoral fins, linear body and oozing milt on applying slight pressure on abdomen.

Rearing of spawn

The yolk sac after absorption, fry are harvested and stocked in ponds for further rearing. The fry are grown to 12 to 15 mm in 18days.

 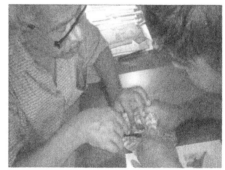

Fig. 61: Hypophysation technique of *P. sarana*

Fig. 62: Spawn of *P. sarana*

The reproductive cycle of *P. sarana* is found to follow the normal pattern, timing of growth phase and maturation stage of germ cells in both male and female and is comparable to that of the other medium carps as indicated earlier. *P. sarana* is a hardy fish and in general diseases are not encountered very often, thus can successfully be introduced in carp polyculture system to increase the fish production.

vi) Culture possibilities of *Oreochromis niloticus*

Tilapia is a symbol of rebirth in Egyptian art and in a country like India , tilapia the *Oreochromis niloticus* **in particular, is said to be the "fish of the future"** because once introduced need not to recruit again and again, since these are prolific breeders and get matured within 3-6 months automatically in tropical shallow water ponds and lakes. Since Male tilapia grows approximately twice as fast as female's, sex reversal has now become a common practice to the commercial fish farmers.Tilapia has become the third most important fish in aquaculture after carps, at least in India and because of their high protein content, large size, rapid growth (6 to7months to grow to harvest size) and palatability, have become the focus of major aquaculture efforts.

Filter-feeding tilapia *Oreochromis niloticus* (Fig. 63) isrightly designated as a core white fish, "every-mans fish", "the aquatic chicken" and "not unlike the culinary versatility of chicken" and is able to convert plant proteins, diatoms, algae, heterotrophic plankton (biofloc), bacteria, zooplankton and low trophic level organisms into a valuable, succulent and mild tasting fish being enjoyed universally. Indeed, it has been argued that tilapia are perhaps the only truly herbivorous fishes since they possess not only a highly efficient digestive system (RLG=3.5 times the length of the fish) but also a highly specialized one, capable of producing pH values < 1 facilitating the digestion of highly refractory plant carbohydrate compounds perfectly commented by Ramon Kourie, Chief Technical Officer, Sust. Aqua Fish Farms (Pvt.) Ltd. At present the expected growth of 2.6 percent in 2018 with average production of 6.5 million metric tonnes (MMT).

Most of the global production of tilapia is produced in freshwater pond systems and consumed in producing countries contributing to food security in the developing world where the sector is concentrated. China is the leading producer country followed by Egypt and Indonesia. Production estimates in 2017 have been pegged at 1.7 MMT for China, almost 900, 000 metric tonnes (MT) for Egypt and 800, 000 MT for Indonesia.

Tilapia is the most widely cultivated of all species with more than 120 countries reporting some commercial activity.

In addition, tilapia are cultivated in the highest number of production environments from rice paddies and simple fertilized earth ponds to cages in lakes, aquaponics systems, biofloc technology (BFT) tanks and Recirculation Aquaculture Systems (RAS) and are considered easy to cultivate.

Systematic position:

Phylum- Chordata.

Class- Actinopterygii.

Order- Cichliformes.

Family- Cichlidae.

Genus- Oreochromis.

Species- Niloticus

Binomial name: *Oreochromis niloticus* (Linnaeus, 1758).

Biological features

Body compressed; caudal peduncle depth equal to length. Scales cycloid. A knob-like protuberance absent on dorsal surface of snout. Upper jaw length showing no sexual dimorphism. First gill arch with 27 to 33 gill rakers. Lateral line interrupted. Spinous and soft ray parts of dorsal fin continuous. Dorsal fin with 16 - 17 spines and 11 to 15 soft rays. Anal fin with 3 spines and 10-11 rays. Caudal fin truncated. Colour in spawning season, pectoral, dorsal and caudal fins becoming reddish; caudal fin with numerous black bars.

Nile tilapia is a tropical species that prefers to live in shallow pond water. All *tilapia* species are nest builders (Fig. 63a); fertilized eggs are guarded in the nest by a brood parent. These are are mouth brooders; eggs are fertilized in the nest but the parents immediately pick up the eggs in their mouths (Fig. 63b) and hold them through incubation (Fig. 63c) for several days even after hatching (Fig. 63d). Only females are mouth brooders.

Rajiv Gandhi Centre for Aquaculture (RGCA), the R&D arm of Marine Products Export Development Authority, has established a facility in Vijayawada to produce mono-sex tilapia in two strains. This project involves the establishment of a satellite nucleus for the GIFT strain of tilapia in India, the design and conduct of a genetic improvement program for this strain, the development of dissemination strategies, and the enhancement of local capacity in the areas of selective breeding and genetics. The development and dissemination of a high yielding tilapia strain possessing desirable production characteristics is expected to bring about notable economic benefits for the country.

Fig. 63: *Oreochromis niloticus*

Fig. 63a: The male tilapia digs nest at
the pond bottom.

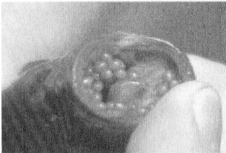

Fig. 63b. Eggs within the mouth.

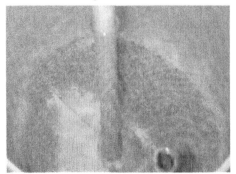

Fig. 63c: Incubation of tilapia eggs.

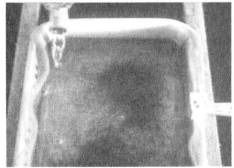

Fig. 63d. Yolk laden early fry. (Clockwise)

The Genetically Improved Farmed Tilapia (GIFT), a faster-growing strain of
Nile tilapia (*Oreochromis niloticus*) is suitable for both small-scale and commercial
aquaculture. Nile tilapia was selected due to its popularity in aquaculture, short
generation time of approximately 6 months, and naturally hardy and high tolerance
to variable water quality, good disease resistance, and ability to adapt many
different farming systems.

Sex Reversal of Tilapia: 17 α-Methyltestosterone (MT) Dose Rate.

Early sexual maturity in tilapia culture is a well-recognized problem. There are a number of ways to control reproduction in mixed sex population. One of these is the culture of all-male tilapia. Sex reversal by oral administration of feed incorporated with 17α -methyl testosterone (MT) is probably the most effective and practical method for the production of all male tilapia. This is the most common method of sex reversal.

Sex reversal of newly hatched tilapia generally is accomplished via oral administration of 17 α –methyl testosterone (MT), which has been incorporated into a starter fish feed @ 5-6 mg/kg feed. Although the use of the 5-6 mg MT/ kg feed dose consistently yields populations comprised of less than 5% females (i.e.> 95% males). Other investigators have reported sex reversal of tilapia at dose rates less than 5-6 mg/kg feed; however, results from some of these studies are inconsistent, and it is difficult to separate treatment environment effects. Thus, it is necessary to identify the optimal dose of MT for consistent, successful sex reversal in a variety of treatment environments.

Although a variety of hormones have been used for sex reversal, 17α - methyl testosterone is the most commonly used androgen. Dose rate and treatment durations vary depending on the environment and the experience of the producer. Tilapia fry <14 mm should be treated for at least 14 day before reaching 18 mm, if growth is slower the duration of treatment should be extended until all fish reach this size or a total treatment period of 28 days are exceeded. A dose rate of 30-60 mg of Methyl testosterone/kg of diet fed at an initial rate of 20%body weight/day should result in successful treatment. The efficacy of treatment should be based on gonadal development.

Food and feeding habit: Nile tilapia, *Oreochromis niloticus* is one of the most known members of the tropical and subtropical freshwater fishes. It is recommended by the FAO as a culture fish species because of its importance in aquaculture and its capability in contributing to the increased production of animal protein in the world. Therefore, it is now globally distributed and has become very popular through the advances in the cultivation techniques. Juvenile and adult Nile tilapias are reported to filter phytoplankton hence indicate the herbivorous behavior. Since Nile tilapia use algal protein raising tilapia for food at lower trophic level can be a cost-efficient culture method.

Table 35: Diseases and control measures of *Oreochromis niloticus*

Disease	Agent	Type	Syndrome	Measures
Aeromonas septicaemia	*Aeromonas hydrophila*	Bacteria	Loss of equilibrium; lethargic swimming; gasping at surface; haemorrhaged or inflamed fins & skin; bulging eyes; opaque corneas; swollen abdomen containing cloudy or bloody fluid; chronic with low daily mortality	$KMnO_4$ @ 400 gms/acre/meter + antibacterial agent in feed at 50 mg/kg. Repeat the same after 5-7 days.
Vibriosis	*Vibrio sp.*	Bacteria	Same as *aeromonas* infection.	Same as before. It is advisable to use Soil and water probiotic @1 litre and 600gms. respectively/acre meter.
Columnaris disease	*Flavobacterium* sp.	Bacteria	Frayed fins &/or irregular whitish to grey patches on skin &/or fins; pale, necrotic lesions on gills	$KMnO_4$ @400gms/acre/meter+ antifungal agent @300gms/acre/meter.
Edwardsiellosis	*Edwardsiella* sp.	Bacteria	Few external symptoms; bloody fluid in body cavity; pale, mottled liver; swollen, dark red spleen; swollen, soft kidney	Antibacterial agent@300gms/acre/metre.
Streptococcosis	*Streptococcus & Enterococcus* sp.	Bacteria	Lethargic, erratic swimming; dark skin & red eyes; and operculum, around mouth, anus & base of fins enlarged, nearly black spleen; high mortality. respectively.	Antibacterial and antifungal agents@300gms/acre/metre. Water and soil probiotics@600gms and 1litre
Saprolegniosis	*Saprolegnia parasitica*	Fungus	Lethargic swimming; white, grey or brown colonies that resemble tufts of cotton; open lesions in muscle	Antifungal agent@300gms/acre/metre.
Ciliates	*Ichthyophthirius multifiliis; Trichodina* & others	Protozoan parasite	Occurs on gills or skin	$KMnO_4$ @400 gms/acre/metre + antifungal agents@300gms/acre/metre.
Monogenetic trematodes	*Dactylogyrus* spp. *Gyrodactylus* sp.	Protozoan parasite	Occurs on body surface, fins or gills	Same as for ciliates

vii) Culture possibilities of *Anabas testudineus*

Culture possibilities of Climbing perch (*Anabas testudineus*) as a candidate species in fresh water ponds.

Systematic position:

Phylum- Chordata.

Class- Osteichthyes.

Order- Perciformes.

Family- Anabantidae.

Genus- Anabas.

Species- Testudineus.

Binomial name: *Anabas testudineus* (Blotch, 1792).

Morphology

The body is laterally compressed. The mouth is anterior and the lower jaw is slightly longer.

Villiform teeth present on the jaw. Long dorsal and anal fin is found. The body colour is dark to pale greenish, fading to pale yellow on belly where as dorsal and caudal fin is dark grey, anal and pectoral fins are pale yellow, pelvic fin pale orange colour.Pectoral and caudal fin is rounded.Dorsal, pelvic and anal fin rays are modified to spine. Scales are cycloid, lateral line is interrupted.

Habits and habitat

The fishes are found both in fresh and brackish waters, mostly in canals, lakes ponds, ditches, floodplains, haors and baors.These are principally bottom dwelling fishes and are insectivorous in nature.

Climbing perch (*Anabas testudineus;* Fig.64) is a very important indigenous fish and are locally named as Koi. This kind of fish is famous due to its high nutrition value as well as for great taste and flavor. The demand of this species is very high because it is generally marketed as alive. It is highly esteemed for it's highly nourishing quality and prolonged freshness out of water because of the presence of accessory respiratory organ in its body and is a valuable diet for the sick persons.

However, abundance of this native species is declining drastically due to the ecological degradation, indiscriminate fishing, use of pesticides and fertilizers, destruction of habitats, obstruction to breeding migration and problems in management. Considering the importance of native Koi, the breeding technology of this species has successfully been developed. However, its slow growth and small size do not favour sustainable production per unit area in a culture system. To solve this problem, another fast growing climbing perch known as Vietnamese Koi *(A. testudineus)* has been introduced in 2010 whose growth rate is 60% higher than the native Koi. It can be cultivable in pond and consume artificial supplementary feed.

Fig. 64: The climbing perch *Anabas testudineus*.

The investigation on breeding biology of Vietnamese strain of climbing perch, *Anabas testudineus* (Bloch) is also successful in commercial hatcheries. Farmers can have high financial return in low investment because of its high production. But the major constraint in the culture of this species in a large scale is the non-availability of quality seeds from the hatchery. To utilize and manage this species wisely in culture system, understanding of breeding biology is very essential. Studies on the breeding biology of any fish is essential for evaluating the commercial potentialities of its stock, life history, cultural practice and actual management of small indigenous fishes.

Maximum standard length: Generally attain 150 – 200 mm, but it appears that some forms may be fully-grown at 100 – 120 mm.

Habitat: Inhabits in the majority of drainage systems across its native range and has been recorded in many different habitat-types including swamps, marshes, lakes, canals, pools, small pits, rice paddies, puddles, oxbows, tributaries and main river channels. Though primarily a lowland, freshwater species it also occurs in brackish coastal environments in some areas.

Water quality requirements:

Temperature: 15 – 30 °C, **pH:** 5.5 – 8.0, **Hardness:** 36 – 447 mg/L.

Food and feeding habit: Chiefly predatory though wild fish apparently feed on some vegetative matter including algae and rice grains as well as smaller fishes, invertebrates and mollusks. However, the Koi under **monoculture** can well be cultured and may be fed on the floating feed of different size ranges.

Sexual dimorphism: Sexually active females are slightly larger and noticeably thicker-bodied than males, while males are darker in coloration and apparently develop tubercles on the pectoral fins during breeding. These observations relate to specimens from West Bengal (Fig.65a).

Predation effects: The effects of the climbing perch on native fish and other fauna can be devastating. Climbing perch are expected to out-compete native freshwater and estuarine species. In addition, the fish has sharp dorsal and opercular spines which are extended when the fish is ingested by predatory species of any types e.g. any other carnivorous fish, snakes or fish eating birds.

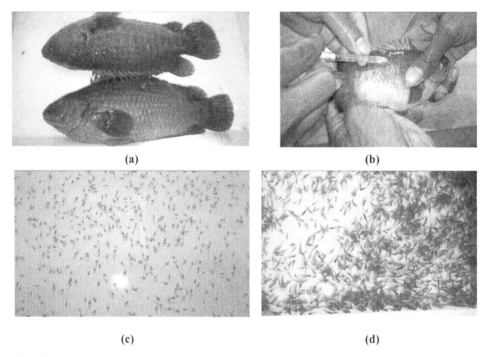

(a)　　　　　　　　　　　　(b)

(c)　　　　　　　　　　　　(d)

Fig. 65: (a) Brooder (b) Hypophysation process (c) Spawn and (d) early fry of *Anabas* sp.

viii) Culture possibilities of *Notopterus notopterus*

Culture possibilities of *Notopterus notopterus*, the Bronz Feather Back (Folui) in fresh water ponds. (Fig. 66).

Systematic position:

Phylum-Chordata

Class- Actinopterygii.

Order- Osteoglossiformes.

Family- Notopteridae.

Genus- Notopterus.

Species- Notopterus.

Binomial name: *Notopterus notopterus* (Pallas, 1769)

Habitat and Ecology

The species inhabits fresh and brackish waters, and appears to thrive well in lentic waters.

Owing to its carnivorous nature, this fish can only be cultivated in wild waters or in fattening ponds in which large fish are present. It breeds both in stagnant and running water in the rainy season. A ripe female bears relatively fewer eggs; they are laid in small clumps on submerged vegetation. Fingerlings are available in upper reaches of Cauvery in July-August. This fish is relished both in fresh and dried state.

Present Status: IUCN listed the species as vulnerable.

Breeding / Reproduction: The fish spawn in the night, dropping eggs to the bottom and on rocks. They were guarded by the male who fanned fresh water over the eggs with his pectoral fins, chasing off all other fishes. The eggs required two weeks to develop and the newly hatched young were sensitive to handling.

Feather fin, Knife fish are mass-produced fish in commercial fish farms. They are commercially spawned and the eggs are normally laid on the substrate or decor and guarded by the male till they hatch, which is usually 2 weeks. Once free swimming these fry will feed on brine shrimp nauplii and micro worms. These fish are intolerant of their own kind, so a huge tank would be need and a large influx of cold water added to the tank to trigger spawning.

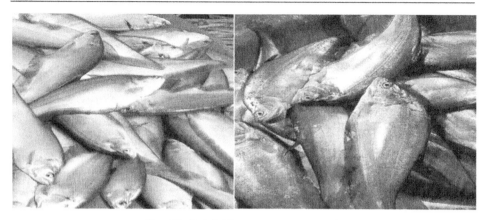

Fig. 66: Folui (*Notopterus notopterus*)

Diseases

The Bronze Feather back does not have scales which make it more prone to disease. Knife fish are normally the first fish in a pond/tank to show signs of ick and will go into spasm and rub around the tank. They respond well to most medication and normally heal quickly. Overall these fish are hardy and disease is not usually a problem in a well maintained tank. That being said there is no guarantee that you won't have to deal with health problems or disease.

Breeding: Preferably in rainy season.

ix) Culture possibilities of *Amblypharyngodon mola*

Systematic position:

Phylum- Chordata.

Class- Actinopterygii.

Order- Cypriniformes.

Family- Cyprinidae.

Genus- Amblypharygodon.

Species-Mola.(Hamilton, 1822).

Binomial name: *Amblypharygodon mola* (Hamilton, 1822).

Morphology

Body of this species is laterally compressed and the dorsal profile is more convex than that of the ventral.Barbels absent. Caudal fin is deeply forked and the lobes are pointed. Dark margins present in dorsal and anal fins. Body colour light greenish on the back and silvery at the sides and beneath.

Amblypharyngodon mola, commonly known as Morala or Mola Carplet or Pale Carplet and is a popular food fish mainly in Indian sub-continent due to its high nutritional value with high protein, vitamin and mineral content and also rich in Fe, Zn and Calcium. *Amblypharyngodon mola* (Cypriniformes: Cyprinidae) is a freshwater fish species (Fig.67); a natural inhabitant of ponds, canals, beels, slow moving streams, ditches, baors, reservoirs and inundated fields.

Feeding biology

Phytoplankton as the basic food and Chlorophyceae as the mostly preferred food class for this fish species. Chlorophyceae followed by Myxophyceae, Bacillariophyceae, Cyanophyceae and Euglenophyceae are dominant in the guts of this fish.The farmers are suggested to apply multi micronutrient mineral mixtures @ 6kg/acre/meter at least once in a couple of months in order to maintain a steady abundance of phytoplankton in the ponds.

Reproductive Biology of Mourala (*Amblypharyngodon mola*)

Males and females are different in color; males are comparatively brighter than females. The color of females is light and they are large in size. In case of mature female, the abdomen is soft and swollen, pelvic fins are smooth and caudal fin is deeply forked. During the spawning season mature females are with distended abdomen by which they can be easily recognized. It is reported that there do exist a significant female dominance over male in *A. mola* (1:1.67).

Experimental works on the breeding biology of *A. mola* conducted by the scientists of CIFE, Kolkata and Comprehensive Area Development Centre of Tamluk Centre suggests that the species can well be reared, propagated both naturally and through hypophysation technique (Fig. 67) along with Indian Major Carps in a pond. For hypophysation technique, the interested farmers may contact Officer-in-Charge Comprehensive Area Development Corporation(CADC), Tamluk, East Midnapur. The stocking density of *A.mola* should be 2/m^2 along with the carps. Since the fish generally prefers phytoplankton as its natural feed stuff which is readily available, hence separate and special attention is not needed (Dr. B.K.Mahapatra, CIFE, and personal communication). The fishes start maturity at the length of 5, 6, 7 and 8 cm. However, it has been noticed that the fish gained first maturation within 4 to 5 months although all of them may not breed. Male attained maturity earlier than female. During this stage it is preferred to keep a provision of keeping some floating weeds at the corner of the ponds which provide the shelter to the fertilized eggs as well as the early fry of the fishes.

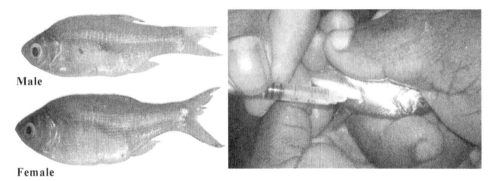

Male

Female

Fig. 67: The matured male and female and hypophysation of *Amblypharyngodon mola*.

Higher percentage of gravid females are found during the month from April to October and fecundity values suggests that the reproductive cycle of *A. Mola* is from April to October with two peaks, one in May and other in September *i.e.,* it breeds twice in a year in both ponds (closed) and beels (open) water environments.

Major threats of *A. mola*: Over fishing of this species is the major threat now-a-days.

x) Culture possibilities of *Mystus vittatus*

Mystus vittatus (Bloch, 1794), an indigenous small fish of Bangladesh, belongs to the family Bagridae, widely distributed in Asian countries including Bangladesh, India, Pakistan, Sri Lanka, Nepal and Myanmar. However, natural populations are seriously declining due to high fishing pressure, loss of habitats, aquatic pollution, natural disasters; reclamation of wetlands and excessive floodplain siltation and it is designated as vulnerable species by IUCN.

Systematic position:

Phylum- Chordata.Class- Actinopterygii.

Order- Siluriformes.

Superfamily- Bagridea.

Family- Bagridae.

Genus- Mystus.

Species- Vittatus

Binomial name: *Mystus vittatus* (Blotch, 1794).

Morphology

Fig. 68: A mature *Mystus vittatus*

Body elongated and slightly compressed.Presence of four pairs of barbels. The maxillary barbells are extending beyond the pelvic fins, often to the end of the anal fin. Dorsal spine is weak and finally serrated on its inner edge. The adipose fin is small, inserted much behind rayed dorsal fin but anterior to the anal fin. The lateral line is present which is straight.

The colour of the fish varies with age, generally delicate grey silvery to shining golden, with about five pale blue or dark brown to deep black longitudinal on side. A narrow dusky spot often present on the shoulder. The fins are with dark tips. The head portion is about 23.3% of the standard length and about 18.2% of the total length. The maximum length is found to be 10.2cm.

Habitat: Generally found in fresh water bodies, in flooded canals, beels, paddy and jute fields as well as streams, haors, oxbow lakes and rivers in rainy seasons.

Mystus vittatus(Fig.68) is an omnivorous fish and they prefer both phyto- and zooplankton along with worms, insects and plant material. It is revealed that the fish is a bottom feeder and the food matter includes algae (18%), higher plants (27%), protozoa (13%), crustacean (24%) and insects (11%).Breeding season is generally during summer and monsoon.

Diseases and control measures: *Mystus vittatus* were more affected showing red spots, deep dermal ulcer in the months of December and January. Marked necrosis, pyknosis, haemorrhage and fungal granuloma were observed in skin and muscle.

Fig. 69a: Mature (♀) and (♂) *M. vittatus.*   Fig. 69b. Male genital papilla   Fig. 69c. Female genital papilla.

 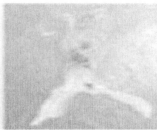 

Fig. 70a: Ovary of *M. vittatus.*   Fig. 70b: Testis of *M. vittatus.*   Fig. 70c: Injecting male *M. vittatus*

  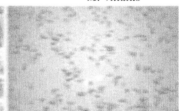

Fig. 70d: Fertilized eggs   Fig. 70e: Fertilized and unfertilized eggs   Fig. 70f: Early fry

Fig. 70g: *M. vittatus* fingerlings   Fig. 70h: Adult *M. vittatus*

Gills of the fishes were found to contain monogenetic trematode, hypertrophy, clubbing and pyknosis in December and January. Fish kidney and liver displayed fungal granuloma, necrosis, haemorrhage and vacuoles. To overcome these diseases it is advisable to apply mycin group of antibiotics @300gms/acre/meter. This application should have to be repeated once in two months.Vankara and Vijayalaksmi (2009) during a survey in River Godavari found that *Mystus vittatus* (Bloch) of family Bagridae serves as an excellent host for a variety of metazoan parasites. These fishes, locally known as 'jella' are highly relished in southern India due to their rich protein content and delicacy.They have described a total of nine parasitic species from four parasitic groups from *M. vittatus*, of which a new species of acanthocephalan, *Raosentis godavarensis* sp. nov.

xi) Culture possibilities of *Nandus nandus*

It inhabits fresh waters. Found in rivers and in agricultural lands. It is commonly collected in the summer months when it is collected from dried-up beds of tanks, beels, bheries, etc. The species changes colour to camouflage against its environment.

Systematic position:

Phylum- Chordata.

Class- Acinopterygii.

Order- Perciformes.

Family- Perciformes.

Gednus- Nandus.

Species- Nandus (Hamilton, 1822)

Binomila name: *Nandus nandus* (Hamilton, 1822).

Nandus nandus, is commonly known as **Gangetic leaf fish**. These are common in slow-moving or stagnant bodies of water, including ponds, lakes, ditches, and flooded fields but have been treated as a threatened one. Attains a length maximum of 20 cm and is very popular in West Bengal, particularly when freshly caught. It is a high-prized fish in spite of its spinous fins and ugly black bands and blotches all over the body. The species changes colour to camouflage against its environment (Fig. 71a&b).The fish matures by the middle of March and its breeding season extends from late March to the middle of July. The males are comparatively smaller than the females (Fig. 72). The sexes can be distinguished externally during the breeding season, by the difference in the general coloration. The mature ovarian eggs measure 0·595 to 0·663 mm in diameter. The fecundity of the fish (length range: 75 to 138 mm) ranged from 1, 573 to 23, 546.

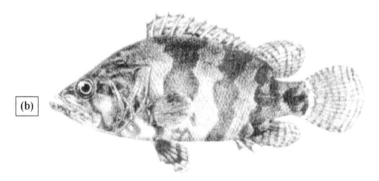

Figs. 71a and 71b: The young and adult *Nandus nandus*.

Fig.72: Sexual dimorphism of *N. nandus*

Mature fishes may be induced to breed (Fig. 73) by the administration of carp pituitary hormones successfully. The fertilized, swollen eggs are adhesive in nature and yellowish in colour with a central, spherical oil globule. The newly hatched larva measuring 2·07 mm in length has three characteristic vertical bands and pigmented eyes.

Fig.73: Artificial insemination of *N. nandus*

The fish feed on plankton at their post-larval stage. It is predominantly insectivorous and piscivorous in dietary habits in the juvenile and adult stages respectively. The predatory habit of the fish warrants its eradication from nursery ponds. Since the species is highly planktivorus sufficient phyto- and zooplankton production in the pond must be there in order to provide balanced nutrition. The information related to diseases are not seriously been documented yet except some helminth infections in the intestines of females of this species predominantly in summer months.

xii) Culture possibilities of *Piaractus brachypomus*

Culture possibilities of *Piaractus brachypomus, the Pacu*as a candidate species along with Indian Major Carp in fresh water ponds.

Common name for this species vary by region, in Brazil the fish is known as Pirapatinga, USA as Cachama, UK as freshwater pompano and in India especially in Tripura, Assam and West Bengal the fish is locally known as Rupchanda or Pacu (Fig. 74.)

Fig. 74. *Piaractus brachypomus*

Systematic position:

Phylum-Chordata.

Class- Actinopterygii.

Order- Characiformes.

Family- Serrasalmidae.

Genus- Piaractus

Species- Brachypomus

Binomial name: *Piaractus brachypomus* (G.Cuvier, 1818).

Pacu (*Piaractus* sp.) (Fig. 74) is a freshwater fish of the order Cypriniformes, suborder Characoidei, and family Serrasalmidae. Pacu is a herbivorous fish, which preferentially feeds on leaves, flowers, fruits, and seeds of superior plants.Introductions of alien fish species are an important part of human activities concerning aquatic ecosystems. In India, over 300 exotic fish species including 291 ornamental species, 31 aquaculture species and 3 larvicidal fishes are recorded. These introductions are human mediated, which may be intentional or unintentional.Pacu the *Piaractus brachypomus* habe been recently included in aquaculture activities in India. It is now understood that over 0.1 million tonnes of pacu are produced in the country.

It has small scales and weighs between 3 and 7 Kg at normal growth. Water temperature is the single most important factor, other than food, affecting fish growth. In most warm water fishes, feeding activity starts when the temperature reaches 17 - 18° C and attains its maximum rate at 28 - 30° C. The water temperature ranging from 23-28°C, the pH ranging between7.5-8.5 are found to be ideal. Pacu is an obligate gill breather which can survive low oxygen concentrations of less than 0.5 mg l^{-1}.Stomach analyses of wild specimens show it to be primarily a herbivorous species, feeding on fruits, nuts and seeds. It is an opportunist, feeder and also takes insects, zooplankton and small fish.

Behavior and Compatibility: Small fish will likely be eaten but in general it's quite peaceful with bigger species. It's a shoaling species when juvenile, becoming more solitary as it matures.

Reproduction: A single 5 Kg female pacu can produce between 0.5 - 1 million larvae under natural conditions. The reason is that the ripening of the gonads requires a synergistic stimulus of several environmental factors, such as conductivity, water level and rainfall. Pacu spawn usually during the months of November - March when the water temperatures are 26°C.

Sexual Dimorphism: Mature females are a little rounder-bellied than males. The fish are often induced to spawn via the use of hormones.

Aquaculture: Pacu have been traditionally grown in earthen ponds however, they are now extensively cultured in floating cages. Pacu fed under extensive systems in earthen ponds, can take up to two years to attain 1.2 Kg. Cultured pacu fed formulated diets reach 1.2 - 1.5 Kg in 12 months. Market size is at least a kilo and farm - gate prices now command Rs.150.00 per kg depending on size and weight. Pacu was introduced in India as an alien species during 2004 from Bangladesh.

Piaractus brachypomus showed higher growth with application of supplementary feed in the culture tanks. The enhanced production in tanks treated with fertilization only can be justified by the fact that the fertilizer contributed to the fertility of the pond. In ponds where the Pacu seeds are stocked, the farmers are advised to apply 6kg/acre/meter multimicronutrient mineral fertilizer and humic acid together @ 5liter /acre/ meter each which will greatly help to generate enormous planktonic communities, of course after a couple of days water probiotic application in the pond will help. Since substantial work on Pacu in fresh water pond culture is yet to be done, occurrence of disease outbreak during culture have not been noticed till now.The overall best growth performance of *P. brachypomus* was in polyculture with Indian major carps, with feeding especially adopting Substrate Based Aquaculture **(SBA)** system. It is concluded that pacu is a promising candidate species in freshwater aquaculture. Monoculture with feeding is a feasible approach, although Polyculture in fertilized systems is equally productive. Pacu is a compatible species for culturing along with Indian major carps.Growth of pacu is higher in fertilized systems with added substrates.

Diseases encountered: Both external and internal parasites are common in Pacu. The gills are very often infected with *Dadaytrema oxycephala,* while *Spectatus spectatus* occurred in the intestine and abdominal cavity, the highest intensity was however, found in the intestine. Ivermectin based products sounds well to get rid of these external parasites.Ivermectin 1.2% administration (in feed treatment) should be done following the advice of aquaculture technicians.

xiii) Culture possibilities of *Ompok bimaculatus*

Systematic position:

Phylum- Chordata.

Class- Actinopterygii.

Order- Siluriformes

Family- Siluridae.

Genus- Ompok.

Spcies-.Bimaculatus (Bloch, 1794)

Binomial name: *Ompok bimaculatus* (Blotch, 1794).

Ompok bimaculatus (Fig.75) is a catfish belongs to the family Siluridae, a freshwater fish used to occur naturally in streams, lakes, ponds, canals and inundated fields. This fish is commonly known as Indian butter cat fish .It is a highly priced and popular food fish due to its good taste and nutritional value with good amount of protein, lipid and mineral content in its flesh. At present in nature this fish species is facing high risk of extinction as over the last 10 years its wild population has undergone a steady decline (>50%)due to over exploitation mainly due to indiscriminate fishing during the breeding season, disease, pollution, siltation, poisoning, loss of habitat etc. It has been listed among the 97 endangered fish species of India and is under near threatened category as per IUCN Red list.

Fig.75: *Adult male and female of Ompok bimaculatus*

Physical Characteristics

Pabda fishes are small sized freshwater fish species. The physical characteristics of the Pabda fishes are described below:

● The body of Pabda fishes is flat in both sides.

● The caudal side is narrower than head side.

● Pabda has similarity in look with Boal fish (*Wallago attu*) when young.

● There are no scales in their body.

- Their chest is silver colored.
- There is a pair of mustache in their mouth.
- Has two pairs of fins.
- Pelvic fin lengths from belly to tail.
- Anal fin is divided into two parts.
- There is a fishbone in the pectoral fin of pabda fish.
- Pabda fish lengths between12 to 30 cm.

Morphology

Body elongated and strongly compressed.Head depressed and the snout is rounded.The mouth is superior. The lower jaw is longer than the upper. Two pairs of barbels are present. The maxillary barbels extends posterior to anal fin base.Nostrils are widely separated with each other. Teeth are found on the jaws and on the vomer.The caudal fin is deeply forked and its upper lobe is long. Dorsal side of the body is grey, a traverse blackish spot is present behind the operculum and on the lateral line. Caudal stripped with black spots.There are purple and yellowish spots throughout the body. The anal fin is with 57 or 58 finrays.

This fish is of great demand with a high market price because of its good taste, flavor and invigorating effects.

Food and Feeding Habit

Ompok bimaculatus prefer to feed on insects of various categories hence insectivorous, but also used to feed on crustaceans, fish and molluscs.

Pabda fish usually live in the upper level of water. They are omnivorous. They generally eat protozoa, aquatic insects, crustacea, moss etc. But they also like to eat cake and fishmeal as a supplementary feed. Feeding with supplementary feed containing 40% rice bran, 30% mastered cake and 30% fishmeal before their breeding period attain sexual maturity very quickly.

Sexual Dimorphism

In male pectoral fin is with a strong internally serrated spine while in female it is feebly serrated. In male, pectoral fin is strong and hard, laterally flattened, broad and somewhat thick along the entire length, abruptly tapering into blunt spine, inner edge of the spine with strong and prominent serration; in female, pectoral fin is weak and flexible, thin and narrow along the entire length, gradually tapering into sharp spine, inner edge of the spine feebly serrated or nearly smooth. They also have reported that in male, genital papilla is a small outgrowth but in female, it is some what fleshy and comparatively larger in size and almost double in size that of the male. In male, it is visible only during the breeding season.

Even after its high demand as a food fish and tolerant nature; *Ompok bimaculatus* has not received that much attention in aquaculture due to unavailability gravid brood stock for experimentation, lack of information on its feeding and breeding biology, larval rearing and culture technology. High mortality rate during the period of larval rearing is the most serious problem as has been reported for commercial production of this catfish species. Success achieved so far in its induced breeding experiments can be considered as a line but further study is needed to get success in larval rearing. Further investigations of this delicious species on its suitability for commercial culture are immediately needed because of its fine adaptation to confinement and good tolerance to captive environment.

xiv) Culture possibilities of *Mystus cavasius*

Mystus cavasius (Fig.76) is native of Bangladesh, India and Nepal, this species inhabits a wide variety of fresh water habitats especially in rivers primarily with sandy or muddy substrate.

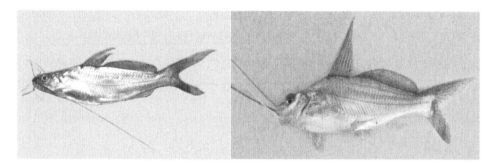

Fig. 76: Male and Female of *Mystus cavassius*

Systematic position:

Phylum- Chordata.

Class-Actinopterygii.

Order- Siluriformes.

Family-Bagridae.

Genus-Mystus

Species-Cavasius.

Binomial name: *Mystus cavasius* (Hamilton, 1822).

Major threats: Although there is a marked decline in the population in southern West Bengal due to overfishing, the threats to this species in other areas of its distribution are unknown. Since sufficient information on the biology is not available, current threats to aquatic biodiversity in all of its known distribution have also not been adequately identified.

Although IUCN has identified habitat loss as a major threat to this species, but this has not been verified by empirical evidence.

Food and feeding habit

It is omnivorous and predatory fish with wide range of food preference. Presence of insect larvae & nymph, insects, molluscs, fish larvae, fish & fish scales, chironomid larvae, algae, higher aquatic plants etc. are documented in the gut content of *Mystus cavasius*. High feeding activity has been documented during monsoon, post-monsoon and winter months while poor feeding activity has been reported during summer months.

Reproductive biology

Sexual dimorphism has been documented and the sexes of *Mystus cavasius* can be identified externally by the presence of genital papilla; which is present only in males. The papilla becomes very prominent during the breeding season.

Sex ratio

It is reported that females dominate over male in *Mystus cavasius* population.

Fecundity and the gonadal maturity

The fecundity of *Mystus cavasius* to be ranged from 4, 026-25, 960 and there are five stages of gonadal maturity namely immature, maturing, mature, ripe and spent. August-September as a period of intense breeding for Mystus *cavasius*.

Scope of culture

Except few scattered works, not much information is available on the scope of culture of *Mystus cavasius*, and it is documented that a stocking density of 2, 00, 000 fry/ha is the best one to get highest growth, production and net benefits. The experiments on the culture of *Mystus cavasius* with Indian Major Carps and *Ompok pabda* have concluded that it can be cultured effectively in low input carp polyculture management while in monoculture management practice, the production to the tune of 1, 370 to 1, 535 kg/ha in six months of culture of *Mystus cavasius* (Fig.76) has been obtained.

Diseases and remedial measures of *Mystus cavasius*

A variety of diseases have been found in *M. cavasius* which include EUS and diseases caused due to protozoa and worms etc. Qualitative maintenance of water is the principal key to abate diseases. Application of Zeolite powder of repute@15kg/Acre/meter at least once in a month coupled with Water and Soil probiotic at their respective doses should have to be applied to get rid of such infestations.

xv) Culture possibilities of *Macrobrachium rosenbergii*

Culture possibilities of Scampi, the *Macrobrachium rosenbergii as* a candidate species along with two major carps in fresh water ponds.

Systematic position:

Phylum- Arthopoda.

Subphylum- Crustacea.

Class-Malacostraca.

Order- Decapoda.

Infra order-Caridea.

Family- Palaemonidae.

Genus- Macrobrachium.

Species- Rosenbergii.

Binomial name: *Macrobrachium rosenbergii* (De Man, 1879).

The genus *Macrobrachium* Bate, 1868 is represented in India by about 40 species. Commonly they are referred to as fresh water prawns, eventhough majority of them migrate down to estuaries for breeding except *Macrobrachium choprai* (Tiwari, 1949) which is reported to complete their life cycle in freshwater itself. Out of 40 species recorded so far in India, about 15 are important from aquaculture point of view. They are either medium sized prawns, abundant in natural water bodies and supporting a fishery or big sized ones with culture potentials, might be available in lesser numbers (Jaychandran and Joseph, 1992).

In India, aquaculture experiments are concentrated principaqlly on *Macrobrachium rosenbergii* (Fig.77)-the giant fresh water prawn, *Macrobrachium malcolmsoni* –the river fresh water prawn, *Macrobrachium birmanicum and Macrobrachium choprai*. In addition to the above mentioned species, *M. dayanum* has also been rendered to as a notable crustacean withstanding good economic potential.

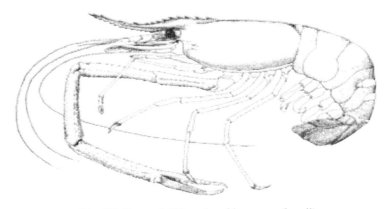

Fig. 77: Scampi *(Macrobrachium rosenbergii)*

Prawn, Shrimp and Lobster

The term 'prawn' is used for freshwater species, which are comparatively larger than shrimps (can grow up to 150gms.max.) on the other hand, shrimps are smaller in size (can grow up to 35-50gms.max) belong to either fresh, brackish or marine waters while the Lobsters are much bigger in size, the cephalic region is laterally compressed and belong to deep sea organisms.

1. Prawns are in general fresh water inhabitants but shrimps live both in fresh and salt waters while lobsters are absolutely marine in habitat.

2. Prawns and shrimps swim while lobsters crawl.

3. Shrimps are generally smaller than both prawns and lobsters.

4. Female lobsters carry their eggs with them while shrimps disperse their eggs to the sea.

5. Lobster types include clawed lobsters, spiny lobsters, and slipper lobsters.

6. Lobsters have 10 legs or five pairs, and the first pair has claws. Lobsters can be as long as 50 centimeters while prawns can be maximum up to 33 centimeters long.

7. Lobsters mostly live in muddy, rocky, or sandy bottoms. Their carapace is harder than the exoskeleton of prawns.

8. Each segment in the body of the prawn overlaps the one after it. That is, the first segment overlaps the second and the second segment overlaps the third.

9. In shrimps, the second segment overlaps the first and the third segment. Also, the legs of prawns are longer than those of shrimp. The front pincers of shrimp are the largest, while the second pincers are larger than those in prawns.

Macrobrachium rosenbergii is an economically important cultured freshwater prawn throughout the world. It is the most favoured species for farming and farmed on a large scale in many countries including India. Considering its high export potential, the giant freshwater prawn enjoys immense potential for culture in India. About 4 million hectares of impounded freshwater bodies in India offer great potential for freshwater prawn culture.

Due to certain constraints such as shortage of quality seeds and differential growth rates, freshwater prawn farming in the country has not spread to the extent it deserves. The virtual collapse of shrimp farming due to viral disease out breaks, the ban imposed by the Supreme Court for shrimp culture in the Costal Regulation Zone (CRZ) and the persistent decline in profitability of carp culture in recent years has forced the aquaculture entrepreneurs to look for an equally remunerative alternative species. The giant freshwater prawn is a profitable and viable alternative species. In India too, it has gained importance as an aquacultural crop. But, problem of differential growth which is also known as heterogenous individual growth (HIG) is still continued to be a major problem in scampi culture. Males of *M. rosenbergii* grow faster and reach higher weights at harvest than females. Hence, culture of all-male population is desirable since sexual dimorphic growth patterns are common among decapod crustaceans. To overcome this problem, many techniques like monosex culture, size grading, production of neomales and different management techniques were adopted in freshwater prawn farming.It has been noticed that the prawns and shrimps show some specialized regional characteristics. Both males and females grow with molting processes. This molting takes place in every three to eight weeks.

Culture techniques

(A) **Extensive Technique:** Extensive culture means rearing in ponds with a production target of less than 500 kg/ha/yr. Stocking density of PL's or juveniles @ 1-4/m2. There is no control of water quality; the growth or mortality of the prawns is not normally monitored.

(B) **Semi-intensive Technique:** Semi-intensive systems involve stocking PL or juveniles @ 4-20/m2 in ponds, with a production target of more than 500 kg/ha/yr. Fertilization of ponds and a balanced feed ration is mandatory. Surveillance on predators and competitors are minimized, controlled and the water quality, prawn health and growth rate are monitored. This form of culture is the most common in India, West Bengal in particular.

(C) **Intensive culture Technique:** Intensive culture refers to small earthen or concrete ponds (up to 0.2 ha) provided with high water exchange and continuous aeration, stocking density @ $20/m^2$ and above with a production target of more than 5 000 kg/ha/yr. Construction and maintenance costs are high and a high degree of management is required, which includes the use of a nutritionally balanced complete feed, the elimination of predators and competitors, and strict control over all aspects of water quality. This technique however, is not popular in India.

Release of Scampi seeds in ponds: It is always necessary to acclimatize the post larval stages of prawn brought from outside. Sudden changes in temperature and pH can cause mortalities when PL's are stocked. Before their release, the bags containing the postlarvae should be floated in the ponds to bring the temperature within them gradually to that of the pond. Any adjustments to the pH of the transport water should have been made in the hatchery, before transport (Fig. 78a, b).

Molting

In general, fresh water prawns molt at regular interval to grow Hard external shell of prawns is incapable of expanding, so to grow, prawns must periodically shed their entire shell.

The shell is composed mainly of protein, calcium carbonate and a stiffening agent, called chitin. Prior to shedding its shell, the prawn re-absorbs most of the protein and chitin from the old shell as the new one forms underneath, and increases its water intake to create a space between the body and the new shell. At molting, the soft shelled prawn emerges from the old shell via a split between the carapace and tail. During this process the entire shell including the part covering the long feelers or antennae, eyes, legs, gut and associated structures is renewed. After molting the new shell hardens and the prawn grows into the space between the body and the shell. The frequency of molting is dependent upon many factors including size, gender and species of prawn and water temperature. The process is controlled by hormones which are released by glands located in the eyestalks. Small prawns grow more rapidly and hence need to molt more frequently than larger ones and these growth rates are usually highest during the warm, summer months. Numbers of soft shelled individuals of some species peak around the time of the full moon. Molting may be inhibited by low water temperatures, whilst higher temperatures may stimulate its onset.

     (a)   (b)   (c)   (d)

Fig. 78: (a) Acclimatization of PL's, (b)Slow release of PL's and (c) Surface based Culture sheets for PL's and (d) Vertical placing of substrates.

Hideouts

During pre -molt stage, the prawns search for places to hide themselves to avoid attacks from other prawns that are not on molting phase since prawns are in general, **cannibalistic** in nature. To provide shelter, broken pipes, unused tiers etc. are unevenly distributed throughout the pond bottom so that during molting when the prawns become weak should not be a victim of cannibalism. Prawns are cannibalistic and display aggressive and territorial behaviour especially when stocking densities are high, such as in nursery systems and food is insufficient. Cannibalistic behaviour causes stress and mortality especially on moulting individuals which are vulnerable to these attacks.

Substrate Based Farming (SBF) of Scampi

Albeit the SBF of Scampi is not yet developed in India, but encouraging results in freshwater prawn culture have been reported from other countries where artificial substrates are placed in the ponds (Fig.78c and d), which makes it feasible to increase stocking rates above the level recommended earlier for ponds without substrates. PVC fencing forms an ideal substrate (Fig.78 c, d and Fig.79).

The advantage of SBF of Scampi

i) Increased production target, ii) Stocking rate of PL's may also be increased to the tune of $6.5/m^2$ from initial $4/m^2$. iii) No extra labour cost is involved. iv) The PVC substrates may be used for several cycles of culture.

This new technology is still being developed but it clear that the use of substrates can markedly increase the productivity of freshwater prawn farming.

Fig. 79: Rearing pond with horizontal substrates

Food and feeding habit of Scampi

The natural food preference of prawns depends on their age. Larvae are, feeding primarily on zooplankton (especially small crustaceans), while the post larvae and adults are omnivorous, feeding on algae, aquatic plants, mollusks, aquatic insects, worms, and other crustaceans. During the larval stages, prawns seize food by their thoracic appendages and since at this stage they are non-active hunters, they seize food items as they encounter them, thus the importance of live prey that remains suspended in the water. Moreover, the size of the food is also important, such that the brine shrimp *Artemia* nauplii have been found to be more suitable than Cladocerans, Moina at the early prawn larval stages.

Environmental factors affecting feeding

There are many factors that relate to the growth and feeding activity of prawns. Ideal ranges for water quality variables of *Macrobrachium rosenbergii*:

Table 36: Ideal ranges of physico-chemical parameters suitable for rearing *M. rosenbergii*

Parameters	Ideal range
Temperature	25-32°C
Transparency	25-40 cm
Alkalinity	20-60 mg/L
Hardness	30-150 mg/L
Un-ionized ammonia	0.1-0.3 mg/L
Dissolved Oxygen	3-7 mg/L
pH	7.5-8.5

Prawns become stressed at dissolved oxygen levels below 2 ppm. pH should not be more than 9.5. However, the most critical parameter is temperature which means that in conducting tests with freshwater prawns, temperature has to be within the acceptable ranges.

Water stable pellets with moderate nutritive value are the preferred food of *Macrobrachium rosenbergii*. However, it is necessary to maintain an adequate phytoplankton density, to provide cover and control the growth of weeds in freshwater prawn ponds. This is done by encouraging the growth of phytoplankton. *Benthic fauna are very important features in the ecosystem of freshwater prawn ponds, forming part of the food chain for prawns. Fertilization to encourage the development of benthic fauna is therefore necessary.* Organic manures have been used for this purpose (e.g. 1000-3000 kg/ha).

Freshwater prawns are omnivores and, their nutritional requirements are not very demanding. Some farmers utilize commercial feeds designed for marine shrimp in freshwater prawn nurseries or during the first few weeks of the grow-out phase when prawns are stocked as PL. Marine shrimp feeds have a much higher protein content than is needed for freshwater prawns, so cheaper commercial feeds that have either been specifically designed for freshwater prawns or for a species of fish (e.g. catfish) must be used in grow-out ponds stocked with nursery-reared juveniles, or substituted as soon as possible in those stocked with PL. The prawns are fed daily with formulated pellet diet (2-3 mm size) @ 10% of the biomass initially and then reduced to 3% of the biomass towards the end of the culture period.

Feed management

There can be no exact general recommendation for daily feeding rates, because these depend on the size and number of prawns (and, in a polyculture system, fish) in the pond, the water quality, and the nature of the feed. The feed should be broadcasted in the pond as mentioned above. Spread the feed around the periphery of the pond in the shallows, which are good feeding zones. Check trays 3-4 nos. may be kept in different corners of the pond to check the consumption of food in the gut of the prawn.

Fig. 80.a and b: Check tray and its lifting. (It is always necessary using Check tray for periodical evaluation of consumption of feed and health monitoring).

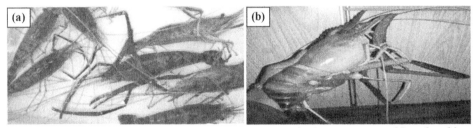

Fig. 81a and b: Young and adult maturing and matured fresh water prawn *Macrobrachium rosenbergii.*

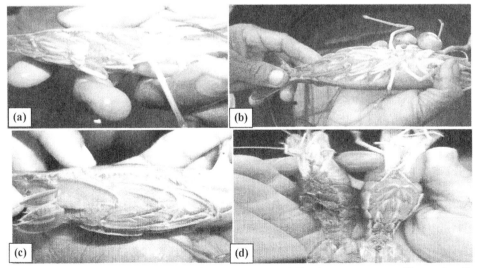

Fig. 82a, b, c and d: Gravid (maturing) and matured Fresh water prawn *M.rosenbergii.* (Clock wise).

Considerations for fresh water prawn hatchery

The larval stages of freshwater prawns require brackish water for growth and survival the reason, the hatcheries do not have to be located on coastal sites. Prawn hatcheries can be sited on inland sites. The necessary brackish water can be obtained by mixing locally available freshwater with seawater or brine (and sometimes artificial seawater) which may be transported to the site. Inland hatcheries have the advantage that they can be sited wherever suitable freshwater is available and their market (namely outdoor nurseries and grow-out facilities) is close by. Where to site a hatchery is therefore, not only a technical but also an economic consideration.This involves balancing the costs of transporting seawater and brine, or using recirculation, against the advantages of an inland site. Prawn hatcheries, regardless of type, require an abundant source of freshwater as well as seawater or brine.

The quality of intake water, whether it is saline or fresh, is of paramount importance for efficient hatchery operation. Water quality is thus a critical factor in site selection. Hatchery sites should preferably be far away from cities, harbors and industrial centers, or other activities which may pollute the water supply.The brackish water derived from the mixture of seawater, brine or artificial sea salts with freshwater for use in *M. rosenbergii* hatcheries should be 12-16 ppt, have a pH of 7.0 to 8.5, and contain a minimum dissolved oxygen level of 5 ppm. Both freshwater and seawater must be free from heavy metals (from industrial sources), marine pollution, and herbicide and insecticide residues (from agricultural sources), as well as biological contamination (e.g. as indicated by the presence of faecal coliforms, which can be common in residential and agricultural areas). The analyses of water found suitable for use in freshwater prawn hatcheries are given in Table: 37.

Other requirements for hatchery sites

In addition to having sufficient supplies of good quality water, a good hatchery site should also:

- Have a secure power supply which is not subject to lengthy power failures. An on-site emergency generator is essential for any hatchery - this should be sized in a manner, so that it has the output necessary to ensure that the most critical components of the hatchery (e.g. aeration, water flow), can continue to function; ensure that the most critical components of the hatchery (e.g. aeration, water flow), can continue to function;

- Have good all-weather road access for incoming materials and outgoing PL;

- Be on a plot of land with an area appropriate to the scale of the hatchery that has access to the quantity of seawater and freshwater supplies required without excessive pumping. The cost of pumping water to a site elevated high above sea level, for example, may be an important factor in the economics of the project;

- Not be close to cities, harbors, mines and industrial centers, or to other activities that may pollute the water supply;

- Be situated in a climate which will maintain water in the optimum range of 28-31°C, without costly environmental manipulation;

- Have access to food supplies for larvae;

- Employ a high level of technical and managerial skills;

- Have access to professional biological assistance from government or other sources;

- Have its own indoor/outdoor nursery facilities, or be close to other nursery facilities; and

- Be as close as possible to the market for its PL. In the extreme case, it should not more than 16 hours total transport time from the furthest farm it will be supplying.

Outdoor nurseries and grow-out facilities

The success of any nursery facility or grow-out farm depends on its access to good markets for its output. Its products may be sold to other farms (in the case of nurseries), directly to the public, to local markets and catering facilities, or to processors or exporters. The needs and potential of each type of market need to be considered. For example, more income may result if you can sell your market-sized prawns alive. The scale, nature and locality of the market is the first topic that you should consider and the results of your evaluation will determine whether the site is satisfactory and, if so, the way in which the farm should be designed and operated.

Despite the obvious importance of the market, it is surprising how often that this topic is the last criterion to be investigated.

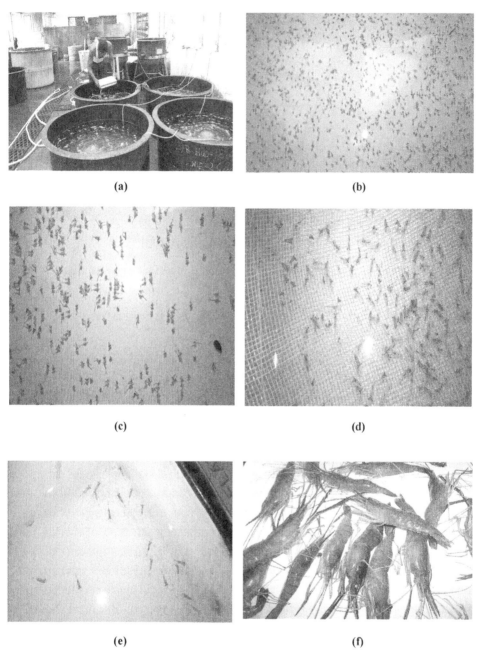

Fig.83 (a-f): Fresh water prawn hatchery and their developmental stages. [Early spawn (3-days, 7-days, 10-days, 15-days and 50-days old Fresh water prawn *Macrobrachium rosenbergii*].

Table 37: Characteristics of water suitable for freshwater prawn hatcheries.

Parameters	Fresh water (ppm)	Sea water (ppm)	Brackish water (ppm)
Total hardness (as CaCO3)	<120	-	2 325-2 715
Calcium (Ca)	12-24	390-450	175-195
Sodium (Na)	28-100	5 950-10 500	3 500-4 000
Potassium (K)	2-42	400-525	175-220
Magnesium (Mg)	10-27	1 250-1 345	460-540
Silicon (SiO2)	41-53	3-14	5-30
Iron (Fe)	<0.02	0.05-0.15	<0.03
Copper (Cu)	<0.02	<0.03	<0.06
Manganese (Mn)	<0.02	<0.4	<0.03
Zinc (Zn)	0.2-4.0	0.03-4.6	<3
Chromium (Cr)	<0.01	<0.005	<0.01
Lead (Pb)	<0.02	<0.03	<0.03
Chloride (Cl)	40-225	19 000-19 600	6 600-7 900
Chlorine (Cl2)	nil	-	nil
Sulphate (SO4)	3-8	-	-
Phosphate (PO4)	<0.2	-	-
Hydrogen sulphide (H2S)	nil	nil	nil
Total dissolved solids (TDS)	217	-	-
Turbidity (JTU)	nil	nil	nil
Dissolved oxygen (DO2)	>4	>5	>5
Free carbon dioxide (CO2)	nil	-	nil
Ammonia (NH3-N)	-	-	<0.1
Nitrite (NO2-N)	-	-	<0.1
Nitrate (NO3-N)	-	-	<20
pH	6.5-8.5 units	7.0-8.5 units	7.0-8.5 units
Temperature	-	-	28-31($^{\circ}$C)

It is also important to consider other factors to ensure success, including the:

- Suitability of the climatic conditions;
- Suitability of the topography;
- Availability of adequate supplies of good quality water;
- Availability of suitable soil for pond construction;
- Maximum protection from agricultural and industrial pollution;

- Availability of adequate physical access to the site for the provision of supplies and the movement of harvested animals;

- Availability of supplies of other necessary inputs, including post larval and/or juvenile prawns, equipment, aqua feeds or feed ingredients, and power supplies;

- Availability of good skilled (managerial) and unskilled labor;

- Presence of favorable legislation; and

- Availability of adequate investment.

M. rosenbergii is a freshwater prawn, but since its larvae require brackishwater for survival, hatchery operations are usually conducted using 12 parts per thousand (12 ppt) salinity. This species regulates its hemolymph osmolarity to be equivalent to that of brackishwater (12 ppt or 450 mOsm). In both freshwater and brackishwater, *M. rosenbergii* is able to maintain hemolymph osmolarity (based on ionic and free amino acid concentrations on the hemolymph), but at higher salinities, the prawn loses this ability (Huong *et al.*, 2001).

Table 38: Water quality requirements for freshwater prawn nursery and grow-out facilities.

Parameters	Recommended (ideal)range for fresh water prawn	Levels known to be lethal (L) or stress ful(s) to juvenile prawns
Temperature(^{0}C)	29-31	<12(L)
		<19(s)
		>35(L)
pH(units)	7.0-8.5	>9.5 (S)
Dissolved Oxygen(ppm)	3-7	2 (S)
		1 (L)
Salinity(ppt)	<10	-
Transparency (cm)	25-40	-
Total hardness (ppm CaCO3)	30-150	-
Alkalinity (ppm CaCO3)	20-60	-
Non-ionized ammonia(ppm NH_3-N)	<0.3	>0.5 at pH 9.5 (S)
		>1.0 at pH 9.0 (S)
		>2.0 at pH 8.5 (S)
Nitrite nitrogen (ppm NO_2-N)	<2.0	-
Nitrate nitrogen (ppm NO_3-N)	<10	-
Hydrogen sulphide (ppm H_2S)	nil	-

Among various environmental parameters, salinity plays a vital role for hatching and survival of hatchlings in *Macrobrachium rosenbergii*. The brooders of *Macrobrachium* sp. which are usually from freshwater in habitat they show higher rate of hatching in 0 ppt. The brooders acclimatized in freshwater did not show any stress during hatching (Jayalakshmy and Natarajan, 1996).The water quality parameters for brooder prawns and larval rearing are depicted in the Tables-38 and 39.

Larvae hatched very well from *M. rosenbergii* brooders kept in slightly brackish (0.5 ppt) rather than brooders kept in 2-10 ppt. brackish water (Rao, 1986 and John Samuel *et al.*, 1997). Whereas hatching took two to four batches from the brooders kept in 10, 15 and 20 ppt salinities.

Hatching rhythm: Complete hatching of eggs in a single batch was noted from the brooders kept in 0 ppt and 5 ppt.

Hatching rate: The hatching rate was significantly higher in the brooders kept in 0 ppt (98.66 %) and 5 ppt (97.33 %), than that of 10 ppt (95.00 %), 15 ppt (93.66 %) and 20 ppt (91.00 %).

Table 39: Water quality parameters for the brooders (Saundarapandian *et al.*, 2009).

Sl.nos.	Water quality parameters	Optimum range
01.	Temperature	29 to 31° C
02.	Dissolved oxygen	5.0 to 6.0 ml/L
03.	pH	7.0-8.5
04.	Photo period	12/12hL/D
05.	Nitrate-nitrogen	<20ppm.
06.	Nitrite-nitrogen	<0.1ppm.
07.	Ammonia	<0.1ppm.

Post-larval appearance: Post larvae become visible from the brooders kept in different salinities (Soundarapandian *et al.*, 2009) appeared significantly in shorter period where brooders, maintained in 0 ppt (22.66 days) and 5 ppt (24.66 days) than brooders kept in other salinities viz., 10 ppt (27.33 days), 15 ppt (29.66 days) and 20 ppt (31.33 days). However, the brooders kept in 0 ppt (22.66 days) and 5 ppt (24.66 days) and also in 15 ppt (29.66 days) and 20 ppt (31.33 days) salinities did not show any significant difference between their first post larval appearances.

Table 40: Water quality parameters of the larval rearing (Saundarapandian *et al.*, 2009).

Sl.nos.	Water quality parameters	Optimum range
01.	Temperature	28 ± 2^0C
02.	Salinity	12 ± 1 ppt.
02.	Dissolved oxygen	5.0 to 6.0 ml/L
03.	pH	7.0-8.5
04.	Photo period	12/12hL/D
05.	Nitrate-nitrogen	<20ppm.
06.	Nitrite-nitrogen	<0.1ppm.
07.	Ammonia	<0.1 ppm.
08.	Hardness	100 ppm.

Rearing periods

The larval cycle was completed significantly quicker in brooders kept in 0 ppt salinity (35.33 days) rather than the brooders kept in other salinities viz., 5 ppt (37.66 days), 10 ppt (41.33days), 15 ppt (44.66 days) and 20 ppt (47.66 days). The brooders kept in 20 ppt salinity took 47.66 days to complete their larval cycle.

Survival rate

The survival rate of hatchlings was significantly higher from the brooders placed in 0 ppt (82.66 %). It shows significant discrepancy with other brooders kept in different salinities viz., 5 ppt (79.33 %), 10 ppt (70.66 %), 15 ppt (68.33 %) and 20 ppt (59.66 %). The lowest survival rate of hatchlings was seen in the brooders kept in 20 ppt (59.66 %).

Brackish water preparation and treatment:

Filtered seawater and freshwater were mixed to prepare 12 ppt salinity water. Sodium hypochloride solution was added to the prepared 12 ppt water, which was then aerated for 24 h. Excess chlorine was removed by treating the water with sodium thiosulphate (Soundarapandian *et al.*, 1997).

Water exchange

On every morning, left over feed, detritus and dead larvae were removed by turning off the aeration and siphoning the settled particles from the tank bottoms. Fifty percent of the water was exchanged each day.

Feeding

Both live *Artemia* nauplii and formulated feed (Table 41) may be fed to the larvae. The ingredients were dried and powdered separately. The pellets were prepared by mixing together; the required quantities of the finely powered materials and the mixture were kneaded or moulded well by adding minimum quantity of water to form dough. The dough was cooked in a pressure cooker for 30 minutes and the cooked material was extruded through a hand pelletizer with required perforation in the form of noodles (Fig. 94) on a filter paper and oven dried at 60 °C. The dried pellets were broken into pieces of required length and stored in polythene bags for future use (John Samuel *et al.*, 1997).

Table 41: Composition and nutritive value of the formulated feed. (Saundarapandian *et al.*, 2009).

Sl.Nos.	Ingredients	Quality	Nutrient content	Percentage
01.	Hen's egg	4.0%	Protein	29.0%
02.	Oyster meat	15.0%	Carbohydrate	17.0%
03.	Shrimp meal	10.0%	Fat	8.1%
04.	Water	50.0%	Moisture	11.9%
05.	Beef meat	15%	Ash	6.2%
06.	Milk powder	3%	-	-
07.	Amino acid & Vitamin	1%	-	-
08.	Minerals	1.5%	-	-
09.	Antibiotic powder	0.5%	-	-

Diseases encountered in *Macrobrachium rosenbergii*

Disease has been considered as one of the important constraints to limit the production of freshwater prawn world wide. Generally, *M. rosenbergii* is considered to be moderately disease-resistant in comparison to penaeid shrimps. The major disease problems affecting *Macrobrachium rosenbergii* generally occur because of poor water quality, poor husbandry, overcrowding, poor sanitation, and non-existent or inadequate quarantine procedures.

Table 42: Diseases encountered and their possible remedial measures of *M.rosenbergii*.

Sl.nos	Diseases	Agent	Type	Syndrome	Rectification
01.	*Macrobrachium* Muscle Virus(MMV)	Parvo-like virus	Virus	Opaqueness of infected tissue, with progressive necrosis; affects juveniles.	Application of 115 kg/acre zeolite/week+Water and soil probiotics as per specified dose /fortnight is a must.
02.	White spot Syndrome Baculo Virus (WSSV)	Baculovirus	Virus	White spots; affects juveniles	Same as above.
03.	White tail Virus	Nodavirus	Virus	Whitish tail; affects larvae	Same as above.
04.	Black spot; brown spot; shell disease	*Vibrio*; *Pseudomonas*; *Aeromonas*	Bacteria	Melanised lesions; affects all life stages, but more frequently observed in juveniles & adults.	Same as above +100 gms.of antibiotic and antifungal powder each in the morning hours only.
05.	Bacterial necrosis	*Pseudomonas*; *Leucothrix*	Bacteria	Like Black spot but affects larval stages only.	Same as above.
06.	Luminescent larval syndrome	*Vibrio harveyi*	Bacteria	Moribund & dead larvae luminescent	Same as above.
07.	White post larval disease	Rickettsia	Bacteria	Larvae become whitish	Same as above+ heavy liming
08.	Protozoan infestations	*Zoothamnium*, *Epistylis*, *Vorticela* etc.	Protozoans	External parasites that inhibit swimming, feeding and moulting; affect all life stages	Liming+ soil and water probiotic application, if necessary bath in formalin

xvi) Culture possibilities of *Cyprinus carpio haematopterusas*

The Amur Carp (*Cyprinus carpio haematopterus*)

The original Common carp was native to the inland delta of the river Danube down to the Black sea. Since then it has been introduced on purpose for food and sport or by accident to every continent except Antarctica and near every countries in the world. The Romans bought *Cyprinus carpio* to Western Europe as a food fish and later spread all the way across Europe to Britain by monks (Balon, 1995). Cross breeding has produced variants of the species with domesticated sub-species such as the Koi carp (*Cyprinus carpio haematopterus* Fig. 84), Mirror carp (*Cyprinus carpio carpio)* or Leather carp.

Fig. 84: Slim bellied *Cyprinus carpio haematopterus* (New strain).

Common carp (*Cyprinus carpio*) is one of the highly domesticated and extensively cultured aquaculture fish species in India especially in southern and north eastern parts of the country, contributing significantly to enhance inland fish production in the world. Common carp belongs to Cyprinidae, the largest family among freshwater teleosts.

Carps have become the main stay of aquaculture in Asia and contribute significantly to inland aquaculture. Aquaculture in India is essentially a polyculture system with Indian major carps and exotics forming major components. The flattened bellied common carp (*Cyprinus carpio,* Fig.85), an introduced species is an important and contribute significantly to increase overall production. It is preferred to grow along with grass carp and silver carp. The common carp presently grown in India originated from two introductions, in 1939 (German strain) and 1957 (Bangkok strain).

Cyprinus carpio **(German strain):** These have become mixed over many generations to give the current stock. This stock of common carp is characterized by early sexual maturation (at an age of approximately six months and sometimes at a weight below 100 g) and slow growth rate. Further the damaging nature of the dykes is also a serious problem of stocking this species (Mandal, 2017). The morphological appearance as well as the taste of the fish flesh of this flattened bellied *Cyprinus carpio* (Fig.85), usually refuge to accept in their food dish by majority of the fish lovers if other carps are available in the market. This is considered as the serious problem in the culture of this species (Mahanta *et al.*, (2010).For faster growth and successful aquaculture of this species, it is imperative to replace the stock with improved strain e.g., Amur-China type of wild carp, *Cyprinus carpio haematopterus* (Fig. 84).

Fig. 85: Fat bellied *Cyprinus carpio* (Old German strain)

Systematic position of *Cyprinus carpio haematopterus*

Phylum- Chordata.

Class: Actinopterygii

Order: Cypriniformes

Family: Cyprinidae

Genus: Cyprinus

Species: Carpio (Linnaeus, 1758).

Subspecies: Haematopterus.

Name: Cyprinus carpio haematopterus

The Amur wild carp is an ancient form that originated from the Asian carp centre (Amur-China type of wild carp, *Cyprinius carpio haematopterus*) and spread to the water bodies of Asia. During the centuries, after settling in the river Amur, this carp adapted to the local environmental conditions. It was brought into the gene bank of Fisheries College and Research Institute (FCRI), Thoothukudi, Tamil Nadu in 1982 from the Russian National Fisheries Research Institute. It has no breeding history as it is an ancient wild fish. Carp is native to areas from the Black Sea to Manchuria in China, but has been spread by humans to many parts of the world (Pethon, 1994). It was introduced to Europe by the Romans. The common carp (*Cyprinus carpio* Linnaeus, 1758) is an introduced fish species in India and today has a scattered distribution throughout the country. The species is well being adopted in Indian waters and encouraged in the polyculture especially in composite fish culture, so as to utilize the food as produced in all the entire ecological niche of the ecosystem.

Food and feeding habit of this species differs with the situations of pond water and pond bottom sediment qualities. The ponds after due fertilization usually prefer artificial feed over benthic macro invertebrates, followed by zooplankton. The species prefer to graze individually and never graze on phytoplankton. They are mainly benthic in habitat; feed on available macro invertebrates in the system. In the absence of benthic macro invertebrates, their feeding niche shifted from near the bottom of the tanks to the water column and fed principally on zooplankton. The species readily switched to artificial feed when available, which led to better growth.However, the principal disadvantage of culture *Cyprinus* in an earthen pond is that they use to dig the bottom soil especially the lower portion of the dykes resulting serious damage to the entire pond. The pond become useless for pisciculture until and unless it is repaired (Mandal, 2017).Almost everywhere else the common carp is considered invasive and destructive of both native fish (also other wildlife effected by the links in the entire ecosystem) and native habitats with its tendency to stir up sediments, eat, destroy and uproot vegetation thereby severely disrupt the natural balances of water nutrition which in turn creates algae covered water sucking out oxygen, killing fish and water plants and having a knock on effect to ducks and water fowl and their predators. The constant grubbing around of the carp schools churning up the sediment can lead to permanent turbidity in waters that otherwise would be clear. This turbidity of the water and eventual blocking of light leads to dead vegetation and loss of habitat then in-turn loss of native fauna.

The Amur strain (*Cyprinus carpio haematopterus*) showed about 40% faster growth over earlier local strain in mono and polyculture systems, respectively. There was no significant difference in the survival rate of Amur and existing strain under monoculture (74.47 and 70.85%) and polyculture systems (74.16 and 75.30%). Although, the production performance varied with type of water body, Amur strain consistently showed its superiority over existing strain in all the trials. It may be inferred that Amur strain of common carp *(Cyprinus carpio haematopterus)* has greater potential in low-input aquaculture systems due to its better growth than the existing strain.

xvii) Culture possibilities of *Osteobrama belangeri* - *Pengba*

Culture possibilities of *Osteobrama belangeri* – *Pengba* the famous state fish of Manipur state, North-eastern India as a candidate species along with Indian Major Carp in fresh water ponds.

Osteobrama belangeri (Pengba Fig. 86a and b) is one of the most important minor carp fish emerging as potential candidate species for diversification of freshwater aquaculture in India, and have high market demand in Manipur and other North East states (Fig. 86a and b). *Osteobrama belangeri* (Val.) locally known as Pengba in Manipur and 'Nga-hpeh-oung'and 'Nga-net-hua'in Myanmar is a medium carp (Family-Cyprinidae) endemic to the eastern part of Manipur (India), Myanmar and Yunan province of China.

It has slow growth (38.0 cm standard length), and due to its delicious flavor (very much to those of *Tenuolosa ilisha*-the Hilsa fish of Bengal) it fetches very good market price in North-East India. Moreover, the fish has almost vanished from the *Loktak Lake* (A Ramsar site) and other water bodies of the Manipur. Furthermore, this fish species is categorized as threatened by the IUCN and it faces a high risk of extinction, which raises the issue of sustainable exploitation of this biological resource. Pollution, habitat loss, over-exploitation besides species invasion are major threats in Manipur. Though, no species was categorised as globally extinct or extinct in wild especially in the Eastern Himalaya assessment region, Manipur state fish Pengba (*Osteobrama belangeri*) was reported to the regionally endangered in wild as the root of this Myanmar origin minor carp has been disturbed with the construction of Thai barrage across Manipur river for the operation of Loktak hydro electricity project about 30 years ago. Therefore, induced breeding and larval rearing should be considered as a high priority for culture intensification, diversification, conservation and restocking aspect (Kumar et.al, 2017).

Systematic position of *Osteobrama balengeri*

Phylum- Chordata.

Class: Actinopterygii

Order: Cypriniformes

Family: Cyprinidae

Genus: Osteobrama

Species: Belangeri

Binomial name: *Osteobrama belangeri* (Valenciennes, 1844).

It is one of the most important food fishes and has a ready demand not only in Manipur but also in the entire NE region of India. However, the fish has almost disappeared from the Loktak Lake (the species is reported to have represented up to 40% of the total fishery of Loktak Lake) and other water bodies of the central plain of the state. Hence intensification of the induced breeding and attempt to culture this particular fish species has been given high priority (Basudha and Viswanath, 1993).The species is originally a riverine in habitat, naturally breed and spawn in the running waters and is very difficult to bred in captivity without using synthetic hormone (Devi *et al.*, 2009).

Very recently, Central Institute of Fresh Water Aquaculture (CIFA), Kausalyaganga, Bhubaneswar after a couple of years of study could have successfully bred the species in captivity and started distributing the spawns and fry/fingerlings of the species to the interested farmers to introduce the same as a candidate species and also a suitable alternative of Indian Major Carps.

Fig. 86a: The solitary photo of Pengba

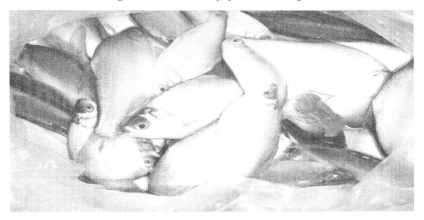

Fig. 86b: *Osteobrama belangeri* -the Pengba

CHAPTER

4

MANAGEMENT OF NURSERY, REARING AND STOCKING PONDS

1. INTRODUCTION

Aquaculture also known as aqua farming, which evidently means "the farming of fish, and brackish water populations under controlled conditions". India is blessed with large number of rivers, lakes and many natural water resources. Along with good pond environment to the fish starting from spawn, fry to fingerling and advanced fingerlings to adult form, fishes do need good quality of nutritionally balanced quality feed and a bit of care and management. The test of pH level (7.5 to 8.5), adequate dissolved oxygen as well as minimal ammonia content in water and soil beneath the water column is necessary for suitable fish farming. The prevention of entrance of various types of predators are necessary, so as the steps to be taken against various types of fish diseases. It is said that preventing diseases is better than curing it. Various management practices involved in nursery, rearing and stocking ponds have been described in the following paragraphs.

2. MANAGEMENT OF NURSERY POND

The principal purpose of all fish farmers is the rearing of fish spawn to a marketable size to the earliest possible time, wherein some inputs of technology is of utmost importance. Spawn rearing in nursery ponds is one of the most significant aspects where the aim of each farmer is to raise maximum number of fish fry in minimum water areas. It is no exaggeration that quite a good number of fishermen are involved in rearing fish spawns to a marketable fry and fingerling stages in nursery ponds. This necessitates enlightening them the proper technology of the nursery pond management practices. The tender spawn start external feeding after 3-days of hatching and therefore, they need congenial environment with enough natural food from their environment for faster growth and survival.

Selection of nursery ponds

These are usually small, ranging from 200 to 500 sq. meters in area with water depth of about 0.8–1.2 meter. These ponds may be seasonal or perennial in nature. However, seasonal ponds are quite preferable and favorable too. The reason, because soil strata of the pond bed get exposed to direct sunlight to a measurable time which helps improvement of the pond water when filled. Generally, nursery ponds are prepared in late spring or summer (February-March). In case of perennial water bodies, where water remains throughout the year, adequate attention has to be given for weed clearance, improvement of soil and water qualities and productivity so also the eradication of weed fishes, mollusks, amphibians, reptiles etc., if any. The ideal nursery ponds are shown in Fig.87.

Fig. 87: Batteries of nursery ponds.

Weed clearance

Weed plants and grasses not only absorb nutrients from the soil but also provide shelter to the predatory insects and serve as their breeding ground and impose difficulty in fry harvesting.

Therefore, they should be eradicated carefully. Since nurseries are small and shallow, the clearance of the same may be done by employing manual labour preferably during late spring or early summer when there is no water or at minimum water level. Marginal grasses also give a low survival rate of spawn.

Eradication of predatory and trash fishes

In perennial nursery ponds, predatory and weed fishes will take a direct toll to the tender spawn while trash and weed fishes would also compete the natural food along with supplemental feed besides space and oxygen. Therefore, total eradication of these unwanted fishes is very much essential. Now-a-days, the tendency of maximum number of fishermen to use different types of insecticides to eradicate weed fishes, which is of course not ethical and the farmers not to go for such type of activities, since we never know the long term effect of residual effects of such insecticides/pesticides in water or soil. The information on the pathway of transformations, bio magnifications of such insecticides and pesticides are still lacking.

It is always advisable to use piscicides of plant origin e.g., Mahua oil cake (MOC), Tea Seed Cake (TSC), Rotenone Powder (RP), Croton etc.

The modus operandi of Mahua oilcake (MOC) is dual i.e.

1) Immediately after application of MOC @2000-2500kg /ha/metre mortality of unhauled fishes of importance including undesirable weed fishes starts. If fresh material (MOC) is available, less than 2000kg/ha/m may be enough.

2) After 3-weeks (21days) the toxicity of saponin (the plant poison present in MOC) get reduced and the nutrients present therein (the NPK) are released which in turn helps generating the desired natural food items e.g., phyto- and zooplankton which make the water colour green. However, an *in situ* test should be performed to make sure about the survival of the spawn before their release.

The availability of Tea Seed Cake (TSC) and Rotenone Powder (RP) are still a matter of great concern, hence the subject is not considered here.

Chemical method of eradication of predatory, trash fishes and other aquatic animals

Alternatively, the farmer may use bleaching powder@350kg/ha/meter or combination of bleaching powder and urea @175kg and 100kg/ha/meter respectively for complete eradication of all unwanted fish. The toxicity lasts for 7-8 days in the pond. It also possesses disinfecting effect besides oxidizing the decomposing matter on the pond bottom. In view of limited supply of Mahua Oil Cake (MOC), bleaching powder is an effective substitute with easy availability and lower cost. Dissolve required quantity of bleaching powder with Urea in water, the slurry thus produced is to be sprayed on the water surface (Pers.com: Dr. Utpal Bhoumik, ex-officio of CIFRI, Barrackpore).The fishes will face discomfort and start surfacing followed by gradual mortality. These fishes are to be removed immediately by repeated netting and are also found to be fit for human consumption. Instead of 3 weeks as in the case of MOC, only after 7-10 days the *in situ* test for examining the survival of fish spawn may be performed. However, the pathway of activity of bleaching powder and urea is still to be investigated.

Immediately after eradication of all the weed and unwanted fishes from the nursery pond it is necessary to apply Zeolite powder @ 30-35kg/ha /meter respectively to keep the pond water free from any obnoxious gases due to decomposition and to maintain alkaline state of water.

Application of lime in nursery ponds

Liming is the first step in fertilization owing to the fact that it supplies calcium, one of the essential macro nutrients. The major functions of lime are the following:

It corrects the acidity of soils and water. *Speeds up the decomposition of organic matter, releasing carbon-di-oxide from bottom sediments.* *It raises the bicarbonate content and lack of CO_2 does not become a limiting factor.* *It counteracts the poisonous effects of excess magnesium, potassium and sodium ions and fixes inorganic acid like sulphuric acids.* *Keep the fishes, disease free at least to some extent.*

If the soil of the pond is found to be acidic, it has to be treated with quick lime to bring the soil pH to alkaline condition, which is ideal for fish production. Lime is applied when all the water is drained out and the bottom is well dried, exposed to the hot sun for about 15 days.

During this period, the lime will help to kill all the pathogenic bacteria present in soil which may affect adversely on the fishes stocked in pond after filling the water. The doses of quick lime depend on the pH of the pond soil as given in Table-23.

The quicklime is either sprinkled or spread on the pond water in the form of a paste or to act as an antiseptic and also to neutralize the toxic effect of old organic deposits at the bottom. It also stabilizes the pH of water at a slightly alkaline level, which enhances the growth of phytoplankton and fish, apart from increasing the calcium content of water. Further, it increases not only the bicarbonate content of the pond but also counteracts poisonous effects of ions like magnesium and sodium.

From the useful effects of liming as given above, it is concluded that liming is essential preliminary to successful pond manuring. A pond containing lime is likely to be more fertile than the one without it. The doses of lime depend upon the characteristic of water.In the lateritic zones (semi-arid zones in particular) however, where both the water and soil beneath it, is found to be acidic in nature. In those cases, the amount of lime should be applied as depicted in Table-24.In nursery ponds following fertilization, an increase in planktonic population are also accompanied by an increase of the aquatic insect population. Some of them are of course, very carnivorous and predatory viz. backswimmers, water bugs, water scorpion, dragon fly nymph etc. A nursery pond, even with a little marginal vegetation, provides support to the insects for multiplication. Even there should not be any vegetation in the pond, these insects being able to fly, would come from the surroundings and inhabit the pond. Complete eradication of these insects as well as frogs, tadpoles and snakes if any, is however, impossible. An earlier recommendation in such case, as it is learnt from the literature is to spray vegetable oil and soap emulsion@ 56:18 kg ratio/hectare/meter is now become obsolete due to high price of oil. Even though in the rural and urban areas where the fish hatcheries do exist they sometime prefer this only to avoid pesticides.

The soap-oil emulsion: It is necessary to check whether the nursery pond has already been infested by backswimmers, water bugs, water scorpion, dragon fly nymph etc. If so, following is the procedure of preparing the emulsion (Table-43).

A day earlier from the date of purchase of spawn, following the above mentioned process, the emulsion is prepared and subsequently beating up in froth vigorously and the same is slowly released in the nursery pond in favour of wind direction. Gradually the emulsion will spread over the pond water as a layer of film. Care must be taken that it should not be a rainy day since rain drops may damage the film cover.

Table 43: Preparation technique of soap-oil emulsion.

Sl.No	Ingredients	Quantity/33Deci.	Quantity/1acre
01.	Vegetable oil	7400 gm.	22kg200gm.
02.	Low cost bath soap	2400gm.	7kg200gm.

The film will create an obstacle in the respiratory processes to the back swimmers, water bugs; water scorpion, dragon fly nymph etc. and majority of them will die. Some of them will try to jump to the pond dyke, hence it is necessary to spray the emulsion also at the water and dyke surface.To overcome this dilemma, Phenitrathion 50EC (Sumithion) @300ml per acre may be sprayed over the pond surface at least 48 hours before stocking of spawn.

Pond fertilization: Manuring is done with a view to increase the plankton volumes particularly the zooplankton (rotifers and crustaceans) which form the natural food of the spawn. For this purpose, seasonal nurseries are manured with only dried cow dung as organic manure @ 10, 000kg/ha about 15 days before the anticipated stocking date. In the perennial ponds where MOC has been used earlier do not require cow dung as manure instead, good quality organic manure@ 150kg/Acre should be applied. At these stage i.e., before stocking of spawn sometimes both in soil and water there is a probability of generating some undesirable pathogenic bacteria and fungi which may cause loss of appetite followed by mortality as well. It may be mentioned in this context that during this period application of Humic grannules@ 4kg/Acre of pond water or soil area followed by Water probiotics @600gms.not only turn the pond water congenial for raising fish spawn stage to fry and fingerling stage but also help to increase the survival rate considerably augmenting appetite and growth of spawn.

Stocking of Spawn: Stocking of spawn should always be done in the early morning or late evening when the water temperature remains low (preferably below 28^0C). It is always advisable to acclimatize the spawn in the poly bag with the pond water temperature. Stocking of spawn may vary pond to pond. But in general the moderate stocking densities ranges from 3-5 million/ha/meter for carps and 6-9 million /ha/meter for cat fishes.

Natural and Supplementary feeding to spawn and fry stages: Spawn starts external feeding from the 3rd day. Even after proper manuring of the pond, sometime it is difficult to maintain the usual development of natural food items (Plankton) for growing the fishs pawn/fry. Hence it is absolutely essential to provide extra feed. A mixture of finely powdered ground nut or mustered oil cake and rice bran or rice polish in equal proportion by weight is to be supplied to the fish spawn fry.

Table 44: Stocking density of spawn in nursery ponds.

Type of spawn	Density/33decimal	Density/100decimal	Density/Ha
Indian and Exotic major Carps(fam.Cyprinidae)	2.5lakhs	7-8 lakhs	3-5 million
Cat fishes of diff. types (fam. Clariidae, Heteropneustidae & Serrasalmidae: the Pacu)	2.5-4lakhs	8-12lakhs	6-9million

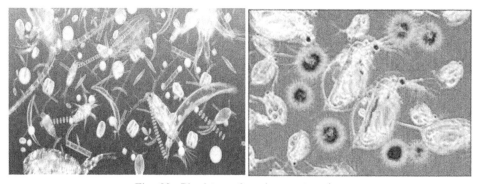

Fig. 88: Plankton of various categories

Table 45: Natural food preferences of IMC and Exotic carps at different stages of their life cycle.

Species	Stages of life cycle			
	Larvae	Fry	Fingerlings	Adult
Catla (*Catla catla*)	Protozoans, rotifers unicellular algae, etc.	Protozoans, rotifers and crustaceans.	Crustaceans, algae, rotifers and some vegetable debris	Crustaceans, algae, rotifers, plant matters, etc.
Rohu (*Labeo rohita*)	- do -	Protozoans, rotifers, crustaceans, unicellular algae.	Vegetable debris, phytoplankton crustaceans, detritus, etc.	Vegetable debris, microscopic plants, detritus and mud.
Mrigal (*Cirrhinus mrigala*)	- do -	Crustaceans, rotifers, planktonic algae.	Vegetable debris, unicellular algae detritus and mud.	Blue-green and filamentous algae, diatoms, pieces of macrophytes, decayed vegetable matters, mud & detritus.
Grass carp (*Ctenopharyngodon idella*)	Protozoans, rotifers, copepod nauplii.	Protozoans, rotifers, crustaceans, microzoobenthos, detritus, microalgae, plant fragments.	Detritus and aquatic plants.	Aquatic plants such as wolffia, lemna, spirodela, hydrilla, najas, ceratophyllum, chara, etc.
Silver carp (*Hypophthalmichthys molitrix*)	Unicellular planktonic organisms, nauplii and rotifers.	Copepods, cladocerans and phytoplankton.	Falagellata, dinoflagellata, myxophyceae, bacillariophyceae etc.	Mainly phytoplankton.
Common carp (*Cyprinus carpio*) Var: *Communis*	Protozoans, rotifers, cereodaphnia, moina, nauplii, etc.	Rotifers, cyclops, cereodaphnia, moina, nauplii, euglena, oscillatoria, etc.	Diaptomus, cyclops, moina, cereodaphnia, ostracods, insects including chironomid larvae.	Decayed vegetable matter, worms, molluscs, chironomids, ephemerids and trichopterans.

Magnesium sulfate and yeast together @ 0.5mg/liter water area boost up plankton of various categories (Fig. 88) after broadcasting over the pond surface once in a week during morning hours from the day of stocking. In general, mixture of mustered oil cake or groundnut oil cake and wheat bran or rice polish @ 1:1 is used as supplementary feed which gives fair result.

To obtain appreciably good results with negligible spawn mortality, the farmers are recommended to use finely meshed crumble of nutritionally balanced formulated (Fig.89) feed.

This feed should be broad casted over the pond surface twice daily in the morning and evening (half of the total quantity at a given time) in the dry powder form.

Fig. 89: Crumble feed

Harvesting and transportation

The fry is reared to about two weeks which generally grow between 20-25mm sizes. Feed is provided up to 12th day of stocking and stopped on the 13th day. Harvesting is done on the 14th day by fine meshed cotton fry net. Fry however, should not be harvested on a bright sunny or cloudy days as higher air and water temperature or dissolved oxygen depletion may cause heavy mortality.

The fry should be kept in hapa for some time prior to transport where they will release faecal matters, emptying the stomach. Without conditioning, if fry are transported, faecal matters which are released due to decomposition of the same in the container may cause mortality. The fry are transported by traditional methods of hundies or by plastic bags filled with oxygen. The following table provides information about the number of seed to be packed in plastic bag under oxygen pressure (Table 48 and Fig. 90):

Table 46: Nutritionally balanced nursery fish feed.

Sl.no.	Fish Feed	Size	Protein	Fat
01.	Nursery Mash Feed	0-mm	36%	6%
02.	Nursery Mash Feed	1-mm	34%	6%

Table. 47: The generally recommended feeding schedule is as follows.

Period	Rate of Feeding /Day	Qty/1lakh Spawn/Day
1st to 5th day after stocking	4 times the initial body weight of spawn stocked (single spawn weighs 0.0014gms)	0.56kg
6th to 12th day after stocking	8 times the initial body weight of spawn stocked	1.2kg
13th day	No feeding	—————
14th day	Harvesting	—————

Fig. 90: Oxygen packing of fish spawns in plastic bag.

Table 48: Number of seed to be packed in plastic bag.

Size of seed	No. of seed per bag in 6 liters of water			
(mm)	Indian major carp	Silver carp	Grass carp	Common carp
20	2, 200	300-500	400-600	750-1000
30	600	200-300	300-400	500-750
40	330	160-200	200-300	300-500
50	225	100-150	150-200	200-300
60	80	—————	—————	—————
70	70	—————	—————	—————
80	40	—————	—————	—————

3. MANAGEMENT OF REARING PONDS

Characteristic of pond soil and water: The characteristics of soil beneath the water in rearing ponds play an important role for success of aqua farming. The nature of pond soil, its water holding capacity, its acidity & alkalinity and other physico-chemical properties play a great role.

Nature of pond soil: For successful aquaculture enterprise silty loam, sandy loam and alluvial soil is ideal. Generally such type of soil, decomposition of organic matter does not generate undesirable gases like ammonia and others. Soil with reddish gravels found generally in arid and semi-arid zones, acid-sulfate soils and sandy soils, aquaculture seems to be meaningless. Sandy soil does not have water retention capacity while the water turns turbid easily in gluey mud soil. Generally, the productivity of pond water is largely dependent on the quality & characteristics of pond soil and thus very often it is rightly said that, *the soil beneath the water is the chemical laboratory of a pond.*

Technological guideline for composite fish farming

In general, the composite fish culture is preferred in perennial ponds. Of course, in India there still do remain a lot of areas where temporary ponds do exist. Most of the months in a year ponds remain dry. In such impoundments the technology of fish culture is also different.

Nutritional status of pond soil

The rearing ponds are comparatively bigger in size than those of nursery ponds and should be rectangular in shape. The ideal size of a rearing pond may be less than an acre (100 decimal) with a water depth of 1.0-1.5 meter. The fry stage (20-25 mm) of fishes are generally transferred from nursery pond and allowed to grow in the rearing ponds. Similar to those of nurseries the rearing ponds are also treated in a manner so that there should not be any predators and competitors. The fish fry in rearing ponds also need an environment which is free from the predator fishes e.g. *Sal, Sol, Boal, Chital, Lata, Chang, Koi, Magur, singhi* etc. On the other hand, the small trash and weed fishes like, *Punti, Chanda, Mourala, Kholse and Tilapia* etc. have also found to play foul role in pond. Mostly they are prolific breeders and produce lot of offspring's causing problems as well as competitions for both food and space with the desirable quality fish. Therefore, it is necessary to eradicate these fishes from the rearing ponds before the fry is stocked (Fig.31).

Sometime it is observed that farmers use Aluminum Phosphide tablets which are a pesticide of chemical origin, generally stored by the FCI warehouses for eradicating various pests therein, but it is not desirable to use in aquaculture ponds. In our country however, Rotenone powder is not available which is a piscicide of plant origin although are in use throughout the world at least in aquaculture arena. Mahua Oil Cake or **bleaching powder** at the dose already mentioned in earlier (page-239) chapter also seems to hold good for rearing pond management. Ponds will be ready for stocking the fry stages after 3-weeks of Mahua Oil Cake application. However, while in the latter case (**Bleaching powder**), the ponds can be stocked within a period of 7-10 days of application of the same. Albeit few plants viz. *Croton, Bisloti, Hijol* (which are naturally grown in India, but the availability has become a question right now due to large scale deforestation) etc. are also equally effective like previous ones to kill predatory and weed and trash fishes.

The principal aim of rearing pond management is to raise maximum number of fingerlings (100-150mm) in minimum area in a period of 3months or so. In order to fully utilize the natural food resources of a rearing pond it is always advisable to stock the same with fry of different species.

Pond fertilization

The ponds already treated with Mahua Oil Cake (MOC) need no further fertilization since the same acts as organic manure after detoxification of the saponin (the toxicant) present in it. However, the following fertilization module is framed before stocking of fry in rearing ponds. In case of seasonal perennial ponds fertilization of water and pond bed, may be done by superior quality organic manure@ 500kg/Acre, dried cow dung@200kg./Acre may be applied in 4 equal installments fortified with 75kg of Mustered Oil Cake, 30kg of ground nut oilcake and 10kg of Humic granules in the rearing pond before stocking fry/ fingerlings. This mixture is to be done at the pond site in a rectangular cemented tank measuring about 10'x6'x3' (Fig. 91). Where provision of cement tank is not available by simple excavation of soil up to desired depth, a HDPE sheet has to be laid on the soil on which the entire mixture is dumped to avoid seepage in the soil (Fig. 91a).These ingredients together, should be mixed thoroughly and kept under bright sunlight for a period of 7-8 days. It is necessary to agitate this mixture at least once or twice a day manually. This compost together with organic manure will enhance the abundance of quantity of plankton, the natural food for fish fry and fingerlings. It is necessary to apply zeolite powder at this stage @15kg/acre since sometimes some undesirable gases like ammonia, sulfur-di-oxide, hydrogen sulphide may be produced due to decomposition of organic debris present on the pond soil surface. To adsorb these gases zeolite powder at appropriate dose (30-35kg/Ha) keep the soil surface free from all these obnoxious gases and helps in the mineralization process at the soil-water interface.

It is sometime also necessary that if the water color of the pond become dark bluish green due to presence of blue-green algae and to avoid this situation before stocking the fry of fishes apply copper oxy- chloride 50%@750gms/Acre/meter. This will make the water lighter. If it is found that planktonic organisms (Phyto-and Zooplankton) are less than what it should be, immediately apply multimicronutrient mineral fertilizer@ 6Kg/ Acre/meter. This will help accelerating both phyto- and zooplankton within 2-3days (Fig.88).

Fig. 91 and 91a: Processing of organic manure at the pond site

Fry Stocking

After complete detoxification and fertilization, the rearing ponds are stocked with 15-day old fry (20-25mm) at a density of 2, 00, 000-3, 00, 000/ha. (Table-49).

Table 49: Stocking density of fry in rearing ponds.

Indian and Exotic Major carps	2-3 lakhs/hectare	80,000-1,20,000/Acre
Catfishes (Pangas, Pabda, Tangra, Magur, Koi etc.)	1-1.5 lakhs/hectare	40,000-60,000/Acre

If the provision of additional oxygen may be supplied in the rearing ponds the stocking of fry might be increased to the tune of the following (Table 50):

Table 50. Additional oxygen and fry stocking.

Indian and Exotic Major carps	5-6 lakhs/hectare	2 lakhs/Acre
Catfishes like Pangas, Pabda, Tangra, Magur, Koi etc.)	2-3 lakhs/hectare	80,000-1,20,000/Acre

The rearing ponds whose depth of water is about a meter only, it is advisable not to stock exotic carps instead, stock only Indian Major Carps. After stocking of fry and fingerlings in the rearing pond it is strictly advised that, to the stocked fishes the farmer should not apply any external food immediately (i.e., supplementary feed) in order to allow them to acclimatize with the new environment for a couple of days.

Table 51: Stocking may be done at any of the following ratios.

	Species	Ratio
1.	Catla+Rohu+Mrigal	2:4:4
2.	Silver Carp+Grass Carp	1:1
3.	Silver carp+GrassCarp+Common Carp	4:3:3 or 5:1.25:3.75
4.	Catla+Rohu+Mrigal+Common Carp	3:4:1:2
5.	Catla+Rohu+Mrigal+Grass Carp	3:3:3:1
6.	SilverC+GrassC+CommonC+Rohu	4:2:2:2 or 3:1.5:2.5:3

Similar to those of the Nursery pond management, in the management practices of rearing ponds also proper liming followed by the application of zeolite powder powder at the specified quantity should be broadcasted in pond water. It is noticed that during acclimatization of fry/fingerlings, bacterial infections followed by the secondary attack of fungus causes some diseases like tail rot, fin rot, red scars on the operculum etc. In order to overcome this problem, the farmers are advised to use antibacterial and antifungal agents as per reccomendations of the aquaculture consultants as preventive dose. Further, after releasing the fishes the farmers should use water and soil probiotics for healthy growth. The use of the probiotics will alleviate the pathogenic microbes of water and soil and improve the productivity status of the pond significantly.

The technique of using the probiotics

Take desirable quantities of probiotics (i.e., 600gms. of water and 1 liter of Soil probiotic/acre/meter of depth) in a container and mix with pond water. Mix these two in approximately 25liters of water and with it add molasses (Gur) about 2kg, shake the solution well by a wooden rod thoroughly and allow to mature the same for a period of 4-5 hours under shade (not under bright sunlight). The bacterial spores will automatically bloom in the mixture. The process may be done at night and preferably during early morning next day, broadcast the same mixture over the pond surface continuously by further mixing with pond water. This application of probiotics not only keeps the pond water clean and free from pathogenic bacteria but also the pond muck will be healthy enough for better mineralization at the soil-water interface. These probiotics should have to be applied at least twice in three months.These probiotics will correct the water and soil conditions and improve the productivity of the pond. There is another way of increase planktonic development in the pond: Take 30 kg. rice bran or DOB, 9-10 kg mustered oil cake, 600gm.of yeast and 9-10 kg of molasses (Gur). This mixture is then thoroughly blended with water and kept in shadow and kept for 72 hours. Then the mixture is to be further mix with sufficient water for applying in one (1) acre pond. The pond will be full of plankton within a couple of days. In every 2-3 months repeat the same process.

In general, the length of the released fry in the rearing pond is about an inch or bit less and is stocked at a higher density. The reason, there is a chance of decreasing the values of dissolved oxygen (DO) in the pond. For the instant supply of oxygen provision of a pump or an aerator at the pond site (Fig. 43 and 44a & b) is kept. However, recently to increase dissolved oxygen content in pond waters both oxygen releasing powder and in tablet forms are available. This oxygen releasing products whether in powder form or in tablets, these are to be mixed with sufficient dry sand which are being broad casted over the pond surface to combat this situation. At this stage nutritional support to the fingerlings is a pre requisite and continued to be applying at a regular basis@ 7-8% twice a day (Table 52).

Table 52: Fry/fingerling feed for rearing ponds.

Sl.no.	Type of fish feed	Size	Proein%	Fat%
01.	floating type	1.2mm.	36.0	6.0
02.	floating type	1.2mm.	32.0	6.0
03.	floating type	2.0mm.	30.0	5.0

The rate of application of crumble 7-8% i.e., per day requirement is 7-8kg for 100kg of the stock. It is always better if the farmer use multivitamin powder (Vitamin mixture) with it.The purpose of stocking higher quantity of fry in the rearing pond is to keep them with restricted growth. The fishes are not allowed to grow beyond 100-150 gms of weight. The water is kept lighter by the application of Zeolite with sufficient oxygen and nutritional support. The growth of fishes will remain stunted even after 7-8 months.

4. COMPOSITE FISH CULTURE IN STOCKING PONDS

Judicious and proper utilization of various food organisms grown naturally in different niches of a pond and transformation of the same in animal protein is the basics of composite fish culture. The purpose of composite fish culture in a pond (otherwise referred to as mixed fish farming or polyculture) at least in India, is to culture both Indian and exotic carps together along with smaller quantity of minor carps e.g., *Labeo bata* etc., so as to the proper utilization of autochthonous plankton and benthic production in the pond. The species of different types of fishes to be stocked in the stocking pond is segregated on the basis of the following:

i)   Market demand,

ii) The types of the niches the fish generally inhabit,

iii) The preference of natural feed, and

iv) Maintainance of the usual growth rate without any competition among the fishes stocked.

Table 53:   Selection of advanced fingerlings to be stocked in the stocking ponds (Percentage wise).

Types of fishes to be stocked	Rohu	Catla	Mrigal	Cyprinus	Silver carp	Grass carp
Three species culture	30	40	30	—	—	—
Four species culture	30-35	30-40	15-20	10-20	—	—
Six species culture	20-30	10-15	15-20	15-20	20-30	5
Local farmers of Dt.Hoogly/Howrah/Midnapur (E) and Burdwan	80	10	7	2	—	1

Market survey on the demand of cultured fish in West Bengal reveals that the demand of Rohu (*Labeo rohita*) is maximum followed by Catla (*Catla*

catla). It is estimated that if the culture practice is maintained following the above recommendations as made supported by the judicious management practice of both water and soil quality and application of nutritionally balanced supplementary feed, it is easy to produce 2, 400-3, 000 kg or bit more of fish/acre/2meter depth pond.

Fig. 92: Technique of releasing advanced fingerlings in stocking pond.

In general the stocking ponds are comparatively bigger (Fig. 32) where the technique of releasing advanced fingerlings are also different as depicted in the figure (Fig. 92).The stocking density is also strictly maintained at a lower level to avoid inter and intra specific competition for both food and space. Fishes ranged from 6-8 inches are generally released from the rearing ponds where they were kept yearlong under stress and naturally the growth remains stunted. Depending on the fertility status and the depth, the release of fish fingerlings maximum to the tune of 3000 pieces/acre/6ft depth of water should be stocked in the stocking ponds. Before releasing, a pinch of $KMnO_4$ is desirable to be added in the container (hundi) so as to avoid pathogenic infections, if any. It will be advantageous to keep a time gap in releasing the catla and silver carp, since both the species inhabit at the same ecological niche and prefer similar type of feed. This time gap will help to reduce the interspecific competition for food. Generally, after release of Catla there should be a minimum 40-50 days' time gap for silver carp dispense in the pond.

The ponds where fresh water giant prawn (*Macrobrachium rosenbergii*) is cultured along with carps, under no circumstances Mrigal (*Cirrhinus mrigala)* and common carp *(Cyprinus carpio)* should be released since all these three species are the inhabitants of the bottom of the pond. The reason, if stocked together there will be every possibility of competition for both food and space resulting extensive decline in the production of the giant prawn. It is advisable that before release of the PL's of prawn in the pond multimicronutrient mineral fertilizer @6kg/acre/meter and Humic grannules @ 6kg / acre/ meter together should have to be broadcasted after thorough mixing over the pond water. The chemical constituents which are present in these two items will boost up comprehensive development of both planktonic and benthic populations. Whatsoever, supplementary feed may be applied to the fishes and prawns, without the planktonic and benthic natural food items neither the colour nor the desired weight will be achieved (Fig.93 and 93a). It is observed that if the stocking density is high or due to higher rate of the decomposition of muck, the fishes and prawns come up to the surface water engulfing air during late night and early morning. It is due to the shortage of dissolved oxygen in water and soon after the sunrise, the fishes and prawns gradually come down to their respective ecological niche. The fishes and prawns will suffer from the loss of appetite and starts exhibiting various types of diseases. In such situation the farmer must take an immediate action and should use oxygen releasing powder or tablets (750-900gms/acre/meter depth) with about 5kg dry sand to the entire pond, of course, without premixing the same in water. This will help the cultured fishes to go down due to the instant supply of oxygen.

Figs. 93 and 93a: Full grown Catla and Rohu

However, if it is found that the surfacing of fishes has become a recurring problem and there is no other way than to replace the existing stock to some other ponds at least in part. The reason behind the surfacing of fishes is the depletion of dissolved oxygen in water. Oxygen in water is primarily a by-product of "Photosynthesis" generally performed by the phytoplankton in the pond and in presence of sunlight they produce their own food. All the living organisms of the pond including phytoplankton during respiration consume this oxygen dissolved in water. **Respiration is a continuous process but photosynthesis is light dependent.** In absence of sunlight there is no photosynthesis. During day time in presence of sunlight what so ever oxygen is produced, when gets exhausted there is a massive respiratory trouble to the animals especially the stocked fishes. They come up to the surface waters (Fig. 45) and starts engulfing air in late night and early morning, so long the sun is not visible on the sky. Hence there are two possible ways to solve this problem:

1) Either supplementary support by aerators or pumps,

2) To reduce the stocking density of the fishes.

5. FOOD AND FEEDING HABITS OF FISHES

In general fishes thrive on the natural food organisms i.e. planktonic, benthic organisms those develop in the aquasystem itself.The classes of food substances represented include proteins, carbohydrates, fats, lipids and vitamins. They are of animal, vegetable and mineral origin (Lagler, 1956).Primary feeding of most fishes in nature consists of bacteria, diatoms, desmids, unicellular protozoans, rotifers, microcrustaceans and microscopic plankters. Periphyton and associated tiny animalcules, forming more or less a slimy coating on bottom materials, on debries, and on plant stem leaves are also browsed upon.

Schaperclaus (1933) classified the natural food items under three groups:

(a) 'Main food', or the natural food which the fish prefers under favourable conditions and on which it thrives best: Certain microscopic planktonic crustacean groups and rolifers form the 'main food' of spawn and fry (15 to 20mm size range) of the Indian major carps and majority of other culturable species,

(b) 'Occasional food' or the natural food that is well liked and consumed as and when available, and

(c) 'Emergency food' which is ingested when the preferred food items are not available and on which the fish is just able to survive: Phytoplankton forming the 'emergency food' (Alikunhi, 1952). Spawn and fry with a small and short straight intestine appear to digest rotifers and cladocerans fairly rapidly and thrive well on zooplankton. Phytoplanktonic algae are not so easily digested and, at least, some algal genera (*Euglena, Phacus, Eudorina, Oscillatoria, Microcystis*, some filamentous green algae, etc.) remain undigested, and ejected intact along with faecal matter.

Nikolskii (1963) further divided food of fishes into four categories acording to the relationship between the fishes and their food. These categories are:

a) 'Basic food' which the fish usually consumes, comprising the main part of the gut contents,

b) 'Secondary food' which is frequently found in the guts of fishes, but in small amounts,

c) 'Incidental food' which only rarely enters the gut unintentionally and

d) 'Obligatory food' which the fish consumes in the absence of basic food.

Adult fishes, according to the character of diet they thrive on, have been classified into herbivores if they feed on vegetable matter, carnivores, if their food comprises animal matters and omnivores, if' they subsist on a mixed diet composed of both vegetable and animal matter. Hora and Pillay (1962) put plankton and detritus feeding fish into a separate group, including therein fishes like *Catla catla, Hypopthalonichthys molirrix, Labeo fimbriatus, Cirrhinus mrigala, C. reba* etc., which consume phyto - and zooplankton, decayed microvegetation and detritus. Nikolskii (1963) grouped fishes into:

1. Herbivores and detrito-phagic, including in the group, species which feed on vegetable matter and detritus,

2. Carnivores, which feed on invertebrates, and

3. Predators which prey on fish.

Most of the culturable fishes are omnivorous in their feeding habit. Carnivorous species often behave as predators and vice-versa. Nikolskii (1963) categorised fishes according to the extent of variation in the types of food consumed by them, such as:

1. Euryphagic, feeding on a variety of foods;

2. Stenophagic, feeding on a few selected types of food, and

3. Monophagic, feeding on a single type of food.

The feeding behaviour is species - characteristic. Cultured fishes are often classified according to the trophic niche they occupy in a water body. Following this system, fishes have been grouped into:

1. Plankton eating surface feeders, such as *Catla catla and Hypopthalmichthys molitrix,*

2. Column or mid-feeders, such as *Labeo rohita* and

3. Bottom feeders, such as *Cirrhinus mrigala, C, reba, Labeo calbasu* and others.

In the bottom feeder category, sand and mud are very common in the guts along with the detritus. Fishes belonging to surface, column and bottom feeding categories have been sub-grouped according to character of food they consume into herbivores, carnivores and omnivores (Das and Moitra, 1955). Feeding habits of adult fishes vary according to the amount and type of food present in a particular environment. Food spectrum of fish varies in different seasons, depending on maturity stages and 'the quantum of food supply. Herbivores and carnivores are recorded to show always definite peak periods in feeding, while omnivores show little variation throughout the year (Das and Moitra, 1956). For any species, food habits may change seasonally with the type of food available and vary with life history stages. Most fishes are omnivorous even in early life, ingesting and digesting both plant and animal tissues. As fish grow towards adult hood however, specific adaptations develop and the diet varieties become slightly or highly restricted. Many fishes remain largely omnivorous throughout life (bluegill sunfishes and others). A few become plankton feeders at an early stage and remain so throughout their life, (carps, gizzard shad and paddle fish). *Pangasius pangasius* is an omnivore, feeding on a variety of food such as insects, molluscs, crustaceans, offal etc. (David, 1963) but according to Ramakrishnaiah (1986) the fish showed a preference to molluscs when they are available. Some become highly herbivorous (grass carp,), others carnivorous (Perches, Bhekti etc.are Piscivorous while the trouts are insectivorous).Cannibalistic (Murrels, Pike, Large mouth bass) fishes are also common.

In addition to the natural food items what ever is available in the culture ponds only a small fragment of the farmers use supplementary feed for the fishes.

6. SUPPLEMENTARY FEEDING

Throughout India, at present there is a tremendous incitement for aquaculture. It is evident from a survey that about 80% of the fish farmers in our country are not aware of water and soil quality management so also the feeding management. It is no denying that who rear cattle in his house do not hesitate twice to serve food to them, so also in case of poultry and ducks but miraculously enough majority of fish farmers in our country remain reluctant to feed the fish in his pond. Nevertheless, to produce quality fish at least in commercial segment, supplementary feed for the fishes is inevitable.It is needless to mention, up to the stage of fingerlings state, application of sinking feed does not hold good and to avoid the wastage, it is recommended to broadcast only mash type of feed with comparatively higher protein, fat content and at least 3000 kcal energy can support good growth. It has been estimated that the state West Bengal has more than 2.80 lakhs ha of fresh water ponds and tanks. Out of these, not even 70% is utilized for commercial pisciculture. Majority of the fish farmers even now, are reluctant to understand the necessity of using feed for the farmed animals, since the fishes are aquatic and do not shout for the food, even they are hungry. Facts remain; no one will deny that cost of feed in the working capital cost in any farming operation is about 60% of the total culture cost. Even today, fish farmers who are not using the processed food due to higher cost and only with the locally available feed ingredients they prepared the feed what so ever the quantity they require. This effort even remains to be better and they are also achieving much more production than those who do not. Earlier, the practice of broadcasting the feed ingredients over the ponds, which again very often deteriorate the environment resulting additional operation cost for rectification of water and soil quality, is also a question to point out.

The farmers who prefer to feed the fishes with the farm made food, generally use pelletizer (Fig. 94). Before pelletizing a thorough grinding and pulverizing of the food ingredients in a fine manner is to be made (Pulverizing), then after pelletizing, instead of broad casting, use of duly perforated polythene bags which is otherwise named as feed bags (Fig. 95) are to be hanged at a definite distance (preferably 6-7meters). To restore the quality of soil and the environment as a whole, the technique of feeding fishes in bags is scientific since this does not damage the pond system. The technique of feeding the fishes in feed bags does not damage the environment at least the soil frequently. In prawns and shrimp culture however, the importance of supplementary feed has opened a new dimension in feed quality improvement and marketing.

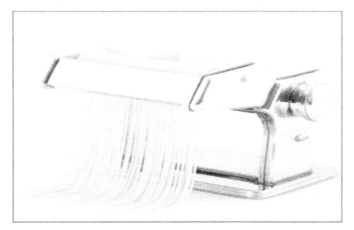

Fig. 94: Domestic type of pelletizer.

Fig. 95: Feed bags hanging at a definite distance

Ingredients used in formulated aqua feed

Rice bran, Maize Powder, Soya bean, Ground nut oil cake, Flour, Chick-Pea, Fish Meal and oil, essential vitamins and minerals etc. are the principal components of the supplemental feed.

To obtain significant growth, irrespective of types (Sinking or floating) application of formulated feed to be made @ 3-4% from April to September i.e., summer to post monsoon months. However, there are many farmers who cannot afford to buy floating type of feed due to higher cost. The following are the proximate composition of the ingredints recommended to prepare the feed at the farm site is given hereunder (Table 54).

Table 54: Carp Feed formulation with locally available ingredients (Quantity for 1MTfeed):

Sl.no	Ingredients	Kg	Protein%	Fat%	Fibre%	EnergyKcal
01	De Oiled Rice Bran	700	14.00	5.4	20.00	2415
02	Ground Nut	100	40.1	12.2	14.0	3018
03	Soyabean Powder	70	44.0	1.5	6.5	3060
04	Maize Powder	75	5.1	8.7	3.9	3326
05	Wheat flour	50	13.9	8.3	13.1	2995
06	Vitamin premix	02				
07	Common Salt	03				
	Total	**1000**	**23.42**	**7.22**	**11.5**	**2962.8**

*Proximate values given may vary from place to place.

Fig. 96: Pelletized sinking and globular floating feed for the fishes

It has been observed that the food conversion ratio (FCR) of this farm made fish is good enough and economically feasible. FCR evidently means, the extent and amount of feed convert into fish flesh. It is experimentally proved in many parts of the state. The FCR of the feed prepared following above, ranged from 1:1.5 to 1:2.This indicates to produce 1kg of fish the total feed requirement is only one and half kilogram to two kilogram. This result obviously depends on the management of the water and soil quality and the good farm management practices.

The importance of binding material of ingredients in pelletizing fish and prawn feeds

Aquaculture diets can either be pelleted or extruded and these should have particles of high durability to withstand handling, transportation and be of good water stability to minimize disintegration and loss of nutrients upon exposure to water. Farm made fish feed requires good quality binding agents that would go a long way in stabilizing feed in water and furthermore enhance prolonged feed floatation time when the floater is trapped or coated within the nutrients. When a water stable pellet is achieved, there will be almost wholesome delivery and utilization by the fish. Also, there will be minimum wastage which is of immense benefit to aquaculture operations in terms of nutrient utilization as compared with broadcast method. Water stability of feeds can be improved through the use of binders. Binders are very important since they have a strong effect on the physical integrity of pelleted feed and the biological availability of nutrients, as it is evident from ARCL'S AQUA STRONG BOND.

Amount of feed to be given and their application pattern

Aquaculture feed, irrespective of their quality and quantity must have to be given to the fishes/prawns, not to the water. For this reason, to maintain the congenial quality of soil beneath the water and the water itself it is imperative to apply feed in the bag, or to the HDPE plastic sheet measurably lower to the water surface level.However, it is equally true that hatchlings of fishes and prawns do not require external feed. After exhausting the yolk food from the fourth day they start external food, preferably small phyto- and zooplankton. From the early fish fry and in case of fresh water giant prawn, PL-5 stage they start feeding on the external food. At this stage broadcasting of Nursery Grade Fish Feed, preferably the 0-mm and 1.0 mm feed to be broadcasted over the water surface.

Fig. 97: Floating feed application within enclosure

Formulation of feed

Easy availability, low cost, high digestibility and high nutrient contents are the major considerations in selecting the fish feed ingredients for feed formulation (Table 54). Feed constitute the major, about 60% of the operating cost in un drainable ponds fish culture and therefore, the objective is to supply essential nutrients at the minimum possible cost. Formulated feeds may be either a complete feed with optimum level of all the essential nutrients and energy to provide complete nutrition or a supplementary feed - a diet basically to supplement energy and a portion of protein and other essential nutrients. In undrainable pond culture systems where natural feed are made available by pond fertilization, feed is required only to supplement the natural feed. The initial step involves surveying market prices of the locally available feed ingredients (Table-55).

Table 55:   A directive of on farm feed preparation technology with locally available ingredients (Kg/MT of feed) in poly culture.

Sl.no	Ingredients**	Quantity(Kg)	Protein%	Fat%	Fibre%	Energy Kcal
01.	De-oiled rice bran	390.00	15.0	1.0	14.4	2174
02.	Oiled rice bran	100.00	13.7	5.4	20.0	2146
03.	G.N. Cake	150.00	40.1	12.2	14.0	3018
04.	Soya powder	100.00	44.0	1.5	6.5	3060
05.	Dry fish powder	100.00	44.1	11.0	————	2754
06.	Wheat Flour	100.00	13.9	8.3	13.1	2995
07.	Maize powder	50.00	5.1	8.7	3.9	3326
08.	Vitamin premix	03.00	————	————	————	————
09.	Aqua Strong Bond*	03.00	————	————	————	————
10.	Common salt	04.00	————	————	————	————
11.	TOTAL	1000Kg	31.16	7.68	13.4	2849

* Aqua Strong Bond: it is a certified feed binder from the house of ARCL, Kolkata. **Proximate values given may vary from place to place and the product quality.

Table 56: Locally available ingredients from plant and animal origin

SL No	Ingredients from Plant origin	CP	CL	CF	Ash	Ca	P	DL-Methionin	L-Lysin	Digestable Energy K Cal/Kg
1	Musur Dahl	20.2	1.9	6.2	4	.....	.....	0.54	1.4	2273
2	Mung Dahl	26.8	0.9	5.3	5.6	0.22	0.39	0.45	2	2318
3	Rice bran	13.7	5.4	20	18.1	.....	.....	0.52	0.6	2416
4	De-oiled Rice bran(DORB)	15	1	14.4	18.8	0.64	1.69	0.57	0.6	2174
5	Rice Polish	12.4	17	12	14.1	.....	.....	0.73	0.8	3154
6	Wheat	13.9	8.3	13.1	4.6	.....	.....	0.42	0.5	2995
7	Maize Powder	5.1	8.7	3.9	1.1	.....	.....	0.1	0.1	3326
8	Maize Grain	10.9	5	2.9	3.4	0.02	0.26	0.22	0.3	3118
9	Ground Nut Oil Cake	40.1	12	14	7.8	.....	.....	0.52	1.4	3018
10	Coconut Oil Cake	18.1	8.9	16.4	4.6	0.21	0.58	0.34	0.5	2960
11	Soybean Oil Cake	47.5	6.4	5.1	6.4	0.13	0.69	1.42	2.9	3009
12	Cotton Seed Oil Cake	47.7	5.4	12.5	6.6	0.22	1.34	1.33	2	3078
13	Sunflower Oil Cake	34.1	14	13.2	6.6	0.3	1.3	1.36	1.2	3394
14	Lin Seed Oil Cake	30.5	6.6	9.5	10.2	0.37	0.96	1.34	1.1	2983
15	Sesame Oil Cake	32.3	14	20.3	11.1	.....	.....	1.64	0.9	3055
Ingredients of Animal Origin										
1	Bone Meal	36	4	3	49	22	10	0.25	1.7	2000
2	Fish Meal	55.6	12	2.9	21.3	.....	.....	.....	.....	3569
3	Fish Waste	44.1	11	.....	44.9	.....	.....	.....	.....	2754
4	Silk worm pupae (Solvent Extract)	77.6	1	4.3	7.3	0.1	1.5	2.95	7.8	3672
5	Yeast	49.9	1.3	1.5	8.5	0.12	0.51	0.52	0.9	3215

Usually the crude protein level of the supplementary feed is fixed at about at about 5-10% below the dietary protein requirement of the fish to be fed. Vitamin, minerals and trace elements are added as and when it is required. (CP=Crude protein; CL=Crude lipid; CF=Crude fibre).

Food conversion raio of supplementary feed: The ingredients used to prepare the supplementary feed can easily be enumerated by the feed conversion ratio (FCR) of the given feed during culture period. Without any fertilization and application of supplementary feed the production of a fish pond may achieved maximum upto 200 – 400kg/ha/year while judicious management of water and soil quality vis-à-vis application of good quality supplementary feed the production may boost upto7, 000-11, 000 (or bit more) kg/ha/year.In this circumstances however, the higher unit productivity though possible but the total cost e.g. aeration in the cultured pond, cost of the advanced fingerlings, fertlization as and when required, periodical netting expenses and prophylactics in addition to feed cost also increases thus increasing productivity of the fishes which have a positive effect on increased income of the farmer.

The efficiency of a feed is normally measured by the amount necessary to produce a unit weight of fish. This is referred to as Feed Conversion Ratio (FCR).The feed conversion ratio is the unit weight of the feed given, divided by the live weight (or the wet weight) of the fish produced.

$$FCR = \frac{\text{Weight of the food given to fishes}}{\text{Weight of live (or, wet fish) produced.}}$$

For example, to produce, 5, 000(5 metric tonnes) of fishes in a pond where total amount of supplementary feed as given 7, 000 kg during the entire culture period, the FCR will be:

$$FCR = \frac{7,000}{5,000} = 1.4$$

It is a convention that the FCR is written as 1.4 which means that 1.4 units of feed has been utilized by 5000 metric tonnes of fish together. The higher the value of the FCR, the less efficient the feed is. To explain this, suppose the feed which has aFCR of 2.5:1 is considered as less efficient than which shows 1.4.

CHAPTER

5

SEWAGE-FED FRESHWATER
AQUACULTURE:
(WITH SPECIAL REFERENCE TO EAST KOLKATA
WETLANDS- A RAMSAR SITE)

Nalban-the sewage fed pond of East Kolkata Wetland

1. INTRODUCTION

It was Lord Bentinck in the year 1830, who first visualized the need of reclamation of urban development. Disposal of wastes from a city like Kolkata, was a chronic problem since its establishment.Initially sewage from Kolkata metropolis was used to be drained into the river Hoogly but subsequently found to be defective both from the civilian's general health and the river Hoogly as well. It was felt necessary to construct a system where from sewage as produced in Kolkata area should have been carried through a disposal drain since there do exist much naturally slope towards the east. But the idea did not come to fruition because of the calculated cost involved in the process. Finally in 1857, the drainage committee considered Mr. William Clarke's recommendations for conveying the city sewage to the eastern part of the city with some modifications in the levels and escalating the number of pumping stations. In 1891, the Kolkata Municipal Corporation compelled to extend the drainage system to the far suburbs to alleviate the sewage and rain water disposal due to high boom of population vis-à-vis increasing rate of the water supply so as the wastes.

Vegetable production continues in the East Kolkata Wetland area and is centred in and around Dhapa, with an estimated 320 ha under horticulture producing 370mt ha^{-1} yr^{-1} in intensively cropped plots (Bunting *et al.*, 2002). Further downstream, enhanced productivity of 18, 260 ha of integrated brackish water rice, fish and shrimp farming over 50 km from the city in the Kulti-Minakhan areas of the Sundarbans Delta was attributed to irrigation with untreated wastewater discharged to the Kulti estuary otherwise named as Kulti gong (Naskar 1985; Bunting *et al.*, 2010).

During 1860's, sewage fed fish farming was tried to be introduced but the attempt was not successful in the area.However, it was Mr.Bidhu Bhusan Sarkar who was very serious and undertook the first formal effort of sewage-fed aqua farming in 1918.

Later after a century of years in 1945 since 1857 of recommendations of Mr.Clarke, a decision was taken by the then Government of West Bengal in 1953 to reclaim a part of the northern salt lakes for urban expansion (Bunting *et al.*, 2010) to support the development of agriculture, horticulture and aquaculture in the other parts (Ghosh and Sen, 1987).That was probably the first to begin with the organized farming of fishes in sewage-fed ponds although reports are available on unorganized approach and fragmentary basis of fish farming in sewage fed ponds which began since 1930 (Nandeesha, 2002).

This particular system used to carry the city sewage to the south east into river Bidyadhari, from there into river Matla and finally to Bay of Bengal.Over the time, within about twenty years or so River Bidyadhari started silting because of the establishment of the lockgate at Dhapa area which was constructed principally for the release the city sewage at low tide. Finally the river was declared as dead.

East Kolkata Wetlands (EKW) is a unique example of innovative recource reuse system through productive activities.According to Kundu *et al.,* (2008), the area is stretching over two districts 24-Parganas South and North covering 12, 500 hectares of area and includes around 254 sewage-fed fisheries, agriculture land, garbage farming fields and some built up area. The resource recovery system, developed by the local people through ages using waste water from the city, is the largest in the world and unique of its type. Long back the area was used as a buffer zone, later the urban waste, both solid garbage and sewage started to be dumped here. Consequently the sewage fed pisciculture and agriculture made this area a natural waste recycling region.The waste water flows through the fish ponds covering about 4000 hactares and the ponds facilitate a wide range of physical, chemical and biological processes which help improve the quality of the water and congenial for the fishes to thrive in. Consequently the wetland system is named as the *"Kidney of the City of Joy—the Kolkata"* and has been described *as "one of the rare examples of environmental protection and development management where a complex ecological process which has been adopted by the local farmers for mastering the resource recovery activities"* by the Ramsar Convention on Wetlands.In august 2002, the East KolkataWetlands area has been included in the 'list' maintained under the Ramsar Bureau established under the article 8 of the Ramsar convention that has given this wetland the recognition of a *"Wetland of International Importance"*Kundu *et al.,* (op.cit).

2. DEFINITION OF WETLANDS

Wetlands are defined as "areas of marsh, fen, peatland or water whether natural or artificial, permanent or temporary, with water that is static or flowing, fresh, brackish or salt including areas of marine water the depth of which at low tides does not exceed six metres" (This definition is included in the text of Ramsar Convention, Article 1.1).

3. PHYSIOGRAPHY OF EAST KOLKATA WETLANDS AREA

Kolkata is persistent by a unique and friendly water regime. To its west flows the river Hooghly, along the embankment of which the city has grown. About 40 km eastward, there flows the river Kulti-Bidyadhari that carries the drainage to the Bay of Bengal (Fig.98, 102, 103 and 106). Underneath the city their lies an abundant reserve of groundwater. Finally, the central to this regime is the vast wetland area beyond the eastern edge of the city that has been transformed to use city's wastewater in fisheries, vegetables and paddy fields. The river side on west of the Kolkata is still the highest part of the city, sloping gradually away from the river towards the east, the original and natural backyard of the city. The entire drainage and sewage networks of whole Kolkata depends heavily on those natural networks of low lying waterlogged areas, ponds, bheries, ditches, nullahs and tidal creeks connecting with estuarine networks of Hoogly Mathla estuarine complex of Mangrove Ecosystems i.e. the Sunderbans . The uniqueness of East Kolkata - Wetland networks are used for its water recycling system and in the development of sewage fed fisheries on 2500 ha of low lying land supplying 20 tones of fishes daily and employing about thousand of peoples. It's resource recovery system, developed by local peoples through cooperative societies, provided employment for a large number of people by way of producing a significant amount of edible biological components as valuable resources for human consumption in the form of fishes of various kinds (Sanyal *et al.*, 2015).

Fig. 98: The lockgate of river Kulti.

The **East Calcutta Wetlands**, also known as the **East Kolkata Wetlands** geographically situated at 88^0 0'E-$88^0$35'E and $88^0$25'N and $20^0$35'N, are a complex of natural and human-made wetlands lying east of the city of Calcutta (Kolkata) of West Bengal , India. The wetlands cover 125 square kilometers, and include salt marshes and salt meadows, as well as sewage farms and settling ponds. The wetlands are used to treat Kolkata's sewage, and the nutrients contained in the waste water sustaining fish farms and agriculture.

The name **East Calcutta Wetlands** was coined by Dr. Dhrubajyoti Ghosh, Special Advisor (Agricultural Ecosystems), Commission on Ecosystem Management, IUCN.The East Calcutta Wetlands(EKW) were designated a **"wetland of international importance"** under the **Ramsar Convention** on August 19, 2002.

The initiation of aquaculture in the past and the present scenario

Around a century ago, a cultivator named Bidu Bhusan Sarkar accidentally allowed untreated wastewater from Kolkata's sewage pipes into his fish pond.

Realising what had happened, Mr. Sarkar expected disaster. Instead of killing his fish however, the water doubled his yields. When fishermen from the surrounding area came to find out more, they discovered that the combination of sewage in the water and sunshine broke down the effluent and allowed plankton, which fish feed on, to grow exponentially. Soon thousands of fish farmers had set up *bheris,* or fishponds, across 12, 500 hectares on the eastern fringes of the city.

The wetlands to the east of Kolkata comprises of many water bodies from north and south 24 Parganas. The hydrological setup of these wetlands is completely different from any other wetlands in India. There is no catchment for these water bodies and suspended or balanced aquifer is found to occur below these water bodies at depth greater than 400 feet. These wetlands are well known over the world for their multiple uses. The wetlands are manmade and the system of wastewater treatment is the largest in the world. It has saved the city from constructing and maintaining a wastewater treatment plant. The wetland comprises of intertidal marshes including salt marshes, salt meadows with significant waste water treatment areas like sewage farms, settling ponds and oxidation basins. In these wetlands a wise use of ecosystem is done whereby 250 million gallons of sewage of the city and after due treatments via sewage pumping station flows to these water bodies and is used for traditional fishing and agriculture (Figs.99&99a).

Fig. 99 &99a: Sewage pumping station transferring Kolkata city sewage.

It is one of the largest multipart of sewage fed fish ponds in the world. Currently there are about 300 large fish farms and ponds covering a total area of 3, 500 ha. There do exist some very large ponds with an area of as much as 70 ha and about 13, 000 tonnes of fish and 150 tonnes of vegetables per hectare per year are produced in these wetlands.

The total area of the East Kolkata Wetlands is 12, 500 ha of which approximately 45.93 % is the water body and 38.92 % is the agricultural land. The remaining portion is occupied by urban and rural settlements and sites for garbage disposal (Kundu *et al.*, 2008).

4. COMPONENTS OF SEWAGE

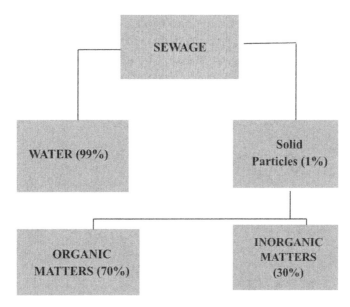

Fig. 100: Components of sewage.

Fig. 101 &101a: The sewage induced ponds after the treatments of sewage water.

Sewage-fed aquaculture was first initiated in Germany and naturally but independantly in Kolkta in late 1930's.Later on, following the model of east Kolkata Wetlands the use of sewage for pisciculture has also been started in the states like Bihar, Madhya Pradesh, and Maharashtra as well as in the countries like Poland, Hungary, Israel, Indonesia and Peru.

5. FLOW OF SEWAGE AT PRESENT

Currently wastewater with an average daily flow of 1100 MLD (1.1 million m3) from the inner city of Kolkata, inhabited by approximately 4.5 million people, is not treated by a conventional sewage treatment plant (STP). An estimated 30–50% of the sewage from central Kolkata is treated and reused by the fishponds of the East Kolkata Wetlands (Edwards 2008a).

Threats

The two most important threats in these wetlands are that of encroachment due to urban development and siltation. The constant change of land use pattern has affected the ecology of these wetlands. Many large pisciculture ponds have been

Fig. 102: Waste water channel in between two grow out ponds.

Fig. 103: Waste water channel.

converted to paddy fields. The industries in the adjacent areas have made unauthorized connection to the sewers to empty their untreated wastewater. The sewers on the other hand empty the water into the channels that later on join the wetlands. This is causing a deposition of the heavy metals in the canals and ultimately the quality of fish and vegetables produced in the wetlands is far below the edible standard.

Fig. 104: Sewage entry point.

6. DEFINITION OF SEWAGE

Sewage is defined as a cloudy fluid arising out of domestic, municipal and industrial waste, containing mineral and organic matter in solution or having particles of solid matter floating, in suspension, or in colloidal and pseudo-colloidal form in a dispersed state. Sludge differs from sewage in that it is the solid portion of waste and does not include fecal matter and urine.

The organic and inorganic constituents of sewage waters also contain various living matters including a variety of bacteria and protozoans.

What does *Sewage* mean?

- *Sewage is waste material that is carried through a sewer from a residence or an industrial workplace to be dumped or converted to a non-toxic form. Sewage is more than 99% water, but the remaining material contains solid material, ions and harmful bacteria. This matter must be extracted from the water with a filtration process before the sewage can be released back into a natural water source.*

Sewage and waste water

Wastes have been rightly referred to as resources out of place (Jana, 1998). Domestic households, industrial and agricultural practices produce waste water that can cause pollution of many lakes and rivers.

- Sewage is the term used for waste water that often contains fecal matters, urine and laundry waste.

- There are billions of people on Earth, so treating sewage is a big priority.

- Sewage disposal is a major problem in developing countries as many people in these areas don't have access to sanitary conditions and clean water.

- Untreated sewage water in such areas can contaminate the environment and cause diseases such as diarrhea.

- Sewage in developed countries is carried away from the home quickly and hygienically through sewage pipes.

- Sewage is treated in water treatment plants and the waste is often disposed into the sea.

- Sewage is mainly biodegradable and most of it is broken down in the environment.

- In developed countries, sewage often causes problems when people flush chemical and pharmaceutical substances down the toilet. When people are ill, sewage often carries harmful viruses and bacteria into the environment causing health problems.

Sewage and its characteristics

The term sewage is used loosely to include the combined liquid waste discharges of domestic and Industrial sources within a given area. It is a cloudy liquid having minerals and organic matter in solution, colloidal form and solids floating as suspension.It contains about 90-99% water. It also contains bacteria and protozoa. It is rich in phosphorus (1-14 mg/l) and nitrogen (18-120%). It contains traces of heavy metals such as zinc, copper, chromium, Manganese, nickel and lead. The BOD and COD of the sewage are very high. The direct use of raw sewage is detrimental to fish because of its high BOD, low DO, High CO_2, high levels of ammonia, hydrogen sulphide and bacterial and organic load.

Composition of sewage

Sewage may vary considerably in composition and strength from place to place owing to marked differences in the dietary habit of the people, composition of trade waste and water consumption. The strength of sewage is determined by the amount of O_2 required to oxidize completely the organic matter and ammonia present in it.

There is also a variation in composition between between domestic and industrial sewage (Fig.100), the later containing more pollutants in terms of heavy metals and bacterial load and other toxic ingredients.While the sewage is very rich in anaerobes when it is raw but gradually transforms to an enriched freshwater when undergoes treatment.Sewage contains living matter especially bacteria and protozoa.The water content of sewage may be 99% the rest being the solid matters.The C:N ratio of domestic sewage is around 3:1, while the industrial sewage may contain more organic carbon and hence may have a higher C:N ratio. Nitrogen in sewage is present partly as organically bound element and partly as ammoniacal nitrogen. Following are the common characteristics of Kolkata City sewage:

Table 57: Common characteristic features of Kolkata sewage waters.

Parameters	Values or Range
pH	6.8-7.8
Carbon-di-Oxide(CO_2)	10-140ppm
Dissolved O_2	Almost zero
Toltal Alkalinity	170-490ppm
Nitrate-Nitrogen(NO_3-N)	0.1-0.6ppm
Nitrite- Nitrogen (NO_2-N)	0 - 0.08 ppm
Free Ammoniacal-Nitrogen(NH_4-N)	12-63.6ppm
Albuminoid ammonium- Nitrogen	1.1-16ppm
Biochemical Oxygen Demand(BOD)	100-500ppm
Phosphate-phosphorus(asP_2O_5)	0.12-14.5ppm
Suspended Solids(SS)	160-420ppm
Settleable solids	1.6 - 2.8 ppm,
Organic Carbon	24-88.8ppm

Gases like CO_2, H_2S, and NH_3 etc. are in dissolved state. *The raw sewage is detrimental to fish and to make it suitable for aquaculture or for usual disposal to the river, treatment is necessary.*

Problems related to sewage fed culture system

● Accumulation of silt and high organic matter at the pond bottom.

● Incidence of parasites and fish diseases.

● Possibilities of pathogens being transferred to humans.

● Accumulation of heavy metals in the system.

7. TREATMENT OF RAW SEWAGE FOR FISH CULTURE

The raw sewage needs to be treated before using in fish ponds.

Mechanical, chemical and biological treatments are the three steps involved in treatment of raw sewage.

Mechanical Treatment

This step is required to remove suspended and floating solids from the raw sewage. The solids are removed first by using screens and then by skimming. Finally they are removed by sedimentation. The mechanical treatment is comprised of:

(a) The screening of sewage water and use of filtration devices

(b) Skimming of floating matters which is lighter than sewage water, and

(c) Sedimentation of suspended solids which are heavier than the sewage water.

Fig. 105: Sedimentation pond

Chemical treatment: Chemical treatment of raw sewage involves steps such as coagulation/chemical precipitation, deodorization, disinfection and sterilization.

Biological treatment: The process of biological treatment of sewage is still not understood very clearly but it is a simple bacterial decomposition process which includes natural bacterial activity for the oxidation of organic matter. The treated sewage can then be used for fish culture after suitable dilution with freshwater. Bacteria may be aerobic, anaerobic or facultative. Aerobic bacteria require oxygen for life support where as anaerobes can sustain life without oxygen. Facultative bacteria have the capability of living either in the presence or in the absence of oxygen. In the typical sewage treatment plant, oxygen is added to improve the functioning of **aerobic bacteria and to** assist them in maintaining superiority over the an anaerobes. Agitation, settling, pH and other controllable are carefully considered as a means of maximizing the potential of bacterial reduction of organic in the wastewater.

8. MANAGEMENT OF SEWAGE

1. **Primary management:** This is mostly the physical removal of solids by mechanical means. The solid material is removed by screening (for larger coarse particles), skimming (for floating solids) and sedimentation (for suspended particles whose density is greater than that of liquid) techniques.

2. **Secondary management:** Soluble organic and inorganic matter, namely the carbohydrates, proteins, fats, hydrocarbons and other nitrogenous materials which are degraded mostly biologically, using microorganisms into the smaller constituents i.e. CO_2, H_2O, NO_3, NO_2, SO_4, PO_4 etc.which can be easily disposed. Sometime chemical and physical removal of substances is combined with this to increase the effectiveness. There are three basic methods for secondary treatments: activated sludge (flocculation), biological filtration and waste stabilization (in oxidation ponds). In the activated sludge or flocculation process, the sewage is aerated by diffused air or by mechanical means. The activated sludge (or biological floc) contains the microorganisms that remove the soluble and insoluble organic matter in the sewage by a combination of adsorption and oxidation or assimilation. Aeration supplies the sludge microorganisms with oxygen and keeps the floc in suspension. After a suitable contact time (1-20hrs) the sludge is separated from the sewage effluent in a settling tank. Some of the settled sludge is returned for aeration along with new sewage but most of it is treated separately in a sludge treatment plant.

9. SEDIMENTATION PONDS

The function of sedimentation is to remove suspended solids from sewage to the maximum possible extent. It is done by letting sewage into a pond/tank at a high velocity of flow (Fig. 99 and 99b). Sedimentation results due to sudden drop in velocity when sewage enters a large pond from sewage channel (DWF, Fig.102)

Sedimentation is best carried out by in two successive stages i.e. primary and secondary.The primary stage is intended to settle down most of the heavier solids while the secondary stages serves two purposes: (a). Provision of additional period to help to mix and homogenize variations in the flow, and (b). Promotion of natural purification process.

It has been estimated that about 33% BOD is got rid of by sedimentation process, which may affect with 90% settlement of suspendedsolids and about 25% reduction in albuminoid ammonia.

Fig. 106: The Dry Weather Flow canal.

10. DILUTION AND STORAGE

Before introduction of sewage into any fishery its dilution by freshwater should be so effected that a positive dissolved oxygen balance (1:1 or 1:2) is maintained and the concentration of unwholesome or objectionable ingredients such as CO_2, H_2S, NH_3 etc. kept below lethal limit. The oxygen required for biochemical reaction is obtained from fresh water used for dilution and through green algae, and other vegetation in the water body. Sewage is stored here for few days. During storage, the biological processes carried out by microorganisms present in the raw sewage oxidize it.

Use of oxidation ponds (waste stabilization ponds) for sewage-fed fish culture has been suggested by several workers. The term waste stabilization ponds is applied to a body of water artificial or natural employed with the intention of retaining sewage or organic waters until wastes are rendered inoffensive for discharge into receiving waters or on land through physical, chemical and biological process (self-purification). This pond is suitable in India because of plentiful of sunshine.These are also cheap to construct and easy to operate. Organic matter contained in the waste is stabilized and converted in the pond into more stable matter in the form of algal cell, which find their way into the effluent. These ponds are of three types:

i) **Anaerobic ponds:** It is pretreatment digester and requires no dissolved oxygen. These are designed to take on higher organic loading so that anaerobic condition prevailed throughout the pond. Such ponds are 2.5 – 3.7 m deep. End products are CH_4, H_2S, and NH_3.

ii) **Aerobic ponds:** These are shallow, having a depth of 0.3m or less, so designed that growth of algae through photosynthetic action is maximized. Waste material is stabilized through microorganisms only and aerobic condition is always maintained. End products are CO_2, H_2O, NO_3, SO_4, PO_4 etc.

iii) **Facultative ponds:** These are 0.9 – 1.5 m deep and are aerobic during day hours as well as for some hours at night. Only for few remaining hours of night, bottom layer become anaerobic. Aerobic, anaerobic and facultative may all be found in a facultative pond. In India, most of the waste stabilization ponds are of facultative type. The village ponds and natural depressions in rural areas are example of waste stabilization ponds.

A conventional oxidation pond retains the settled sewage at a depth of 1 to 2 m (facultative ponds) for a period of 25 to 30 days. This pond contains the algal-bacterial cultures, which oxidizes the organic matter into CO_2, H_2O, H_2S, NH_3 and other decomposition products that are used asnutrients (e.g. NO_3, SO_4, PO_4). If this type of ponds are designed well and operated effectively, well over 90% of the BOD is removed and the micro flora is much reduced.

Conventional methods of fish culture

The fish farmers of Kolkata operating sewage-fed fisheries however, generally use raw sewage, relying on intuition and experience for regulating its application. This practice is not only unhygienic but also harmful since the sedimented organic matter besides raising the bed level of pond being highly oxidisable in character may undergo decomposition and cause negative oxygen balance causes mortality.

But sewage partly or fully decomposed contains a high percentage of nitrogen, phosphorus, Ca, K etc. These nutrients together with adequate alkalinity contribute largely to high productivity in sewage water and for this reason fertilization of fish pond is sometimes carried out with raw sewage. Sewage fed ponds are used for raising seeds of Carps and Tilapia and also culturing them to table size. For raising carp seed, ponds are dewatered completely during summer to remove all the carnivorous and weed fishes. When complete dewatering is not possible treatment with mohua oil cake or other similar fish toxicants is used. Initial fertilization of pond is done with the introduction of fresh sewage effluent, which is taken into the pond up to 90 cm. Following this, the ponds show extreme diurnal fluctuation of dissolved oxygen ranging from super saturation stage at day time to serious depletion in night. However, due to dilution and natural putrefaction process, the wide fluctuation of dissolved oxygen is minimized within a month and the pond rendered suitable for stocking and rearing fish seed. The stocking density in such pond varies from 70.000 to 1,50,000 perha. The density is depended mostly on the size of the spawn or fry. An experimental pond of 0.11 ha at Khardah, West Bengal, having a density of 60, 000 fry/ha with the ratio of the following:

Table 58: The size of the fishes after 25days of stocking.

Types	Stock%	Initial length	Initial wt.	Final length	Final wt.
Catla	40%	72 mm	6gm	133mm	30gm
Rohu	6%	72 mm	5gm	147mm	37gm
Mrigal	45%	74 mm	4gm	126mm	24gm
Common Carp	9%	54 mm	3gm	135mm	50gm

After 25 days when they attained the above-mentioned sizes, were transferred to a bigger stocking pond @ 10000/ha. This pond needed the additional fertilization, which was carried out every month with sewage effluent in small doses. For raising the juveniles, stocking pond was fertilized with raw sewage @ 45, 00, 000 lit/ha. Production in stocking pond was recorded 2500 kg to 3000 kg/ha. It was reported that at a stocking density of 50000 fingerlings/ha at ratio of Rohu, Catla, Mrigal =1:2:1 gave a production of 7076 Kg/ha in 7 months.

For raising *Tilapia* seed more or less the same techniques as that of carp seed are adopted. However, instead of *Tilapia* spawn/fry, adult Tilapia of both sexes is stocked together in the ratio of 6 males: 4 females at about 20, 000/ha. They bred profusely in the pond. The harvesting of fingerlings is initiated two months after stocking of adults and is continued periodically either fortnightly or monthly depending on the density of harvestable size tilapia. Normally 30 – 40 gms tilapia are harvested. Under this system of culture a production as high as 8 – 10 tons/ha/year were obtained by CICFRI.

11. TECHNOLOGIES ADOPTED BY FARMERS

The sewage fed ponds are locally known as bheries. These are the ponds of different sizes, which can be as big as 40 ha. The ponds are shallow with a depth ranging from 0.5 to 1.5 m. Generally the culture practice includes five phases:

1. Pond preparation.
2. Primary fertilization.
3. Fish stocking.
4. Secondary fertilization (Periodic)
5. Harvesting of fish.

Pond Preparation

Pond preparation is undertaken generally in winter (Nov – Feb) when the fish growth is reported slowest. Ponds are drained, desilted, tilled and dried in sun. The pond dikes are consolidated. Silt traps (perimeter canal along the dikes) 2-3 meter wide and 30-40 cm. deep are dug, as they get filled during regular harvesting of fishes. Aquatic weeds as water hyacinth (*Eichhornia*) is grown along the pond dikes, which save the dikes from wave, and give shelter to fishes against high temperature and poaching and above all it extracts heavy metals from the sewage, supplies oxygen by photosynthetic activity. The bamboo sluice gate is repaired which helps to prevent the entry of unwanted fishes and escape of cultured fishes.

Primary fertilization

After pond preparation, sewage is passed in to the pond from the feeder canal through bamboo sluice. It is left to stabilize for 15 – 20 days. The self-purification of sewage takes place inpresence of atmospheric oxygen and sunlight. When the water turns green due to photosynthetic activity, the pond is considered ready for stocking.

Species cultured

Although both Indian and exotic carps are grown, farmers have specific preference for the Indian carps, namely catla (*Catla catla), rohu (Labeo rohita), mrigal (Cirrhinus mrigala)* and *bata (Labeo bata)* with bulk of the stocking consisting of mrigal. Exotic fish like silver carp *(Hypophthalmichthys molitrix), grass carp (Ctenopharyngodon idella)* and common carp *(Cyprinus carpio)* are stocked as a small percentage. However, the popularity of tilapias *(Oreochromis niloticus* and *O. mossambicus)* is increasing and they constitute 5-30% of the species stocked in different farms. There is also a tendency for some farmers to stock *Pangasius hypophthalmus* to control molluscan populations and some are attempting to culture high value species like giant freshwater prawn, *Macrobrachium rosenbergii.*

The following Table 59 depicts the abundance of fishes generally found in the East Kolkata Wetland areas (Partly modified from Kundu *et al.,* 2008).

Rotational cropping system

Farmers have evolved culture systems that are responsive to market demand. Fish are stocked and harvested throughout the culture period leading to periodical

Table 59: The abundance of fishes in East Kolkata Wetlands.

Sl.no.	Scientific name of the fish	Local name of the fish	Abundance
01.	*Catla catla*	Katla	Common
02.	*Labeo rohita*	Rui	Common
03.	*Cirrihinus mrigala*	Mrigel	Common
04.	*Labeo bata*	Bata	Common
05.	*Labeo calbasu*	Kalbaus	Rare
06.	*Hypopthalmichthys molitrix*	Silver carp	Sporadic
07.	*Ctenopharyngodon idella*	Grass carp	Rare
08.	*Aristichthys nobilis*	Bighead carp	Common
09.	*Oreochromis mossambica*	Tilapia	Common
10.	*Oreochromis niloticus*	Nilotica	Common
11.	*Cyprinus carpio*	Cyprinus	Common
12.	*Lates calcarifer*	Bhetki	Rare
13.	*Liza parsia*	Parshey	Rare
14.	*Puntius ticto*	Punti	Rare
15.	*Puntius javonicus*	Japani punti	Rare
17.	*Amblypharyngodon mola*	Mourala	Rare
18.	*Glossogobius guiris*	Beley	Sporadic
19.	*Apocheilus panchax*	Techokha	Common
20.	*Mystus vittatus*	Tyangra	Rare
21.	*M.gulio*	Nona tyangra	Sporadic
22.	*Channa striatus*	Shole	Rare
23.	*C. gachua*	Chang	Rare
24.	*Clarias batrachus*	Desi Magur	Rare
25.	*C.garipinus*	African magur	Rare
26.	*Heteropneustes fossilis*	Singhi	Rare
27.	*Anguiliformis* sp.	Pankal	Sporadic
28.	*M.armatus*	Baan	Sporadic
29.	*Pisodonophis cancrivorus*	Kucho	Rare
30.	*Chanda nama*	Chanda	Rare
31.	*Chanda ranga*	Ranga chanda	Rare
32.	*Notopterus notopterus*	Folui	Rare
33.	*Anabas testudineus*	Koi	Sporadic
34.	*Badis badis*	Banda	Rare

stocking and regular harvest. In larger ponds, harvesting takes place continuously for almost fifteen days in a month. After completion of one cycle of harvest in a large pond, fishes are restocked at the rate of one kg of fingerlings for every five kg of fish harvested. After restocking, fishes are left undisturbed for the subsequent fortnight and harvesting will start again after that period. Drag nets are commonly used for harvesting fishes through an encircling technique. However, for the bottom burrowing and difficult to catch species like common carp and tilapia, encircling with the net and hand picking are adopted as common techniques. There are specialized people to harvest fishes using these strategies.

Fig. 107: Harvesting of fishes.

Fig. 108: Segregation of harvestedfishes.

Fig. 109: Transport of fishes from the bheri to the market.

Fig. 110: *Eichhornia cressipes* lined to protect the dyke from erosion.

12. DYKE PROTECTION

Aquatic weeds like water hyacinth are grown along pond dikes of larger ponds to break waves and prevent damage to dykes (Fig.110). In addition, these weeded areas, provide shelter to fish when the temperature rises, prevent poaching of fishes to some degree and most importantly serve as filters to extract nutrients and metals from the system. When these weeds grow in excess, they are periodically harvested and decomposed in the pond to enhance fertility of water. Surrounding these large ponds, silt traps 2-3 m wide and 30-40 cm deep are dug. These get filled with regular harvesting of fishes. Farmers restrict themselves to cleaning of these silt traps instead of digging the entire pond. Silt rich in nutrients is used for various purposes, including strengthening of dykes.

The uniqueness that the East Kolkata Wetlands deserves proper conservation and management measures including the following:

1. Proper management of the wetlands complying the Ramsar Convention guidelines,

2. Conservation of biodiversity and,

3. Improvement of livelihood of local people

CHAPTER

6

FISH HEALTH MANAGEMENT

1. INTRODUCTION

A successful aquaculture depends on the quality of soil, water, seed, feed and a skillful Fish health (disease) management practice.The term fish health management eventually means the technique of preventing fish disease.It is no exaggeration that once fish get unwell it is very difficult to recover them.

What is a disease any way?

A disease is a particular abnormal condition, a disorder of a structure or function, that affects part or all of an organism.

Aquaculture and fisheries is beset with disease problems resulting from its intensification and commercialization. Fish diseases have been classified according to their cause and general aspects of their etiology, epizootology (*the character, ecology, and causes of outbreaks of animal diseases*)and epidemiology (*analysis of the patterns, causes, and effects of health and disease conditions in a defined population)*.The physiological conditions of fish and their entire environment play an important role is connected with the course of diseases. Fish diseases are broadly classified into pathogenic and nonpathogenic diseases. Pathogenic diseases are virus, bacteria, fungi, protozoans, parasitic, crustaceans, helminthes and other parasites, Non-pathogenic diseases are hereditary, ecological, tumors, environmental and vitamin disturbances. The highest importance is given to fish diseases which are highly contagious (Pathogenic). The host pathogen relationship generally undergoes several stages of development. The incubation period is when the pathogen multiplies, but the host does not yet show any signs of diseases. The incubation period may range from a day or two for virulent pathogens, to prolonged periods of several months.

Aquaculture has a long history, originating at least in the year 475 B.C. in China, but became important in the late nineteen-forties, since the methods of aquaculture could be used to restock the waters as a complement to natural spawning. Nowadays, aquaculture is a lucrative industry. However, the intensification of aquaculture practices requires cultivation at high densities, which has caused significant damage to the environment due to discharges of concentrated organic wastes, that deplete dissolved oxygen in ponds, giving rise to toxic metabolites (such as hydrogen sulfide, methane, ammonia, and nitrites), that often are responsible for mortality. Additionally, aquaculture has appropriated of water bodies used for recreational purpose, and sometimes makes water's waste because this natural resource is not reused in extensive aquaculture systems. Moreover, under these conditions of intensive production, aquatic species are subjected to high-stress conditions, increasing the incidence of diseases and causing a decrease in productivity.

Outbreaks of viral, bacterial, and fungal infections have caused devastating economic losses worldwide, that is, China reported disease-associated losses of $750 million in 1993, while India reported $210 million losses from 1995 to 1996.

Added to this, significant stock mortality has been reported due to poor environmental conditions on farms, unbalanced nutrition, generation of toxins, and genetic factors. In recent decades, prevention and control of animal diseases has focused on the use of chemical additives and veterinary medicines, especially antibiotics, which generate significant risks to public health by promoting the selection, propagation, and persistence of bacterial-resistant strains.

Most of the diseases encountered in fishes and prawns are principally due to the quality deterioration (pollution) of water and soil as well as faulty (unscientific) farm management practices. Besides, subject to the fluctuation of air and water temperature there is a marked variation of depth of water in the ponds and tanks. The significant increase in the pathogenic organisms due to the decomposition of organic matters in the perennial ponds especially where the drying of pond water is not possible at least once in a couple of years. As a result fishes do suffer due to some of the infectious diseases. The epidemics sometime incur severe financial loss to the farmers. It is also true in case of Faulty Management Practices (FMP).

Regular monitoring of fish/prawn health is an effective way to identify causes of disease and appropriate treatments. One major cause of serious fish/prawn mortalities is overlooking the contagiousness of fish diseases and thus delaying treatment. As such, adequate care and treatment should be given to infected fish promptly.

It is no denying that successful fish health management starts with the prevention of disease rather than to treat it.Prevention of fish disease is accomplished through good management practice of better water and soil qualities, fish nutrition management and sanitation. These are the foundation related to successful aquaculture without which it is absolutely impossible to prevent various diseases.The fish is constantly merged to its surrounding water and are exposed to various types of potential pathogens which includes bacteria, fungi and variety of parasites.Even use of various sterilization technologies viz. ultraviolet sterilizers, ozonation etc.can not eliminate all potential pathogens from the environment (Francis-Floyd, 2005).Suboptimal water quality, poor nutrition, or immune system suppression generally associated with *stressful conditions* allow these potential pathogens to cause disease.

Visit to the culture systems (ponds, tanks, large water bodies etc.) at least once a day and subsequent observations on the fish behavior, feeding activity etc.allows early detection of problems when they do occur so that a diagnosis can be made before the majority of the population becomes sick. If treatment is indicated, it will be most successful when implemented early in the course of disease at the same time as the fish are still in a good shape.

The poikilothermic nature of the fishes allow them to adjust with the diurnal changes of the environment continually and acclimate themselves with the changing environment. When these changes go beyond toleration fishes suffer from general adative syndrome (GAS) (Pal and Ghosh, 1990).The general adaptive syndrome has three phases:

1) An alarm reaction,

2) A stage of resistance (when the fish tries hard to adapt itself with the changed situation), and

3) Finally a stage of exhaustion (ultimately leading to the death) if it is not recovered from the long lasting stressful environment (Pal and Ghosh, 1990).

It will be wise to discuss and define stress of fishes:

Brett (1958) while defining the stress of fish pointed out that "***Stress is a state produced by environmental or other factors which extends the adaptive responses of an animal beyond the normal range or which disturbs the normal functioning to such an extent that, in either case, the chances of survival are significantly reduced***".Management practices directed at limiting *stress* are likely to be most effective in preventing disease outbreak.

2. CAUSES OF DISEASE

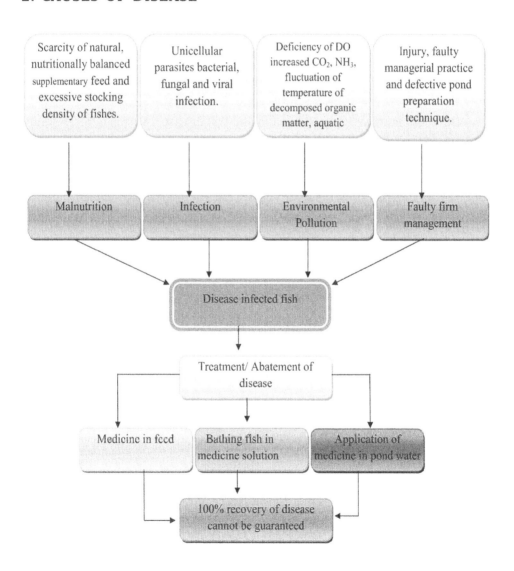

The disease is a simple association between a pathogen and a host fish.

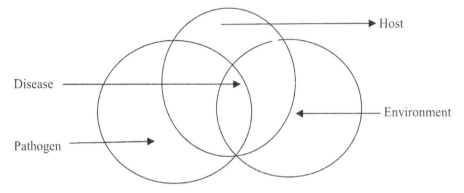

Fig. 111: Host-Pathogen-Environment relation for the outbreak of disease.

Three factors must be considered in any fish disease investigation. These are the susceptible host, the virulent pathogen, and the environment in which they encounter one another. Even though all three may be present, a host and pathogen may interact without resultant disease. However, if a disturbance in any of the three factors disrupts the relationship, disease can appear and spread.

Both fish and its pathogen maintain equilibrium during their existence in a common environment.But sudden a minute change in the environment can break this condition when either fish or pathogen gets an upperhand. If the environment is more congenial for fish then it grows better; when the reverse is true then the pathogen causes harm to fish making it ill.This may be either in subclinical or in clinical level. Most of chronic cases remain in subclinical stages for a considerable period where as the acute cases are easy to detect by clinical signs (Pal and Ghosh, 1990).

3. HOST-PATHOGEN-ENVIRONMENT RELATIONSHIP.

a. **The consequence of fish disease to aquaculture:** The occurrence of disease in the fishes very often incurr substantial financial loss to the aquaculture farmers. This includes the loss in the dead fish, cost of treatment and decreased growth during convalescence.Disease is infrequently a straight forward relationship between a pathogen and a host fish (Fig.111).

b. **Classification of diseases of fishes**

Fish disease can be grouped into five different categories:

1. Diseases of fish and the environmental stress

2. Diseases of fish caused by varios pathogen

3. Nutritional problems also are responsible for various fish diseases

4. Faulty managerial Practices may cause various fish diseases and,

5. Congenital fish diseases.

4. DETERMINATION OF FISH DISEASES

The most noticeable indication of sickness is the presence of dead or dying of the fishes in water.However, the careful observer can usually tell that fish are sick before they start dying because sick fish often stop feeding and may appear lethargic.Healthy fish should eat aggressively if fed regularly at scheduled times. Pond fish should not be visible except at feeding time.Fish those are hanging lethargically in shallow water, gasping at the surface, or rubbing any hard objects, splasing water whirling on the water surface, loss of apetite(avoid feed), vertical hanging etc. indicate some thing has gone wrong. These behavioral abnormalities indicate that the fish are not feeling well or that something is irritating them. The above mentioned symptoms or abnormal behaviours are the fore-runners of any disease outbreak (Pal and Ghose, 1990).

Essentially the stress compels the fish to maintain its homeostasis as defined by Claud Bernard (1813-1878) as maintenance of physiologiocal condition within narrow limits of an organism. The maintenance of homeostasis under stressful condition can be well recognized by two adaptive reactions:

I.   **Reactions during primary stress:** Fishes suffering from primary stress exhibit increasing plasma and increased level of corticosteroids in blood circulation.

II. **Reactions during secondary stress:** During secondary stress fishes exhibit the following physiological changes:

 a. Increased blood pressure and heart beat rate which may be ascertained by the increased movement of the operculum.

 b. Increased level of blood sugar and lactate.

 c. Increased number of thrombocytes in the blood.

 d. Increased level of liver glycogen in liver.

 e. Decreased number of white blood corpuscles.

 f. Decreased level of serum protein and blood chloride.

 g. Decreased inflammatory response.

h. Immuno-suppression and changes in mucus productionwhic is reflected by almost loosing defence mechanism.

What is mass mortality of fish?

An event where large numbers of fish die, indicating a problem in the body of water. Fish kills can be caused by a variety of factors including dissolved oxygen depletion, extreme water temperatures, fish diseases or introduction of pollutants. Most fish kills are natural events.

What to do if mass mortality of fish is observed

Once dying fish are observed it is usually too late to stop a fish kill. The concerned farmers may observe abnormally high fish mortalities after excessive atmospheric and water temperature followed by sudden fall might be due to high undesirable gas which has accumulated under water column, force the fishes surfacing and gulping air. During summer (2-3 months, April, May and June), which is prevalent in Indian condition the farmer should apply Allumino-silicate powder (Zeolite) at least once in every fortnight @15-20kg/acre which will help adsorbing the gases like ammonia, hydrogen sulphide and sulfer di oxide etc. which accumulate at the soil-water inreface.

During winter months various pathogens viz. bacteria and fungi etc. become very active and starts proliferating over the fish causing infections ensuing diseases like gill rot, fin rot, tail rot, eye diseases, epizootic ulcerative disease syndrome etc. and the resultant is heavy mortality of fishes. Judicious stocking management of the fishes, application of sterilized feed followed by quality anti antimicrobials like Gentamycin based medicines coupled with antifungal agents like chelated forms of copper atleast once in a month (as per the suggestions of aquaculture technicians) must be made to avoid mass scale mortality.

Immediately after monsoon there remains every possibility of increasing the population of exoparasites in aquatic systems atleast where there are organic enrichments. The parasites possess a definite position in animal kingdom, due to their adaptation and damaging activities. The normal growth is affected by parasite that lives on the fish if highly infested. Ectoparasites not only harm the fish directly but also render the fish for grown, reduce host population and induce mortalities (Piasecki et al., 2004).

High stocking density demands higher nutritional supply in the form of supplementary feed either pelleted sinking or floating types. The emergence of crustacean exoparasites viz. Argulus, Ergasilus and Lernea are very common in tropical water bodies.

5. DISEASES OF FISH AND THE ENVIRONMENTAL STRESS

The environmental factors play a crucial role in disrupting the balance between the host and the pathogen (Pillay, 1990).Following are the environmental variables which sometimes incur serious stress among the fishes, and these are as indicated by Pal and Ghosh (1990):

Table 57: Environmental stress and symptoms of fishes.

Environmental Stress	Symptoms
1. Raising or drop of water temperature.	1. Indian Major Carps can acclimatize themselves between 10 and 40^0C. But when it cross the limit the fishes suffer from general Gas Adaptive Syndrome (GAS).
2. Super saturation of oxygen in water. Phytoplankton bloom may cause super saturation of oxygen (Gas Bubble Disease, GBD), generally in the nursery ponds. GBD are noticed under the skin, in the fins, tail and mouth and inturn gas embolism in the circulatory syste	2. Accumulation of gas bubbles within the body cavity of fish spawn. Smaller fishes mainly fry stage die during mid-day or late afternoon.
3. Depletion of oxygen in pond waters due to raising of temperature.	3. Surfacing and gasping air on the water surface are the symptoms of depletion of oxygen. During this phase the dead fish's showshows their mouth wideopen and the gills look pale.
4. Excess carbon-di-oxide followed by raising of pH of water.	4. Fishes secrete more mucous from their gills and body surface when both carbon-di-oxide and pH is high in pond waters.
5. Raising of nitrogenous wastes and ammonia in aquatic environs.	5. Gills look dark red due to formation of methaloglobin.
6. Excessive generation of Hydrogen sulphide.	6. Rotten egg smell from the pond muck. The fishes generally found in the benthic region shows early death.
7. Eutrophication due to organic enrichment.	7. Instead of light green colour of the pond water when the same turns pea soup green or, excessive green due to the development of blue-green algae.

[Table Contd.

Contd. Table]

Environmental Stress	Symptoms
8. Organochlorine or organophoshate pesticide contamination in pond waters.	8. Blood vascular system of fish generally rupture. Oozing out of blood from eyes and or other outlets. Dropping of scales, pectoral fins followed by erratic movement though very sluggishly.

Pillay (1993) pointed out that the minimum water quality standards necessary to maintain congenial fish health are:

Table 58: Minimum water quality standars for maintaining good fish health.

Parameters	Values / range / tolerance level(ppm) except pH
Dissolved oxygen	5.0
pH range	6.7-8.6 (extremes6.0-9.0)
Free total CO_2	3.0(or less)
Ammonia	0.02 (or less)
Alkalinity	At least 20.0 (as $CaCO_3$)
Chlorine	0.003
Hydrogen sulphide (H_2S)	0.001
Nitrite(NO_2)	100ppb (in soft water), 200ppb (in hard water)
TSS(Total suspended solid)	80.0 (or less)

Pesticides, insecticides, herbicides and also there are many botanicals exert profound adverse effect even mortality of fishes. The maximum permissible limits of tolerance has been worked out by EPA (Environmental Protection Agency, USA and is depicted in Table-59:

Geosmin produced by actinomycetes and a number of blue-green algae belonging to various species of the genus *Oscillatoria* very often generate off flavor in fishes and the water body as well. All these organism grow on the soil-water interface, decompose, causing reduction of the mud. This off flavor phenomenon may seriously affect the economics of culture.

Another source of off-flavour in fish is industrial wastes. The odour and taste of these wastes are usually concentrated in the fat deposits of the fish's body. The most important chemicals that impart off-flavour are phenols, tars and mineral oils. Chlorinated phenols, such as o-chlorophenol and p-chlorophenol, impart a distinct flavor to carp even in low concentrations of 0.015 and 0.06 mg/l. respectively (Pillay, 1993).

Table 59: Maximum permissible pesticides concentrations which may be tolerated by fish

Pesticide	Concentration
Organochlorine Pesticides:	
Aldrin	0.01
DDT	0.003
Dieldrin	0.005
Chlordane	0.004
Endrin	0.003
Lindane	0.02
Tozaphene	0.01
Organophosphate insecticides	
Diszinon	0.002
Dursban	0.001
Malathion	0.008
Parathion	0.001
TEPP	0.3
Carbamate insecticides	
Caebary	0.02
Zectran	0.1
Herbicides, fungicides and defoliants	
Aminotriazole	300.0
Diquat	0.5
Diuron	1.5
2.4-D	4.0
Silvex	2.0
Simazine	10.0
Botanicals	
Pyrethrum	0.01
Rotenone	10.0

6. DISEASES OF FISH DUE TO PATHOGENS

Biological agents are probably the most common cause of disease initiation and are the primary focus of attention in infectious diseases. Potential pathogens coexist in the same environment which includes bacteria, fungus, viruses, protozoans and parasites in the same environment and so also the worms.. In fact, the combination of one or more of these pathogens is the most common cause of disease outbreak and can increase disease dispersion among fresh water prawn, shrimp and fish.Fresh water prawns and fish become ill and die from diseases when they are exposed to pathogens in an aggressive way, in which their natural defenses, potentially already weakened by intensive rearing conditions and/or abrupt changes in environmental factors, are not able to cope with the aggressiveness of this exposure.

Penetration into the host is the first step for a microbial agent to multiply and invade the vital organs of the host fish. This normally happens through the ingestion by rupturing the skin, gill lamellae and play a decisive role on the virulence of the microbe. The skin when infected invites fungus as secondary invadors and may become numerous enough to weaken the fish (Pillay, 1993).

However, it is needless to say like all other animals, the fishes too, shows a distinct defensive mechanism against its pathogens. Firstly, the body of a fish is covered with the scales and the skin secrets a lot of mucous to drive away the pathgens surrounding it and even when the immune system may, including the phage cells can also acts against them.Secondly, the anterior part of the alimentary canal is acidic in nature while the rest is alkaline and in this course the pathogen has to thrive against these two type of environment, and it is very uncertain whether they can thrive.

a. Bacterial diseases

There are several bacterial diseases of cultured fin fishes and sometimes they appear in association with fungal diseases, as secondary infections. Bacterial diseases have worldwide distribution and occur in both tropical and temperate conditions (Pillay, 1990).

Myxobacters cause tail and fin rot disease. *Aeromonas* spp.and *Pseudomonas* spp. very often cause the following diseases:

i)   Cataract in both the eyes, in any kind of fish,

ii) Staphyllococcus causes blindness especially in the snake head fishes,

iii) Putrification and loss of barbels together with tail rot and ulcers, especially in the catfishes,

iv) Accumulation of water in the body cavity or scale pockets, otherwise named as Dropsy in fishes, this may happen in variety of fishes,

v)   Tumors have been identified in *Anabas testudineus* caused by *Clostridium* sp.

vi) Epizootic ulcerative disease syndrome was a major issue even a few years back caused by *Micrococcus* sp (EUDS).

b. Viral diseases

Fresh water fishes and prawns are seldom being affected by viral infections albeit trouts, salmon and eels are found to be infested with pancreatic necrosis (Infectious pancreatic necrosis, IPN) especially in fry and fingerlings. When the mortality rate is high the infectected individuals swim in a rotating manner and the death occurs within an hour or two. There is no proven effective treatment when the fishes suffer from IPN.

Infectious haematopoietic necrosis (IHN) is another viral disease encounterd in the fries of trout and salmon manifested by the occurrence of dark colouration followed by weakness, abdominal swelling and pale gills. A thereapeutic treatment of this disease is not known so far.

Again Viral Haemorrgic Septicaemia (VHS) is an acute and chronic viral disease of the salmonids under culture. It is reported that the stress during transportation or mishandling may be one of the major issue for outburstanding the disease followed by high rate of mortality.

c. Protozoan disease

Trichodinosis cause by the ciliateprotozoa is one of the parasitic diseases encountered among the fin fishes. This is represented by the family Trichodinidae in which the most common of 6 genera is *Trichodina* sp. This protozoon is probably the most frequently encountered external obligate parasite in cultured fresh water fish's world wide. Some species in this family also parasitize fish and when abundant, irritate the skin and gill surfaces causing hyperplasia of the epithelium. Fish parasitized by *Trichodina* often have a white patches and/or moulting of the skin and fins. Excessive mucus is produced causing a white to bluish shed of the skin. Fins are generally frayed and the infected fish exhibit flashing behavoiur of water and scrapping their bodies against the hard surfaces. If the gills are heavily infested the opercular movement becomes rapid.

Myxobolus bengalensis frequently cause gill spot disease with noticeable white spot on gills. Pale gills followed by excessive secretion of mucus are the other symptoms. Scale spot disease is generally caused by *Myxobolus shericum* and *Myxobolus rohitae* causing reformations on the scale which is followed by the excessive secretion of mucus by the epidermal cells. The resultant is loosening of the scales and drop off from the body.

d. Ichthyopthriasis disease (ICH)

Ichthyophthirius multifiliis is an ectoparasite of freshwater fish which causes a disease commonly known as white spot disease, or Ich. Ich is one of the most common and persistent diseases in fish. It appears on the body, fins and gills of fish as white nodules of up to 1 mm, that look like white grains of salt. Each white spot is an encysted parasite.

The protozoa damage the gills and skin as it enters the tissues, leading to ulceration and loss of skin. Severe infections rapidly lead to loss of condition and death. Damage to the gills reduces the respiratory efficiency of the fish, reducing its oxygen intake from the water. This causes the fish to become less tolerant to low oxygen concentrations in the water.

e. Whirling disease

Whirling disease caused by the protozoan *Myxosoma cerebralis*. A common sign of the disease is rapid, tail chasing behaviour when the fish are frightened or trying to feed.This caused by the parasite feeding on the cartilage of young host fish. In advanced stages of the disease, skeletal deformation, including deforme heads, jaws, operculum as well as spinal curvature, can be observed. Infected fish contaminate water and mud associated with the fish are known to be the reservoirs of infection.

f. Helminth disease

Monogenetic and Digenetic trematodes are common parasites in fishes. Among the Monogenetic trematodes *Dactylogyrus* spp. and *Gyrodactylus elegans indicus* are very frequent resulting fading of colours, dropping of scales and excessive secretion of mucus from the epidermal cells.

Among the Digeneans, *Diplostomum, Neodiplostomum, Clinostum* etc. are very common. Eye disease is very frequent upon infestation of digenian trematodes.

Cestode tapeworms as for example, *Ligula intestinalis* when infected result bulging abdomen and in exreme conditions, burrow through stomach into body cavity causing death.

g. Copepod infections

Copepod parasites of the family Argulidae, Ergasilidae and Larnaedae very often infect several species of cultured fishes especially both Indian major, minor and the Chinese carps. Argulosis caused by the species of Argulus (Popularly known as fish lice) are represented by *A. foliaceous, A.japonicus, A.bengalensis* etc. is a branchiuran parasite may easily be recognized by the naked eye. These are transparent but have two dark spots at the anterior end, round shaped and an attaching organ.

Lernaeosis, another anchor worm disease in fresh water ponds with high organic load. The parasite is not found in brackish waters since they can not tolerate even minute concentration of salt in the medium. These are easily recognized by their thin thread like small structures and the females generally carries two small egg sacs.

Ergasilosis caused by the parasite *Ergasilus* and are found in both fresh and brackish waters. The parasite is found mainly attaching at the gills of the host. When the host is severely infected, symptom of erratic movement and sluggishness is a common phenomenon followed by death.

At the beginning, a symbolic question was asked regarding whether or not the presence of *"the characteristic microbe of a disease might be a symptom instead of the cause."* In many situations, cultured fish live healthy, normal lives in the continuous presence of pathogens. However, when environmental stresses occur and the balance tips in favor of disease, the characteristic microbes flourish. If the fish cannot adequately adjust or, if fish cultural corrections are not made, disease may occur. If losses increase in typical patterns, the fish culturist must act. By resolving environmental problems and applying effective therapeutants, a balance between the host and the pathogen can be restored. The question still remains; was the disease caused by the microbe or were the microbes and the fish merely players in a larger environmental scenario? A microbial infection can often be the symptom of environmental failure and an urgent signal that conditions must be changed. Successful fish culture often hinges on whether correction of adverse environmental conditions can be achieved in time to prevent losses (Snieszko, 1973; Snieszko and Bullock 1975). The skills related to fish culture which are required to maintain the balance between the host and the pathogen in the face of changing environmental conditions indicate that there is still a great deal of "art" in the "science" of fish culture (Warren, 1983).

7. DISEASES OF FISH DUE TO NURITIONAL DISORDERS

In culturing fish in captivity, nothing is more important than proper nutrition and adequate feeding. If there is no utilizable feed intake by the fish, there can be no growth but results death. Under-nourished or malnourished animals cannot maintain health and growth, regardless of the quality of the environment.

Nutrient requirements and deficiencies

i. **Energy:** Energy is not a nutrient. It is rather an end-product of absorbed macro-organic nutrients when they are oxidized and metabolized. All organic compounds in fish feed release heat upon combustion, and thus are potential sources of energy.

ii. **Amino acids and Proteins:** Dietary proteins are the source of essential amino acids and provide nitrogen for the synthesis of non-essential amino acids. Proteins in the body tissues are built using about 23 amino acids. Of these, 10 are essential amino acids which must be supplied in the fish diet. Proteins or amino acids are necessary for maintenance, growth, reproduction and for the replacement of depleted tissues. In addition, certain amino acids are readily converted to glucose to provide an essential energy source for some critical body organs and tissues such as brain and red blood cells. Since carbohydrate is not prevalent in their natural diet, fish are more dependent upon amino acids as precursors to glucose than most other animals. Therefore, a portion of the dietary protein is always used as an energy source in fish. Not all dietary proteins are identical in their nutritional value. To a large extent, the bio-availability of a protein source is a function of its digestibility and amino acid makeup. Some protein feedstuffs which contain a high level of crude protein are low in amino-nitrogen and do not contribute toward the requirement of amino acids. As a result, such materials may merely increase ammonia production into the water environment. A deficiency of essential amino acids may lead to poor utilization of dietary protein, and may result in growth retardation, poor live weight gain, and low feed efficiency.

    Amino acds include Arginine, Histidine, Isoleucine, Leucine, Lysin, Cystine and Methionine, Tyrosine + Phenylalmin, Threonine, Tryptophan and Valine.

iii. **Fatty acids and lipids:** Lipids are a group of fat-soluble compounds occurring in the tissues of plants and animals and broadly consist of fats, phospholipids; sphingomyelins, waxes and sterols.Fats are the fatty acid esters of glycerol and are the principal form of energy storage.The nutritionally active components of dietary lipids are fatty acids. Fish and mammals appear to be unable to synthesize fatty acids that are unsaturated in the Ω-3 or Ω-6 positions unless a suitable precursor is supplied in the diet. Thus, the lipid component of the diet must provide an adequate amount of essential fatty acids for growth as well as for required dietary fuel. In contrast to mammals which have a major requirement for Ω-6 fatty acids, many coldwater and marine fishes require Ω-3 fatty acids. Therefore, sufficient amounts of essential fatty acids (Ω-3 longer chain members of these series) must be included in the dietary lipids. One percent linolenic acid (18:3w3) in the diet is required by rainbow trout to avoid such deficiency signs as loss of pigmentation, fin erosion, cardiac myopathy, fatty infiltration of the liver, and shock syndrome (Castel *et al.*, 1972). Salmonids utilize lipids as a major source of energy and digest complex carbohydrates very poorly. Diets for salmonids therefore, should contain very high levels of lipids (10-18%) in comparison to diets for other animals. Because of the high level of use, lipid quality is critical since marine fish oil is very susceptible to oxidation. In all circumstances, rancid oil must be avoided in fish feed. Fish suffering from lipoid liver disease have extreme anemia, a bronzed, rounded heart and a swollen liver with rounded edges. Histologically, the main feature is the extreme lipid infiltration of hepatocytes and associated loss of cytoplasmic staining and distortion of hepatic muralia (Cowey and Roberts 1978). All salmonids are suceptible to lipoid degeneration of the liver, but it is a particularly significant problem in rainbow trout. Slightly affected fish are usually capable of recovery, but if severe anemia and hepatic ceroidosis have developed, the fish are rarely capable of recovery to an acceptable feed efficiency (Cowey and Roberts 1978).

iv. **Vitamins:** Vitamins are a chemically diverse group of organic substances that are required for normal growth, but either not synthesized by organisms or is synthesized at rates insufficient to meet the organism's needs. They constitute a minute part of the diet and also play a catalytic role in their function. Further more vitamins are more vital for the maintenance of normal metabolic and physiological functions. Vitamin deficiency in fish which may cause various diseases and the symptoms there of is given in Table 60. Vitamins may be classified in two groups, viz:

(a) Water soluble vitamins; principally include eight members of the Vitamin B-complex, Thiamin, Riboflavin, and Pyridoxine, Pantothenic acid, Niacicin, Biotin, Folic acid and Vitamin B_{12}. These vitamins take an active participation in nutritional factors choline, inositol, ascorbic acid and,

(b) Fat soluble vitamins principally include VitaminA, D_3, E and K.

In order to avoid the diseases and the concerned symptoms among the fishes under culture it is imperative to provide them the proper nutritionally balanced feed at different stages, so that optimum production at minimum cost may be achieved. Supplementation of Vitamins in the fish diet may be used by commercially available Vitamin premixes. Algal meal and brewers yeast are excellent source of vitamins. Table-61 gives a fairly good idea of the vitamin premixes made specifically for fish feeds (Hastings, 1979). Flavour, colour,

Table 60: Vitamin deficiency symptoms in fish (after Halver, 1979)

Vitamins	Disease	Symptom
Thiamin	Anaemia, Anorexia	Corneal opacities, degeneration of nerves, fatty liver, hemorrhage, loss of equilibrium, paralysis of dorsal and pectoral fins, whirling motion, weakness.
Riboflavin	Anorexia	Cloudy lens, darkened skin, hemorrhage in eyes, photophobia
Pyridoxine	Anorexia	Convulsions, rapid jerky breathing, spasms, weight loss, rapid onset of rigor mortis
Folic acid	Anaemia, Anorexia	Dark colouration, fragility of caudal fin, lethargy, pale gills, reduction in growth.
Pantothenic acid	Anorexia	Clubbed gills, flared opercula, exudated gills, necrosis of jaw, barbells and fins, poor weight gain.
Inositol	Anaemia	Bloated stomach, poor growth, lesions in skin.
Biotin	Anorexia, Anaemia	Muscle atrophy, dark colouration, contracted caudal fins, poor growth,
Choline	Anaemia	Poor food conversion, poor growth, Hemorrage in kidney and intestine
Nicotinic acid(Niacin)	Anaemia, Anorexia	Colonic leisions, muscle spasm, lethargy, skin hemorrhage, high mortality.
Cobalamin(B_{12})	Anorexia	Erratic haemoglobin and erythrocyte counts. fragmentation.
Ascorbic Acid	Anorexia	Impaired wound healing, lordosis, scoliosis
Vitamin A	Ascites	Hemorrage in kidneys
Vitamin D	-	Reduced conversion
Vitamin E(Tocopherol)	Anaemia	Fragile RBC, poor growth , mortality
Vitamin K	Anaemia	Prolong coagulation time

odour, texture and water stability are the important characteristics related to the acceptance and consumption in case of fish diets. It is needless to mention in this contexct that the basic composition of fish feed, depends on the levels of crude protein, energy, specific amino acids, minerals and vitamins, crude fibre and ashes (Pillay, 1993 and Hastings, 1979).

Table 61: Vitamin and mineral premixes for fish diet. (Modified from Hastings, 1979).

Ingredients	Vitamin and mineral premix in feed/kg for pond culture
Thiamin (B1)	2.4gm.
Riboflavin (B2)	1.5gm.
Pantothenic acid (B3)	2.5gm.
Nicotinic acid (B5)	12.0gm.
Piridoxine (B6)	0.5gm.
Cobalamin (B12)	1.0gm.
Choline chloride	50gm.
Folic acid	0.4gm
Biotin	0.06gm.
Ascorbic acid	20.0gm
Vitamin A-acetate	20, 00, 000IU
VitaminD3	1, 00, 000IU
VitaminE	7.0gm.
VitaminK3(Menadion-bisulphate)	2.5gm.
Butyl-hydroxytoluene	5.0gm.
Iron(carbonate)	2.5gm.
Manganese	3.5gm.
Zinc	1.4gm.
Copper sulphate	0.1gm.
Cobalt	0.4gm.
Iodine	0.1gm.

v. **Minerals:** Minerals are required by all animals either in their elemental form or incorporated into specific compounds for various biological functions such as formation of skeletal tissue formation, respiration, digestion, osmoregulation etc.Of the 26-naturally occurring essential elements which are said to be important for the animals, only nine have been shown to be required by the finfish (Pillay, 1993) (Table-62).

In view of the fact that, water generally in enriched with different types of minerals, supplementation of diets may not be necessary, except in the case of those that are required in relatively high concentrations, especially in fresh water fish.

In fish, minerals perform important roles in osmoregulation, intermediary metabolism, and in formation of the skeleton and scales (Lall 1981). Mineral requirements of fish are difficult to study because many minerals are required in only trace amounts and others are absorbed from water in significant quantities through the gills as well as from the diet. It is also very difficult to obtain mineral free feed ingredients for experimental diets. Most practical diets for salmonids provide the major mineral requirements through fish meal which is also a major source of protein. However, diets which rely heavily on plant protein sources must be supplemented with carefully balanced mineral premixes. The minerals required in finfish diets include calcium, zinc, manganese, cobalt, selenium, iodone and fluorine. The functions of some of these have been described in detail (Nutrition Research Council, NRC 1977). The recommended dietary levels of minerals and related deficiency signs are shown in Table: 62. The potential for toxicity of minerals must also be carefully assessed since fish are very sensitive to excess amounts of minerals.

Table 62: Mineral deficiency syndrome in fishes (after Castell *et al.*, 1986)

Mineral	Deficiency syndrome
Calcium	Poor growth and feed efficiency.
Phosphrus	Skeletal abnormalities, poor growth.
Magnesium	Loss of apetite, poor growth, high mortality, skeletal abnormalities and sluggishness.
Iron	Anaemia.
Copper	Poor growth.
Manganese	Poor growth, short and compact body, abnormal tail growth.
Iodine	Thyroid hyperplasia.
Zinc	Cataract, caudal fin and skin erosion and growth depression.
Selenium	Muscular dystrophy.

All nutrients required for the well-being and normal growth of the fish must be supplied in formulated diets as available (digestible) nutrients. Otherwise, the fish cannot utilize the nutrients present in the feed ingredients. The formulated diets also must be pelletized and processed in such a manner that they are durable and water stable.Proper feeding of a quality diet should be considered as a high priority in the daily routine on fish culture stations. Wasted feed depletes oxygen levels, causes gill damage, and supports fungal and bacterial growth, all of which can lead to disease problems. Because it is necessary to transfer dietary nutrients into the fish through a water medium, problems occur which are unknown in terrestrial animal feeding practices. Recently, in addition to pelletized sinking feed floating feed has been introduced which helps minimizing wastage. The main factors influencing feed intake of fish are water temperature, the energy content of the diet, and expected growth. Therefore, an estimation of feed intake needed must be based on these fundamental factors. If a group of fish is not feeding actively or growing as expected, diagnostic work is needed to determine the cause. Lack of appetite or retarded growth is often early signs of stress and disease (Cho, 1983).

vi. **Prevention is better than cure (Borba, 2018):** Fresh water prawn and fish become ill and die from diseases when they are exposed to pathogens in an aggressive way, in which their natural defenses, already weakened by intensive or semi-intesive rearing conditions and/or abrupt changes in environmental factors, are not able to cope with the aggressiveness of this exposure.

The adoption of standards like the Global Aquaculture Alliance (GAA), Best Aquaculture Practices (BAP) is the main key to maintain a more balanced and safe environment for the aquatic animals in productive systems. Even when they are applied, in some cases their effects are not so efficiently measured, leading aquaculturists to doubt about their effectiveness.

It's a known fact, that fish and prawn/shrimp are still cultured in a reactive way in terms of their immunity and health protection. This means that it is much more common to see aqua farmers trying to cure diseases after they have emerged and spread the pathogens into the cultured environment, rather than adopting some systematic measures aiming at preventing disease outbreaks. However, prophylactic actions exist to prevent or reduce the risk and the level of transmission of a disease, protecting the populations from its occurrence or evolution.

In a coutry like India, most of thew aquaculture specialists, highly recommends a series of measures to avoid disease's dissemination. One of them is the systematic and continuous use of probiotics and paraprobiotics to avoid the pathogenic interventions, which are:

vii. **Probiotics and paraprobiotics:** The use of probiotics in Aquaculture at least in India is a new concept and their use has been significantly growing due to the efficiency and proven benefits so far but mainly restricted to ponds' bioremediation and water treatment. The common use of these live bacteria helps to promote a more efficient mineralization of the organic matter, and thus a better balance of the physico-chemical and hydrobiological parameters of the water.

Now-a-days, besides their proven benefits for treating pond bottoms and helping to maintain water quality at acceptable levels, there is a consensus that probiotics (bacteria and yeast) can also promote the competitive exclusion of harmful bacteria in animal bodies, especially in the gut. Unfortunately, probiotic bacteria alone are not always able to prevent (neither treat) disease outbreaks (Borba, 2018).

Hence, the aquaculture industry has started paraprobiotics utilization. They are prophylactic effective compounds with accessible and easy to incorporate in industrial feed, cost-effective, but still incomprehensible is the low adoption of this technology in India.Such solutions are still in the process of implementation and dissemination among aquaculture farmers.

Fortunately, there are already successful cases with the use of yeast cell wall paraprobiotics in highly productive commercial aqua farms. These could serve as an incentive and reference for many other fish and shrimp producers who often are not sure about what paraprobiotics are, how they act and how to use them.

Intestinal bacteria pathogens cause damage to the host by binding and colonizing the intestinal epithelium and even sometimes crossing the intestinal barrier and spreading to inner organs and tissues. Paraprobiotics, such as yeast cell wall, contain complex carbohydrate molecules that interfere directly with the binding ability of pathogenic bacteria in the gut, thereby reducing the ability of pathogen bacteria to bind to the intestinal epithelium of the animals.

One of these complex carbohydrates is 'Mannan Oligosaccharides', also known as MOS. MOS are non-digestible carbohydrates that when in transit into the intestinal lumen, can bind bacteria due to their strong affinity for lectins (molecules present on the outside of the plasmatic membrane of Gram-negative bacteria cells). Mannans binding to these compounds (lectins) lead to the neutralization of the harmful bacteria and with their subsequent excretion through the feces, thereby decreasing harmful bacteria concentration in the host's body.

Another complex carbohydrate found in the yeast cell walls are β-glucans (1, 3; 1, 6), which present a distinct mode of action generating other benefits. Once "sensed" by the gastrointestinal tract of the fish or shrimp, they are identified as a "warning signal", allowing the activation of macrophages.

Macrophages are highly specialized cells of the immune system that possess the ability to 'eat" microorganisms such as bacteria (phenomena called phagocytosis). After phagocytosis, the pathogen is trapped within the macrophage in a structure called phagosome. The phagosome fuses with a vacuole containing enzymes and peroxides to help digest the pathogen.

In addition to phagocytosis, macrophages are of great importance for immunomodulation, producing and secreting a large number of molecules named chemokines that attract other defense cells to specific sites where inflammatory processes are occurring.

Paraprobiotics therefore, have synergistic effects with probiotics. Probiotic bacteria and yeast promote beneficial effects on animal's gut health, competing with pathogenic bacteria for nutrients and oxygen, for adhesion sites on intestinal epithelium, and producing antimicrobial substances. On the other hand, paraprobiotics can be understood as "traps" to capture, inactivate and eliminate, in an effective way the harmful bacteria in the gut, thus opening space for a better and more effective colonization of the digestive tract of fish and shrimp by beneficial microorganisms.

viii.**Probiotics-there use in Aquaculture:** The term "probiotic" comes from Greek *pro* and *bios* meaning "prolife" and the term probiotics was first used by Lilly & Stillwell in 1965 and has been defined as the microbiological origin factor that stimulates the growth of other organisms. In 1989, Roy Fuller introduced the idea that probiotics generate a beneficial effect to the host (in this case it is Fish). Probiotics are defined by Food and Agriculture Organization/World Health Organization as *"live microorganisms which when administered in adequate amounts confer a health benefit on the host".*

ix. **Effects of application of** *probiotics* **in aquaculture**: Probiotc microorganisms have an antimicrobial effect through modifying the intestinal microbiota, secreting antibacterial substances (***bacteriocins and organic acids***), competing with pathogens to prevent their adhesion to the intestine, competing for nutrients necessary for pathogen survival, and producing an antitoxin effect. Probiotics are also capable of modulating the immune system, regulating allergic response of the body, and reducing proliferation of cancer in mammals. Because of this, when provided at certain concentration and viability, probiotics favorably affect host health. In fact, terms such as "friendly bacteria, " "friendly, " or "healthy" are commonly used to describe probiotics.

x. **Mechanism of action of the probiotic bacteria:**

1. Probiotic bacteria may competitively exclude the pathogenic bacteria or produce substances that inhibit the growth of the pathogenic bacteria.

2. Provide essential nutrients to enhance the nutrition of the cultured animals.

3. Provide digestive enzymes to enhance the digestion of the cultured animals.

4. Probiotic bacteria directly uptake or decompose the organic matter or toxic material in the water improving the quality of the water.

xi. Probiotic organisms: The requirements that a probiotic organism must meet are:

i. Resistance to the acid stomach environment, bile and pancreatic enzymes;

ii. Accession to the cells of the intestinal mucosa;

iii. Capacity for colonization;

iv. Staying alive for a long period of time, during the transport, storage, so that they can colonize the host efficiently;

v. Production of antimicrobial substances against the pathogenic bacteria; and

vi. Absence of translocation.

The species normally used as probiotics in animal nutrition are usually non-pathogenic normal microflora, such as lactic-acid bacteria (*Bifidobacterium, Lactobacillus, Lacto coccus, Streptococcus* and *Enterococcus)* and yeasts as *Saccharomyces spp.*

a. **Competition for binding sites:** Also known as "**competitive exclusion**", where probiotic bacteria bind with the binding sites in the intestinal mucosa, forming a physical barrier, preventing the connection by pathogenic bacteria;

b. **Production of antibacterial substances:** Probiotic bacteria synthesize compounds like hydrogen peroxide and bacteriocins, which have antibacterial action, mainly in relation to pathogenic bacteria. They also produce organic acids that lower the environment's pH of the gastrointestinal tract, preventing the growth of various pathogens and development of certain species of Lactobacillus;

c. **Competition for nutrients:** The lack of nutrients available that may be used by pathogenic bacteria is a limiting factor for their maintenance;

d. **Stimulation of immune system:** Some probiotic bacteria are directly linked to the stimulation of the immune response, by increasing the production of antibodies, activation of macrophages, T-cell proliferation and production.

Table 63: Microorganisms recognized as safe and used as probiotics.

Sl.nos.	Groups	Species involved
01.	Aspergillus:	*A. niger, A. orizae*
02.	Bacillus:	*B. coagulans, B. lentus, B. licheniformis, B. subtilis*
03.	Bifidobacterium:	*B. animalis, B. bifidum, B. longun, B. thermophylum*
04.	Lactobacillus:	*L. acidophillus, L. brevis, L. bulgaricus, L. casei, L. cellobiosis, L. fermentarum, L. curvatus, L. lactis, L. plantarum, L. reuterii, L. delbruekii,*
05.	Pediococcus:	*P. acidilacticii, P. cerevisae, P. pentosaceus, P. damnosus*
06.	Saccharomyces:	*S. cerevisiae, S. boulardii*
07.	Streptococcus:	*S. cremoris, S. faecium, S. lactis, S. intermedius, S. thermophyllus, S. diacetylatis*

e. **Aquaculture probiotics:** The relationship of aquatic organisms with the farming environment is much more complex than the one involving terrestrial animals.It is a microbial supplement with living microorganism with beneficial effects to the host, by modifying its microbial community associated with the host or its farming environment, ensuring better use of artificial food and its nutritional value by improving the host's response to diseases and improving the quality of the farming environment. The microorganisms present in the aquatic environment are in direct contact with the animals, with the gills and with the food supplied, having easy access to the digestive tract of the animal. Among the microorganisms present in the aquatic environment are potentially pathogenic microorganisms, which are opportunists, i.e., they take advantage of some animal's stress situation (high density, poor nutrition) to cause infections, worsening in zootechnical performance and even death. For this reason, the use of probiotics for aquatic organisms aims not only the direct benefit to the animal, but also their effect on the farming environment. The interaction between the environment and the host in an aquatic environment is complex. The microorganisms present in the water influence the microbiota of the host's intestine and vice versa. Changes in salinity, temperature and dissolved oxygen variations, change the conditions that are favorable to different organisms, with consequent changes in dominant species, which could lead to the loss of effectiveness of the product. Accordingly, the addition of a given probiotic in the farming water of aquatic organisms must be constant, because the conditions of environment suffer periodic changes. Thus, the variety of microorganisms present must therefore be considered in the choice of probiotic to be used in aquaculture.

Intensive farming systems utilize high stocking densities, among other stressors (e.g. management), which often end up resulting in poor growth and feed efficiency rates, besides of weakness in the immune system, making these animals susceptible to the presence of opportunistic pathogens present in the environment. In this sense, the effect of probiotics on the immune system has led to a large number of researches with beneficial results on the health of aquatic organisms, *although it has not yet been clarified how they act*. In addition, probiotics can also be used to promote the growth of aquatic organisms, whether by direct aid in the absorption of nutrients, or by their supply.

Probiotics which are most used in aquaculture are those belonging to the genus *Bacillus* spp. (*B. subtilis, B. licheniformis and B. circulans*), *Bifidobacterium* spp. (*B. bifidum, B. lactis, and B. thermophilum*), lactic-acid bacteria (*Lactobacillus* spp., *Carnobacterium* spp.) and yeast *Saccharomyces cerevisiae*.

The benefits observed in the supplementation of probiotics in aquaculture include:

1. Improvement of the nutritional value of food;
2. Enzymatic contribution to digestion;
3. Inhibition of pathogens;
4. Growth promoting factors;
5. Improvement in immune response; and
6. Farming water quality.

Among the most recent studies that point to the effect of the use of probiotics for various aquatic organisms stand those for fish, shrimps, mollusks and frogs.

f. Results of probiotics in fish farming:

Immune system: Larvae of some fishes feed on the rotifers, enriched with lactic-acid bacteria increased resistance against infection by *Vibrio* spp. The joint administration of *Lactobacillus fructivorans* and *Lactobacillus plantarum* through dry or live feed promoted the colonization of the intestine resulting decrease in mortality during larviculture in nursery management. Use of *Pseudomonas fluorescens* as probiotics decreased the mortality of juveniles of rainbow trout (*Oncorhynchus* sp.) exposed to *Vibrio anguillarum*. Higher survival rate of carp *Labeo rohita* is witnessed fed with *Bacillus subtilis*, during *Aeromonas hydrophila* infection.

g. **Probiotics and quality of water in aquaculture:** Another aspect of the use of probiotics in aquaculture is the improvement of the quality of the water in the farming ponds. *Increase of organic load, levels of phosphorous and nitrogen compounds are growing concerns in aquaculture. It is also been noted, that the beneficial effect of probiotics on organic matter decomposition and reduction of the levels of phosphate and nitrogen compounds. Aerobic denitrifying bacteria are considered good candidates to reduce ammoniacal nitrogen to nitrite to nitrate to free Nitrogen in aquaculture waters.* To this end some bacteria were isolated in shrimp farming tanks. *Acinetobacter, Arthrobacter, Bacillus, Cellulos imicrobium, Halomonas, Microbacterium, Paracoccus, Pseudomonas, Sphingo-bacterium and Stenotrophomas* are some of the denitrifying bacteria already identified. Reduction in levels of phosphorous and nitrogen compounds in the farming water of shrimp *Litopenaeus vannamei* was also observed when commercial probiotics were added to the water. Similarly, for the shrimp *Penaeus monodon*, an improvement in the quality of farming water was observed with the addition of *Bacillus* spp. as probiotic. Gram-positive bacteria are better converting organic matter into CO_2 than gram-negative bacteria. Thus, during a production cycle, higher levels of these bacteria can reduce the accumulation of particulate organic carbon. Thus, maintaining higher levels of these gram -positive bacteria in production pond, farmers can minimize the buildup of dissolved and particulate organic carbon during the culture cycle while promoting more stable phytoplankton blooms through the increased production of CO_2.

A major challenge now-a-days, affecting fish and shrimp farmers globally is the disposal of accumulated organic matter. Improper treatment and disposal leads to many problems which include:

i)   Poor water quality,

ii) High level of stress,

iii) Run away disease problems,

iv) Excessive costs of production affecting profitability, and

v)   Regional pollution that is in consistent with sustainable production.

The positive benefits of using the Probiotic strains

i) **Cleaner pond bottoms:** Shorter time between cycles and less use of oxidizing agents like lime,

ii) **Improved Feed Conversions:** Healthier fish wastes less feed and use feed more efficiently for growth,

iii) **Less water exchange:** Less pumping rates and costs and in some instances completely closed pumps lessening the risk of inducing problems during the production cycle from the outside.

iv) **Eliminating the use of antibiotics:** Compete against many types of bacterial infestations and the farmers stop using antibiotics.

v) **Reduction in *Vibrio* levels:** Significant reduction in vibrio levels.

vi) **Increased Growth Rate**: Healthy fish/shrimps grow better in cleaner ponds with less organic loads.

The feasibility and future of the application of probiotics in aquaculture

Based on the previous research results on probiotics **it is suggest that the use of probiotic bacteria in aquaculture has tremendous scope and the study of the application of probiotics in aquaculture has a glorious future.** At present, the probiotics are widely applied in United States of America, Japan and European countries, Indonesia, India and Thailand, with commendable results.

Application of Probiotics in Aquaculture

Pathogenic microorganisms causing outbreaks are viruses, bacteria, rickettsia, mycoplasma, algae, fungi and protozoan parasites. For preventing and controlling diseases, a host of antibiotics, pesticides and other chemicals are used possibly creating antibiotic resistant bacteria, persistence of pesticides and other toxic chemicals in aquatic environment and creating human health hazards. Thus, how to improve the ecological environment of aquaculture has become the focus of attention of international aquaculture.

Probiotics as tool of bioremediation

- Cleaner pond bottoms
- Reduction in blue green algae loads (via competition for nutrients-these products are not algaecides)
- Less water exchange needed (savings in pumping costs)
- Reduction in green vibrio counts (on TCBS, the vibrio associated with AHPNS produces green colonies)

- Reduction in overall vibrio loads (the Bacillus compete against the vibrios for nutrients)
- Reduction in ammonia levels (broken down by the bacteria in our products)
- Reduction in disease outbreaks (a result of a change in the microbial flora)
- Better feed efficiency (a result of a cleaner environment)
- Healthier shrimp and fishes.
- Increased profits

Probiotics are defined as live microbial preparations that when fed to an animal colonize the animal's intestinal tract and positively impact the animal's health. This term is widely misused in aquaculture and an analysis of the literature (gray and peer reviewed) suggests that the term may be largely inaccurate. There are very few accounts of bacterial attachment to the intestinal wall of fish or shrimp demonstrating the inhibition of pathogenic bacteria.

Besides the aforesaid indications of fishes in fresh water ponds, there do exist some other problems

1. Due to eutrophication of pond water and the soil beneath it, some time there is a dense growth of filamentous algae. The fishermen usually designate it as 'Hairy algal bloom' or Chool Sheola. This situation may well be avoided by application of Chelated copper @300gms and potassium permanganate ($KMnO_4$) @ 900gm/acre/meter.Both these ingredients after mixing well in a tub or on the boat may further be diluted again and again and broadcasted over the pond surface. If it is found that such algal bloom is due to the decomposition of muck it is advised to apply **Zeolite powder**@ 10kg/acre/ meter depth.

2. Various molluscan species e.g. aquatic snails of various types and bivalves etc., (popularly known as geri, googly, shamuk, jhinuk etc.) (Fig.142), generally inhabit at the peripheral region of organically enriched pond bottom

   These creatures create immense problem to the cultivable fishes in fresh water ponds and tanks. They are the carriers of parasites specially the trematodes, share both space and the natural food allotted to the fishes, these are planktivorous in nature. Hence, are not in favour to the aqua farmers.The massive developments of these creatures are used as a source of low cost animal protein and supports livelihood of rural fisher women at least in part.

However, fish farmers at least who are engaged commercial intensive farming especially Indian major, minor and exotic carps generally do not favour the presence of these molluscan populations due to the fact that:

i)   Excessive presence generally bring down dissolved oxygen content of the pond water,

ii) Molluscan populations are the principal carrier of different types of pathogenic trematodes and nematode larvae and so also the leeches.

iii) Continuously feed on various planktonic millieu of the pond and,

iv) Decrease the overall productivity status of the pond.

Based on these four basic points the farmers express unhappiness throughout the growout phases of carps.

It is now the concern of all aquaculture scientists and technicians to find out the possible measures of eradication of the molluscan population, even if the fishes do exist in the ponds. Opportunities in this direction still remain open.

Application of Mahua Oil Cake mixed with Urea or Bleaching powder ofcourse eradicate the infestation at least in part but complete eradication seems to be impossible.

3. Sometimes it is noted that the fishermen refuse to step in to the pond water due to severe itching on the skin. This itching ultimately results in swellings on the skin. This is simply a case of 'allergy' that develops in the body. Insuch conditions, it is to be understood that there is a massive development of raphides in the culture water. To come out with a solution for such conditions it is advised that 3kg of Bleaching powder and Potassium permanganate each per acre are to be mixed and broadcasted over the pond water. Besides, the fishermen are to be provided with an anti-allergic tablet (viz. Cetzine, Lezynlet, Allegra etc.) before they step in the pond water.

4. Very often it is noticed that in carp culture ponds, massive appearance of fry stages of *Tilapia* spp. These weed fishes occupy both food and space allotted for the carps of economic importance, resulting a drastic decease in the ultimate production.

To come across this situation it is advisable to use Forate powder, which is available in granular form in the market. It is advised that Forate granules @2kg/acre/6ft depth of pond water is to be mixed approximately with 10 liters of water thoroughly. This mixture is then to be filtered through a clean cloth and the filtered water is collected in a tub or in a boat. The waste powder which remains undissolved in the cloth is to be collected carefully and to be merged in a hole sufficiently far away from the pond. The hole to

COMMONLY OCCURRING SOME DISEASES OF FIN FISHES

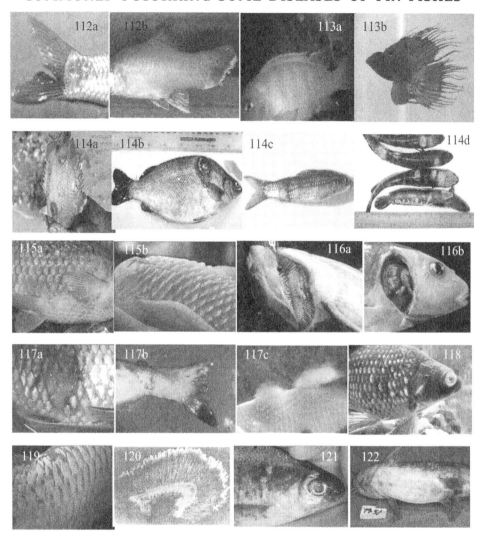

(Clockwise) Figs: 112a, b: Tailrot(a, b), Figs. 113: Finrot(a, b), Figs.114: Epizootic Ulcerative Disease syndrome (a, b, c, d), Figs.115: Dropsy (a, b), Figs.116. Gill rot (a, b), Figs.117: Saprolegniasis (a , b, c), Figs. 118: White spot on body, Fig.119: White spot on scales, Fig.120: Costiasis, Fig.121: Septicemic disease, Fig. 122: Edwardsielosis disease.

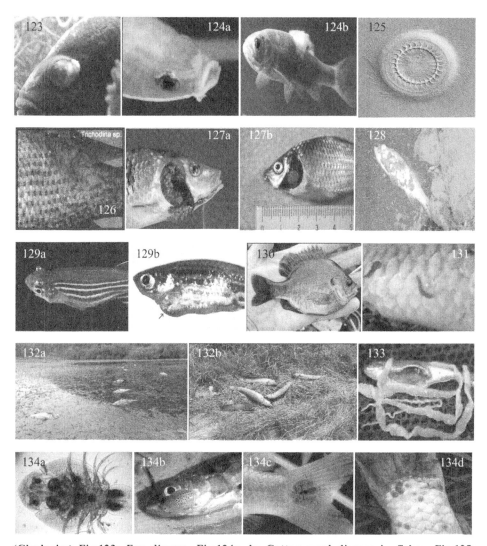

(Clockwise) Fig.123: Eye disease, Fig.124a, b: Cotton wool disease in fishes, Fig.125: *Trichodina* sp. Fig.126: Trichodineasis, Fig.127a, b: *Myxobolus* infection in fish gills, Fig. 128: Algal toxicity, Fig. 129a, b: Gas bubble disease in fishes, Fig.130: Brownish spot disease, Fig.131: Leech infection, Fig.132a, b: Mortality of fishes due to oxygen depletion, Fig. 133: Ligulosis infection, Fig.134: a, b, c, d: *Argulus* sp. and its infection.

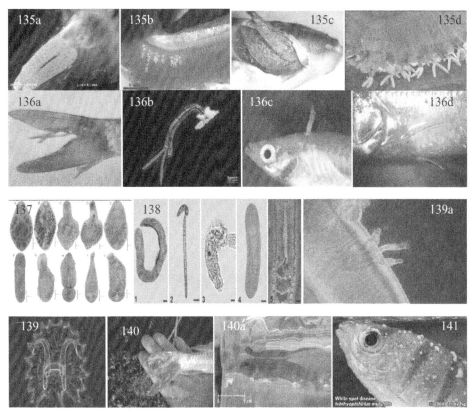

(Clockwise) Fig.135, a, b. c, d: *Ergasilus* sp. and its infections. Figs.136a, b, c, d: *Lernaea* sp. and its infections.Fig.137: Trematodes responsible for intestinal infections. Fig. 138: *Dactylogyrus* infection, Fig. 139 and 139a: *Gyrodactylus* sp. and its infection. Figs.140, 140a: Velvet disease, Fig.141: White spot (ICH).

be carefully covered with soil. The filtered water is to besprayed throughout the pond surface which subsequently kill almost all the fry stages of *Tilapia* sp. while the bigger ones will move down to the benthic region to avoid toxicity. Albeit the procedure of eradication of *Tilapia* fry however, will not affect the commercially important other fish stock of importance. It is advisable to net the pond at least 24 hours after the application of Forate granules to harvest all the dead *Tilapia* fry.

Caution to be taken

i) Avoid feeding to the fishes at least 2 days prior to and 3-days after application of Forate granules.

ii) The waste material to be placed inside the deep hole sufficiently away from the pond to avoid leaching of the toxicity.

Fig. 142: Fresh water snails, bivalves belonging to the Phylum-Mollusca

5. During intensive culture of carps, it is sometimes, small transparent anchor fish (*Chanda ranga* or *Chanda nama*) and some protozoan and crustacean parasites monogenian parasites appear in ponds which adversely affect the natural growth of carps resulting decreased production (Figs.143 & 143a). To avoid such infiltration of chanda, crustaceans, protozoan parasites, it is advocated to use Alkyl phenol ethoxylate (Chloropyrephos 20%) @60-90 ml/acre/5-6ft depth of water. Such application will remove 90-95% of the infestations.

Supplementary feed application must have to be stopped at least for a week from the day of application. Simple netting is advocated for the eradication of such infestations however, repeated netting should not be done at any cost. After application of Chloropyriphos 20% it is recommended to apply antibacterial and antifungal agents@300 gms/acre/meter.

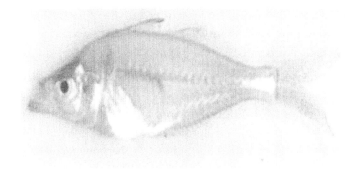

Fig.143: *Chanda nama*

Fig. 143a: *Chanda ranga*

6. Very often cultured fishes do not accept supplementary feed which poses a great concern to the farmers. In such situation Ostovet Forte@200ml per 100kg of supplementary feed for a week is advocated. This will increase the appetite to the stocked fishes.

7. It is sometime observed that the gills of the stocked fishes turn pale instead of reddish. The major reason attributes to the problems in blood circulation. In such situation the natural movement of fishes ceases down. To overcome these situations apply Zinc sulphate 21% @750gms/acre/6feet depth of water in the pond. It is further advocated that vitamin premix @250gms/100kg of supplementary feed is to be mixed together at least for a period of 30-45 days for fry and fingerling stages and 20-30 days in case of adult stage of fishes.

8. In nursery ponds very often it is noticed that the fry stages of IMC are sharply cut into two pieces at a large scale and the pieces start floating on the surface of the pond water. This attributes a massive development of Dytisid larvae, which have sharp mandibular appendages. By the help of these mandibles, fry are being cut in two pieces randomly. To overcome this situation, it is advised to broadcast tobacco dust @ 8-10 kg/acre area of the nursery ponds in dry condition. The toxic effect of tobacco powder will kill

the larvae within 2-3 days. After netting out the semi-decomposed larval remnants thoroughly from the ponds apply Zeolite powder @ 10-15kg/acre/meter within 48-hours.

9. After excavation of the soil in sewage-fed ponds, a thick mat (Fig.144&144a) of algal bloom is found, and even the water is also do not favour stocking of advanced fingerlings for further growing at least for a month. Application of Gypsum (Calcium sulfate) and Zeolite powder@30kg/acre is advocated and after 3-4 days Chelated copper and Potassium permanganate@ 500gms each is to be broadcasted after thorough mixing. The pond water will be lighter and fit for stocking of fish fingerlings.

10. Very often it is reported that the pond water turned blackish or to that of the color of snuff. The abundance of both phyto- and zooplankton are in question. In such situation 16% aqueous solution of humic acid @4-6 liters/acre/meter and Soil probiotic @ 1 liters/acre/meter together with 600gms of Water probiotic respectively is to be applied. The blackish snuffy nature of water color turn light green within a couple of days and even the stocked fishes will enjoy the turned environment.

11. Sometime in eutrophicated (nutrient enriched) tropical ponds free floating and filamentous blue green algae which are otherwise designated as Cyanobacteria pose constant problem. These algal bloom very often compete with fishes stocked in such ponds for:

    i)   Dissolved Oxygen

    ii) Various nutrients present in soil and water

    iii) Dominate over other phytoplankton and compete for available nitrogen

    iv) The blue greens can regulate the cell buoyancy in the water column and gain additional benefit over other organisms for photosynthetic activities.

To control these blue-greens, application of chelated copper in soft water ponds to prevent copper poisoning in fishes stocked. It is further reported that the toxicity of chelated copper compound is very less than those of the equal concentration of copper in Copper sulphate ($CuSO_4$). It is further emphasized that the species with a high susceptibility to copper are not harmed in waters of low alkalinity (Boyd, 1979). Application of chelated copper and Potassium permanganate as oxidizing agent together at the ratio of (3:1) may reduce filamentous algae to a great extent from the pond.

(Water containing little or no dissolved salts of calcium or magnesium, especially water which contain less than about 85 parts per million of calcium carbonate is regarded as 'Soft Water'. On the contrary, the

'hard Waters' are those that contain mineral salts as calcium and magnesium ions, that limit the formation of lather with soap).

Fig. 144a and 144b: Thick mat of algal bloom

12. To control filamentous algae from the pond some of the farmers also use different algaecides of chemical origin@ the quantity specified by the aquaculture consultants. However, to achieve a good result freshly prepared solution of copper sulfate (500gm of $CuSO_4$ in one liter of water) @ 300-400 ml. per 33 decimal of pond water area may be broadcasted.

13. Faulty managerial practice of a fish hatchery also pose serious problem to the farmers, who are raising fish in ponds and tanks. It is noted sometimes that

the seed (Fingerlings) procured from the source (Hatchery) are of the same size and weight ranges (Genetically dwarf). When released, these genetically stunted fishes do not grow equally with who do not, even after maintaining the similar environmental conditions, food and feeding etc. Dwarfism, of the stock creates a severe crisis to the concerned farmer. The actual reason may be attributed to the "*inbreeding of fishes*". It is assumed that F2 and F3 generation from the same parental stock may attaind warfism. The farmers (the fish growers) should always carefully check the up to date" ACCRIDITATION CERTIFICATE" issued to the concerned hatchery by the Department of Fisheries of the state.

14. Fluctuation in weather condition (generally during end of winter and beginning of spring) coupled with foggy morning hours, sometimes there is a heavy toll of fish fry and fingerlings including advanced fingerlings, till the weather is stable. Continuous aeration (Figs. 43 and 44) and bathing of advanced fingerlings with Potassium permanganate for primary disinfection before releasing to stocking ponds during this season is advocated.

15. Another problem in the fish ponds where jute retting, is observed. During retting of Jute, excessive liberation of plant fluid (organic substances) as secreted undergo oxidation followed by putrification, turn the water color blackish and it may so happen, the BOD of water increases followed by massive destruction of planktonic milieu. This natural suffering of fishes is unavoidable. It is justified to apply Zeolite powder@15-20kg/acre/meter at least once in a fort night; so long the retting procedures are in vogue.

16. Mass scale mortality of fishes especially who are engaged in semi-intensive culture of Indian Major and Minor Carps during the onset of monsoon every year.What happens, during summer months (April to last week of June) there remains scorching air and water temperature. The rate of feeding (supplementary feed) is also at a higher rate by the fishes followed by the concurrent higher decomposition of the organic matters present at the soil-water interphase. This decomposition stimulates production of ammonia and other toxic gases which generally tend to remain little above the level of the substratum. Fact remain, immediately after the first or second day of the fall after a long cloudy day, these toxic gases starts coming at the upper level along with the stocked fishes resulting massive suffocation followed by death.It is evident that the ponds with more than 10 feet depth due to oxygen tension die and the ponds which are within 6-7 feet depth they are surfaced through out the day egulfing oxygen from the air. Professional farmer's starts aeration

by operating pumps and paddle wheels but a little success is observed. To avoid such conditions the farmers must take precautions from the summer months. Application of **good quality of zeolite** @ 15-20 kgs/acre/meter depth must have to be applied in the ponds in every month till the monsoon starts during July.Such practice can only save the existing stock of fishes.

To prevent fish disease what a farmer is to do

During culture operations, all the animals do suffer from variety of problems-fishes are no exception. In aquaculture- the most causative agent is water-where the fishes do live. Healthy eco-condition of water can only save the fishes from various contaminations. The most important disease of fishes is caused by the infestations of various pathogenic virus, bacteria, fungus and parasites of various kinds. Environmental degradation or even the nutritional deficiencies also create enormous problem to the fishes. What so ever the cause of the problems, the ultimate result thus appears financial loss to the fish farmer. It is always better to follow the age old verse: *'prevention is better than cure'.*Preventive measures are always necessary rather than to take curative measures. What a farmer is to do to prevent fish diseases are described below:

- It is always necessary to examine the pollution status of the pond water and soil before starting the aquaculture enterprise.
- The pond water should be light enough, devoid of either bloom or swarm of zooplankton.
- The procured fry or fingerlings should be disease free.
- The density of the advanced fingerlings stocked in the stocking pond is to be kept minimum with appropriate species ratio so as to provide them sufficient food and space.
- Raking of the bottom mud is compulsory for easy mineralization and the procedure should be done once in two months.
- The physico-chemical and the biological properties of water and soil are to be kept always at an ideal range.
- To avoid eutrophication of water, organic matter load at the soil-water interphase to be monitored at a regular interval.
- Intrusion of allochthonous wastes from the catchment area is to be stopped during culture period.

- It may be necessary to change the pond water if found extremely polluted.

- Application of lime@30-40kg/acre area is a must at least once in a month together with 15-20kg of Zeolite powder.

- Supplementary feed is to be provided everyday as per the appropriate quantity. However, the same is to be stopped during rainy days or when the weather is cloudy for consecutively 2-3days.

- The disease infected fishes are naturally in distress and suffer from weakness. In such situation netting in the ponds is strictly restricted to avoid mass mortality.

- Application of good quality zeolite powder, Multimicronutrient mineral fertilizer, Chelated copper, Iodine 20% and the Soil and Water Probiotics at the prescribed doses can only prevent the fishes from contaminated diseases.

- Last but not the least, *every farmer should visit his farm site at least once in a day.*

8. THE BASICS OF SUSTAINABLE AQUACULTURE

This has already been illustrated that the aquaculture is still a maturing science so also as a business at least in India. Aquaculture development in one area can change the environment and approaches to its better management. At the present moment aquaculture promises greater growth and diversification because of the advancement of new technologies which safe guards the now vulnerable or critically endangered aquatic species of commercial importance. The uncontrolled aqua cultural development generally cause serious environmental degradations as well as have unanticipated economic, social and cultural impacts on surrounding communities.Lot of evidences are there which indicate poorly planned, and unmanaged aquaculture initiatives those achieve short- term financial achievements but are not sustainable for a long period of time. Herein lies the role of the existing Government agencies or the appropriate authorities to fill in the emerging gap of the cheapest source of animal protein in the form of fishes of various kinds, their systems and environments for sustainable production over a long time to come. The urgency to develop workable, cost-effective, widely applicable to sustainability at least in Indian sub-continent where a major forthcoming aquaculture expansion is likely to be expected.

Aquaculture is an activity producing fish or shellfish mainly for human consumption. It is carried out in ponds, enclosures or in open water bodies and thus involves continuous interaction with the environment. Aquaculture can be a sustainable activity, if it is carried out in socially and environmentally responsible manner, by adopting good aquaculture practices.

Sustainable aquaculture means an aquaculture production system that operates in harmony with the environment and living systems, utilizing renewable resources as far as possible, providing living conditions to the animals as close to that of their natural habitats as well as in tune with the human and social environment of the locality.

Sustainability can be achieved by adopting Better Management Practices (BetMPs). BetMPs involve legal compliance, social responsibility, good site selection and farm construction, good practices in farm management right from pond preparations to harvest and post harvest management activities. Adoption of BetMPs would result in better production, productivity and returns on the one hand and environmental and social responsibilities on the other.

Crop planning in advance of the cropping season in consultation with all farmers involves stocking density based on pond carrying capacity, crop insurance and bank finance and bulk input purchase like good quality seed through contract hatchery system, lime, seed, feed and other commonly used chemicals for entering in to marketing contract with processors, setting up of basic water quality laboratory with easy to test water quality.

a. **Crop calendar:** Preparation of Crop calendar- two months before stocking fix dates for pond preparation, biosecurity, water pumping, contract hatchery, seed stocking and harvest.For analysis of farm inputs like seed, feed and any common ingredient used during culture.

b. **Increase the water holding capacity of the ponds:**To reduce disease risks and to reduce the excessive reliance on water exchange, pond water holding capacity has to be increased maintaining water depth at minimum water level of 1.2m and 2.5m at the centre for both prawn, shrimp and fishes respectively.

c. **Completely drain out the water from the ponds:** Complete draining of water helps in removing the disease carrying fish and crustaceans from previous crops in the pond. If it is not possible to drain the water follow wet preparation method.

d. **Remove the organic wastes from the pond bottom:** Organic matter releases toxic gases like ammonia and hydrogen sulfide in the pond leading to stress or death of the cultured animals. Organic waste is in the form of fine thin black layer on the soil found in feeding area, corners, trenches and in the centre of a pond with aerators and should be checked for the presence of black layer when it is in wet condition. Make sure that the displaced organic waste does not enter the pond again through rain water. The black soil if can not be removed completely from the pond it should be ploughed repeatedly during the summer season.

e. **Dry and plough the pond bottom:** Sun drying kills weed fish/crustacean and their eggs in the pond soil. Drying helps in oxidizing the organic matter thus reducing the sludge. The pond should dry under hot sun for 20 to 30 days till the soil cracks. Plough the pond 2 to 3 times with the gap of 2 to 3 days which will help in oxidizing the organic matter and also help in reducing gastropods. After ploughing compact the pond bottom to reduce the turbidity and seepage. When the ponds cannot be dried farmers can follow wet pond preparation as given below:

- Apply mohua oil cake or chlorine at a dose mentioned earlier before, take out all dead fishes and other animals if any,

- Use the tractor with gauge wheels to plough the pond with 15 to 20 cm water,

- After the ploughing drain out the water from the pond if possible.

b. **Fertilization and liming of the pond bottom:**

- This will help in improving mineral content of the pond bottom especially in ponds with low soil fertility and ponds which are in culture for more than 10 years.

- Apply dry vermicompost 250-1000kg/ha or compost manure

- Spread the compost manure all along the pond bottom.

- Do not use raw cow dung or broiler poultry manure for fertilizing the pond bottom. Application of these may turn the pond soil toxic to fishes.

Application of Lime

Use a soil pH meter to test the soil pH. Soil should be wet while using the equipment. The levels of lime application during pond preparation depend on the pH of the soil. Spread the lime all along the pond bottom and along the slopes of the pond bund. A large proportion of the lime needs to be applied along the feeding areas and on the wet portions of the pond. When applying lime farmers/ workers should wear face mask.

If the soil pH is more than 7, there is no need of Lime application.

Water screening: It is very important in keeping the disease carriers away. If good water screening is followed, no need for further disinfectants. Use double layer of fine mesh filter net (60#) at water inlet point. The filter nets to the delivery pipe and it should be supported properly. Provide additional two layers of 80 mesh hapa below the inlet. Pond should be filled with water within 4 days. Foot valve should be placed in bamboo basket or metal cage and it should be covered with 20# mesh to prevent large animals getting in to inlet mesh.

Water Fertilization

Phytoplankton abundance is essential to successful culture. It shades the pond bottom and prevents the growth of benthic algae; provide the darker environment which the fishes and prawns/shrimps find less stressful. One week after the water is filled, normally development of phytoplankton starts especially where manure is applied.

If the color of the pond water is clear i.e the transparency is higher (exceeds 50 cm or higher), chain dragging once a week helps in stabilizing the planktonic abundance. Add 200kg of Dolomite per ha during sunny period. Apply 2 days fermented mixture of rice bran, jaggery and quality brewers yeast @ of 25kg+10kg+0.25 kg/ ha in doses for three days during the morning period. Spread the fermented mixture across the pond using floating device.When the color of the water is green the pond is ready for stocking. If there are benthic or floating algae in the pond, remove them manually.

i) **Stabilization of the plankton bloom:** Do not heavily fertilize the water to get dark green water. It will reduce the oxygen in water during night time thus suffocating the cultured animals. It is one of the reasons for lower survival rate of fish and prawn seeds. During the first month of culture whenever the water color intensity reduces (Secchi disc reading of more than 50 cm), add fermented mixture of rice bran, jaggery and yeast @ of 25kg+10kg+0.25 kg/ha. Chain dragging from the very first day of seed stocking is not recommended. In case of fish nursery ponds when the spawn attains early fry stage at the end of 7 or 10 days may be harvested for selling to other fish farmers using dense meshed net with a gentle operation. In case of shrimps after 15 DOC follow chain dragging of entire pond phased over 3 to 4 days at least once a week. Wherever possible run the aerators two hours every day at least at late night. Instead of using crumbled, sinking pelletized feed, preference to be given for the floating feed in case of fish culture.

ii) **Biosecurity practices:** Biosecurity in fish and shrimp farming involves stocking disease free seed, pond preparation, and water screening and prevention of entry of disease carriers, personal hygiene and sanitation. General biosecurity precautions that need to be established for each pond to help disease prevention and disease control are given below:

Physical barriers to prevent crabs, birds and other animals: Ensure all farmers in the society implement these biosecurity measures while pond preparation so that it reduces the risk of disease. The objective is to create a barrier which prevents entry of animals, predators and carriers of disease creating pathogens.

iii. **Farm Sanitation and hygiene:**Area surrounding the farm should be kept clean, should have a toilet in good sanitary condition, located 20 to 30 m away from farm area must avoid contamination of domestic sewage , avoid use of animal manure. Each pond should have a separate water sampling container. Farmers/workers should avoid getting in to ponds unless it is necessary.

iv. **Seed selection and procurement:** Although majority of the farmers are in belief that the fish seeds procured from the riverine resources are more susceptible to diseases, attain a good growth and have a high rate of survival but fact remains due to continuos increment of pollution most of the spawning grounds are not in vogue. It is therefore, the farmers are to depend on the fish nurseries. Before fish seed purchase it is mandatory that the farmers (the fish growers) should always carefully check the up to date **"ACCRIDITATION CERTIFICATE"** issued to the concerned hatchery by the Department of Fisheries of the state in order to avoid the problems related to growth of fishes during culture operations.

In case of selection and stocking of good quality and appropriate quantity of fish/frawn seed (PL) in to the pond it is necessary to avoid wild seed and seed from poorly managed commercial nurseries. All the farmers in the area should stock the seed at the same time as per the crop calendar (within a period of one to two weeks) under contract hatchery system. From sources of selling, certified specific pathogen-free (SPF) stock or disease free seed from registered hatcheries should be purchased and stocked in the nurseries.PCR tested healthy seeds and disease free tested brood stock should be procured from approved hatcheries complying norms. Seeds of the same batch form a hatchery are to to be packed with enough oxygen for optimum quantity based on the distance and time required and temperature maintenance.

v. **Seed stocking:** Seed should be released in to pond after proper acclimatization during cool hours of the day, i.e., after 8 PM or before 8 AM. Make sure the plankton is good and stable (green color water). Avoid stocking if pond has transparent water or dark green water.

vi. **Feed Management Practices:** Feed management is one of the most important aspects of successful aquaculture as the feed accounts for 50 to 60% of the operating cost. Feed should be fresh and of good quality and not more than 90 days old form date of manufacture. Ascertain size and quantity of feed to be applied following the chart of the manufacturer. Reduce feeding during periods of low DO, plankton crash, rain fall, molting and extremes of temperature and during disease outbreaks.In case of prawn/shrimp culture, installation of feed trays after 10 days of stocking and monitor the feeding from 20 days. Do regular sampling of prawns once a week after 45 days to determine growth rate and to calculate FCR. Slightly under feeding is better than over feeding. Never mix any antibiotics with the feed. It is preferable to switch off the aerators just before feeding until 2 hrs after feeding, based on stocking density. In case of fish feed, the farmers who prefer to use farm made feed generally depends on De-oiled rice bran, groundnut oil cake, Mustered oil cakes, Soya flakes, Maize powder, Wheat flour, Fish oil in addition to salt, vitamins and minerals.The ingredients generally procured after their harvesting from field following respective processing. Immediately these are transported to the fish farm are being stored in a clean, cool and ventilated area, well protected from sunlight keeping bags stacked neatly on pallets (no more than 10 bags per stack) 30 cm away from walls to prevent feed from being in direct contact with damp floor. Care is taken to that of the store and outside premises and kept clean and use traps to prevent rodents. Do not keep any fuel or liquid items in the feed store. It is important to protect feed bags from sunlight and rain, by storing them off the ground in simple, pond side storage sheds. During the rain proper care should be taken to prevent feed bags getting moist.The empty feed bags are kept properly and recycled them by selling to traders.

vi. **Water quality management practices:** To reduce risk of disease through contamination with the water from outside the farm, follow the minimal water exchange system. If water quality and the pond bottom are good, fishes and the prawns are healthy and growing well there is no need to exchange water. In case of prawns, frequent exchange or in-take water is not recommended. Starting third month, if necessary, water exchange can be done but try to minimize as low as possible. Do not release or in-take more than 10-20 cm (8 to 15%) of water per day. It is recommended to use water for exchange from a reservoir wherever possible. If there is no reservoir, take the water only when there is no disease contamination in water source. Do not exchange water when there is drainage from nearby disease affected pond. Wait for couple of days. Always follow good water screening as mentioned earlier. In early stages of culture (4 to 6 weeks) if the color of the pond water is clear, add fermented mixture of jaggery, rice bran and yeast @ of 25 kg+10kg+025 kg/ha to get bloom. If the water color is too dark, do not use any chemicals to kill the algae; instead change 10cm of top water, if you can during afternoon and intake during the night preferably at high tide. If you cannot, reduce/stop feeding during this period. To control water pH within the optimum range of 7.5-8.3, and limit diurnal pH fluctuation to less than 0.5. If the pH is lower than 7.5 apply shell lime to increase the pH. High pH results from over liming and excess plankton bloom. If the pH is higher than 8.3, apply 50 litres of molasses or fermented juice of rice powder, jaggery and yeast to reduce the pH.

After every water intake / exchange and after rains, use agricultural lime. Agri lime (100 Kg/Ha) should be mixed with water and applied throughout the pond. It acts as a buffering agent for water.

Table 64: List of antibiotics permitted and banned for application in aquaculture (Mohanty *et al.*, 2018).

Permitted antibiotics	Banned antibiotics
Streptomycin	Chloramphenicol
Florfenicol	Nitrofurans including: Furaltadone, Furazolidon, Furylfuramide, Nifuratel, Nifuroxim, Nifurpraine, Nitrofurazone.
Amoxicillin	Neomycin
Enrofloxacin	Nalodixic acid
Erythromycin	Sulphamethoxazole
Furazolidone	Aristolochia spp.and the preparation there of.
Nitrofurantoin	Chloroform
Oxolinic acid	Chlorpromazine
Sulphadiazine	Colchicine
Tetracycline	Dapsone
Oxytetracycline(OTC)	Dimetridazole
	Metronidazole
	Ronidazole
	Ipronidazole
	Various other nitromidazoles
	Clenbuterol
	Diethyestilbestrol(DES)
	Sulfonamide drugs(Except other approved sulfadimethoxine, Sulfabromomethazine and Sulfaethoxypyridazine)
	Fluroquinolones
	Glycopeptides

DIFFERENT UNITS AND MEASUREMENTS:

A. Area:

- 1 decimal = 435.6 sq.ft = 40.48 sq.meter

DIFFERENT UNITS AND MEASUREMENTS:

- 1 bigha = 0.33acre = 33decimal
- 1 Acre = 100decimal = 0.405 Hectare
- 1 Hectare = 2.47 acre = 10,000 meter
- 1 Hactare meter = 10, 000m^{3}water i.e., 10,000,000 liters.
- 1 kilometer = 0.62 mile
- 1 meter = 3.28 foot = 1.094 gauge
- 1 cubic meter = 1.30 cubic gauge
- 1 cubic foot = 0.028 sq.meter
- 1 sq.meter = 10.764 sq.foot

B. Measurementof mass

- 1 metric tone = 1000 kg
- 1 kg = 2.02 pound

C. Measurement of liquid

- 1 gallon water = 8.36 pound = 4.5 liters = 3800 c.c
- 1 Cubic foot of water = 7.5 gallon = 62.4 pound = 28354.3 gram

D. General measurements

- 1% salt = 10 gm/liter of water
- 1:10, 000 = 1 milli liter/40 liter of water

Unit of measure, Calculations, Basic formula

Area = L x B

Volume = L x B x H

Perimeter = 2 (L x B)

REFERENCES CITED

Abowei, J. F. N., 2010. Salinity, Dissolved Oxygen, pH and surface water temperature conditions in Nkoro River, Niger Delta, Nigeria, Advance journal of food science and technology, 2(1), pp 16-21.

Adamu K.M, R. B. Ikomi and F. O. Nwadukwe. 2014. The Design of Prototype Recirculating Aquaculture System and its Use to Examine the Histology of Hybrid Catfish Fed Practical Diets. International Journal of Fisheries and Aquatic Studies. 1(5): 242-249.

Alcover, G.2007.Adaptation of integrated fish–duck–pig farming system in Leyte D. Bagarinao (Ed.), Research Output of the Fisheries Sector Program, vol. 2, Bureau of Agricultural Research, Department of Agriculture. pp. 125-128.

Alferez, V.N., 1977. Engineering aspects and problems in the design and construction of fish pens and fish cages in Laguna Lake, Philippines. In Proceedings from the Joint SCSP/SEAFDEC Regional Workshop on aquaculture engineering. Vol.2. Technical report. Manila, South China Sea Fisheries Development and Coordinating Programme, SCS/GEN/77/15:373–88.

Alikunhi, K.H. 1952. On the food of young carp fry. J. Zool. Soc. India. 4: 77–84.

Allen M.B. and Arnon D.I.1955, Studies on Nitrogen Fixing blue green algae II, The sodium requirement of Anabena, Cylindrica, Physiol Plant, 8, 653-660.

Ayoola, A.A. 2010. Replacement of Fish Meal with Alternative Protein Source in Aquaculture Diets. M.Sc. Thesis, North California State University, North Carolina, USA.

Arboleda C R. 1977. Integrated live stock-fish production system in the Phillipines, FAO, and United Nations.

Anrew A.E. 2001. Fish Processing Technology. University of Ilorin Press, Nigeria. pp. 7-8.

Anderson, E.R., 1952. Energy budget studies. Water loss investigations: Vol. 1, Lake Hefner studies. Tech. Rep., *U.S. Geol. Surv.* Circ., 229: 71 – 119.

Avnimelech Y. 1999. Carbon D nitrogen ratio as a control element in aquaculture systems. Aquaculture 176: 227–235.

Avnimelech Y. 2009. Biofloc Technology — A Practical Guide Book. Baton Rouge, LA: The World Aquaculture Society, p. 18.

Banerjee, B.K.1979. Experiments on fry rearing in floating nurseries (floating cages) in Getalsud reservoir, Ranchi (Bihar). Pages 1-6 In A.V. Natarajan, editor. Proceedings of the Summer Institute on Capture and Culture Fisheries of the Manmade Lakes in India, CIFRI, 7 July-6 August.

Banerjea, S.M. 1967. Water quality and soil condition of fish ponds in some states of India in relation to fish production. *Indian. J. Fish.* 14 (1 and 2): 115 – 144.

Banerjea, S.M. 1995. Role of fish pond soil and water in relation to productivity. PP. 6-10. In: Nutrient *Management in Aquaculture*, G.N. Chattophadhyay (Ed.).

Banerjee, A., G.N. Chattopadhyay and C.E. Boyd. 2009. Determination of critical limits of soil nutrients for use in optimizing fertilizer rates for fish ponds in red, lateritic soil zones. Aqua. Eng. 40 (3):144–148.

Banerjee.R.K and Babulal. 1990. Role of soil and water in fish farming with special reference to Primay production. In.Technologies for Inland Fisheries Development. Edt.V.V. Sugunan and U. Bhowmik.CICFRI, (ICAR), Barrackpore, West Bengal, India.123-129.

Basabaraju, Y and Reddy, A. N.2013. Growth performance of Amur strain of common carp in southern Karnataka. Mysore Journal of Agricultural Sciences . 47 No.(1): 119-123.

Basudha Ch. and W. Vishwanath 1993. Nutritive value and growth responses of formulated aquatic fern Azolla based diets on advanced fry of en-demic medium carp *Osteobrama belangeri* (Val.). J. Freshwater Biol., 5(2) : 159-164.

Basudha.C, Vishwanath W. 1999. Food and feeding habits of an endemic carp, *Osteobrama belangeri* (Val.) in Manipur. Indian J Fish. 46:71–77.

Beveridge, M.1984.Tilapia hatcheries - lake or land based? ICLARM Newsl.7 (1):10–11.

Bhatnagar, A., Jana, S.N., Garg, S.K. Patra, B.C., Singh, G. and Barman, U.K.2004. Water quality management in aquaculture, In: Course Manual of summerschool on development of sustainable aquaculture technology in fresh and saline waters, CCS Haryana Agricultural, Hisar (India), pp 203- 210.

Bhatnagar, A. and Singh, G. 2010. Culture fisheries in village ponds: a multilocation study in Haryana, India. Agriculture and Biology Journal of North America, 1(5), pp 961-968.

Bhatnagar, A., and Devi, P., 2013, Water quality guidelines for the management of pond fish culture. *International Journal of Environmental Sciences*, vol. 3(6), pp. 1980-2009.

Bhowmik, R.M. 1990.Hypophysation of Indian and Exotic Carps. In: Technologies For Inland Fisheries Development.Edt.V.V.Sugunan&U.Bhaumik.Central Inland Capture Fisheries Research Institute.Barrackpore, India.Pp.15-24.

Bhuyan.D, 2014. Paddy-cum-FishFarming- A Sustainable way of Agriculture, practiced in Jorhat district of Assam with special reference to Hatagarh village. J. Aquatic Biol. and Fishs Vol. 2126.

Bijo, P.A.2007.Feasibility study of a RecirculationAquaculture System.Malaysian Fisheries Development Authority. Kuala lampur, Malaysia.

Blancheton, J.P. 2002. *Developments in Recirculation System for Mediterranean Species.* Science Direct, Aquaculture Engineering Volume 22.

Bondad MG, Subasinghe RP, Arthur JR, 2005. Disease and health management in Asian aquaculture. Veterinary Parasitology.132 (3-4):249–272.

Borba, M.2018.Preventive measures in Aquaculture:how to better protect our crops.Aquacuture Brasil Magazine.march/April issue.

Bossier, P and J.Eksari. 2017. Biofloc technology application in aquaculture to support sustainable development goals. Microb. Biotechnol. 10 (5):1012-1016.

Boyd, C. E.1979. *Biology of Fishes*. Philadelphia: W.B. Saunders Co.

Boyd, C. E., 1979. Water Quality in Warm water Fish Ponds, Agriculture Experiment Station, Auburn, Alabama, pp 359.

Boyd, C.E. 1982. Liming fish ponds. Journal of Soil and Water Conservation, 37:86-88.

Boyd, C.E. 1995.Bottom soils, sediment, and pond aquaculture. New York: Chapman and Hall, 348p.

Boyd C.E. and Tucker CS. 1998 Pond Aquaculture Water Quality Management. Norwell, Mass, USA: Kluwer.OJECT RAF/82/009.

Boyd, C.E., C.W. Wood and T. Thunjai, 2002. Pond soil characteristics and dynamics of soil organic matters andnutrients. 19th An. Ann. Tech. Rept. Pond dynamics/Aquaculture CRSP, Oregonstate. Edt. By McEmwco*et al.,*

Boyd, C.E. and J.R. Bowman, 1997. Pond bottom soils. In: H.S. Egna and C.E. Boyd (Editors), Dynamics of Pond Aquaculture. CRC Press, Boca Raton/ New York, pp. 135-162.

Bregnballe, J 2015. A Guide to Recirculation Aquaculture FAO, United Nations.

Brett, J. R. (1958). Implications and assessments of environmental stress. In The Investigation of Fish Power Problems (ed. P. A. Larkin), pp. 69-83. University of British Columbia Press, Vancouver.

Brylinsky, M. 1980. Estimating the productivity of lakes and reservoirs. In E.D. Le Creu and R.H. Lowe-McConnell. Eds. The functioning of Freshwater Ecosystems. Cambridge Univ. Press. PP. 411 – 453.

Bunting SW, Kundu .N, Mukherjee. M.2002. Situation Analysis: Production Systems and Natural Resource Management in PU Kolkata. Institute of Aquaculture, University of Stirling, UK.

Bunting, S. W, Pretty and Edwards P.2010.Wastewater-fed aquaculture in the East Kolkata Wetlands, India: anachronism or archetype for resilient ecocultures? Reviews in Aquaculture 2, 138-153.

Cahu, C., Salen, P. and Lorgeril, M.D. 2004. Review article. Farmed and wild fish in prevention of cardio vascular diseases: assessing possible differences in lipid nutritional values. Nutr. MeTable Cardiovasc. Dis. 14: 34-41.

Castel.J.D.*et al.,* (1986).Aquaculture nutrition. In Realism in Aquaculture: Achievements, Constrains.Perspectives, (Ed.by.M.Bilio, H.Rosenthal and C.J.Sindermann).pp.251-308.European Aquaculture Society, Bredene.Belgium.

Chakma.S, Md.M.Rahman, M.Akter.2015.Composition and Abundance of Benthic Macro-invertebrates in Freshwater Earthen Ponds of Noakhali District, Bangladesh. American journal of Bioscience & bioengineering. 3(5): 50-56.

Charudattan, R.2001. Are we on top of Aquatic weeds? Weed problems, control options and challenges. Talk presented at International symposium on the world's worst weeds.British Crop Protection Council, Brighton, United Kingdom.

Chávez-Crooker, P and J. Obreque-Contreras. 2010. Bioremediation of aquaculture wastes.Curr. Opin. Biotechnol, 21, pp. 313-31.

Chopin.T, A.H. Buschmann, C. Halling, M. Troell, N. Kautsky, A. Neori, G.P.Kraemer, J.A. Zertuche-Gonzalez, C. Yarish, C. Neefus. 2001. Integrating seaweeds into marine aquaculture systems: a key toward sustainability, J. Phycol., 37, pp. 975-986.

Cho C.Young. Nutrition and Fish Health.http://www. glfc.org/pubs/SpecialPub sp832/ pdf/chap8.pdf

Coche. A.G.1967. Fish culture in rice fields: a worldwide synthesis. Hydrobiologia, 30(1): 1–44.

Covich, A. P., Margaret, A. P. and Todd, A. C. 1999. The Role of Benthic Invertebrate Species in Freshwater Ecosystems: Zoobenthic species influence energy flows and nutrient cycling. Bio Sci., 49(2):119-127.

Cressey D. 2009. Aquaculture: future fish. *Nature*. 458(7237):398–400.

Clayton, J.2000.Weed species, station design and operational features contributing to management problems in NewZealand hydroelectric lakes.Abstracts of the Third International Weed Science Congress.3:216.

Cole, G.A. 1983. *Text book of Limnology* III EDn. The C.V. Mosley Company. Torento. 401 PP.

Clerk, R.B. 1986, Marine Pollution. Clarandon Press, Oxford, pp 256.

Cruz.P.M, A.L. Ibáñez, Oscar.AM.Hermosillo, Hugo C.and R. Saad, 2012.Use of probiotics in aquaculture.ISRNMicrobiol .916845. Published online 2012, Oct 16. doi:10.5402/2012/916845

Da Silva, J.B, M. H. T. Mascarehans and J F R Lara, 2000.Large scale cattail (*Typha subulata*) control with imazapyr in restored irrigation drains of Janauba. Abstracts of the Third International Weed Science Congress.3:323.

Das, A.K, D. K. Meena and A. P. Sharma. 2014. Cage Farming in an Indian Reservoir. World Aqua culture. September, 256-59.

Das. Subrata. 2017. Polyculture in fresh water ponds of Risra, Hoogly (Pers.communication).

Das SM, Moitra SK. 1955.On the correlation between fish food and fish gut in food fishes of Uttar Pradesh, India. Proc. 42nd Ind. Sci. 4:10.

Das SM, Moitra SK. 1956.Studies on the food of some common fishes of Uttar Pradesh, India. Part II. Proc. Nat Acad Sci India, 26B (4):213-223.

David A. 1963. Fishery biology of the Schilbeid cat-fish, Pangasius pangasius (Hamilton) and its utility and propagation in culture ponds. Indian Journal of Fisheries 10: 521-600.

Deka P. 2015. A comparative study of the seasonal trend of Biological Oxygen Demand, Chemical Oxygen Demand and Dissolved Organic Matter in two fresh water aquaculture ponds of Assam International Journal of Fisheries and Aquatic Studies 2015; 3(1): 266-268.

Dela Cruz. 1980. Capture and culture fisheries in Chinese lakes. ICLARM Newsl. 3(4):8–9.

Devi, G.A, G.S. Devi, O.B. Singh, S. Munil Kumar and A.K. Reddy. 2009. Induced Spawning and Hatching of *Osteobrama belangeri* (Valenciennes) Using Ovatide, an Ovulating Agent. Asian Fisheries Science 2:1107-1115.

Dela Cruz, C.R., 1982. Fishpen and cage culture development project in Laguna de Bay. Work plan implementation (Working Paper). Manila, Philippines, South China Sea Fisheries Development and Coordinating Programme, SCS/82/WP/102:27 p.

Dey.S, K. Misra and S. Hom Choudhury. 2017.Reviewing nutritional quality of small Fresh water Fish species. American Journ. Of food and Nutrition. 5(1):19-27.

Durborow, R. M., Crosby, D.M. and Brunson, M. W. 1997. Ammonia in fish ponds. SRAC Publication no. 463.

Duning.R, D. Losordo, M. Thomasand H.O. AlexO. 1998. *The Economics of Recirculating Tanks Systems, A Spread sheet for Individual Analysis*, Southern Regional Aquaculture Center, SRAC PublicationNo. 456.

Edwards, P. 1999. Aquaculture and Poverty: Past, Present and Future Prospects of Impact. Asian Institute of Technology. Bangkok.

Edwards P. 2000. Wastewater-fed aquaculture: state-of-the-art. In: Jana BB, Banerjee RD, Guterstam B, Heeb J (eds). Waste Recycling and Resource Management in the Developing World, Ecological Engineering Approach, pp. 37–49. University of Kalyani, India and International Ecological Engineering Society, Switzerland.

Edwards P. 2008a. An increasingly secure future for wastewater-fed aquaculture in Kolkata, India? Aquaculture Asia 13 (4): 3–9.

Ekubo, A.A. and Abowei, J.F.N. 2011. Review of some water quality management principles in culture fisheries, Research Journal of Applied Sciences, Engineering and Technology, 3(2), pp 13.

Ekasari, J.2014. Biofloc technology as an integral approach to enhance production and ecological performance of aquaculture. Dissertation. Ghent University 42-1357.

Ekasari, J., Rivandi, D. R., Firdausi, A. P., Surawidjaja, E. H., Zairin, M., Bossier, P. 2015.Biofloc technology positively affects Nile tilapia (*Oreochromis niloticus*) larvae performance. *Aquaculture* 441: 72–77.

ElagbaM, Rabie Al-Maqbaly and H. Mohamed Mansour, 2010. Proximate composition, amino acid and mineral contents of five commercial Nile fishes in Sudan. African Journal of Food Science Vol. 4(10), pp. 650-654.

Elton, Charles and R.S. Miller, 1954. The ecological survey of animal communities with a practical system of classifying habitats by structural characteristics. *J. Ecol.* 42 (2): P. 460 – 496.

Emerenciano, M., Cuzon, G., Paredes, A., and Gaxiola, G.2013. Evaluation of biofloc technology in pink shrimp *Farfantepenaeusduorarum* culture: growth performance, water quality, microorganisms profile and proximate analysis of biofloc. *Aquacult Int* 21: 1381–1394.

FAO Corporate depository. 2017. Review of recycling of animal wastes as a source of nutrients for freshwater aquaculture.

FAO Corporate depository, 2017. Lecture notes on composite fish culture and its extension in India.

FAO. 2014. The State of World Fisheries and Aquaculture: Opportunities and Challenges. FAO Rome, Italy.

FAO. 2008. Human dimensions of the ecosystem to fisheries: an overview of context, concepts, tools and methods.FAO Fisheries Technical Paper. No.4 89, FAO, Rome.152 pp.

FAO. 2013. The State of Food and Agriculture.Food systems for better nutrition. Pomegranate. 114p. Available at: http: //www .fao.org /docrep/ 018/ i3300e/ i3300e.pdf. Accessed on Aug 10.

FAO, 2012. The state of world fisheries and aquaculture.FAO Rome, Italy.

FAO .2015 .The State of World Fisheries and Aquaculture. Rome. 223 p. 2014. Available in: http://www.fao.org/3/a-i3720e.pdf > Accessed: Jul. 23.

Fast, A.W., 1986. Pond production systems: Water quality management practices. In: J.E. Lannan, R.O. Smitherman, and G. Tchobanoglous (Editors), Principles and Practices of Pond Aquaculture. Oregon State University Press, Corvallis, Oregon, pp. 141-168.

Forbes, S.A.1887. The lake as a microcosm. Bull. Sci. A. Poeria. Reprinted in II. *Nat. Hist. Bull.* 15: 537 – 550.

Forbes, S.A. and R.E. Richardson, 1913. Studies on Biology of the Upper Illinois River. *Bull. III. State Lab. Nat. Hist.*, 9 : 481 – 574.

Forel, F.A. 1892 – 1904. Le Leman. Monographic limnologique. Lausanue, 3 Vols. Cited in text from Ps Weleln's*Limnology.*

Forel, F.A. 1895. LeLeman: Monographic Limnologique. *Tome II. Mecanique, Hydrauliqua, Thermique, Optique, Acoustique, Chemic, Lausanne, F. Rouge.*

Forel, F.A. 1904. Le Leman. Monographic Limnologique. *Tome III. Biologie, Historic, Navigation, Peche, Lausanne, F. Rogue.*

Francis-Floyd, R. 2005. Introduction to Fish Health Management. CIR921. EDIS. http//edis.ifas.ufl.edu

FishHu, Z., Lee, J.W., Chandran, K., Kim, S., Sharma, K., and Khanal, S.K. 2014. Influence of carbohydrate addition on nitrogen transformations and greenhouse gas emissions of intensive aquaculture system. *Sci Total Environ* 470: 193–200.Tech. Pap (14):204 p.

Gaedke, U., S. Hochstadter and D. Straile, 2002. Interplay between energy limitation and nutritional deficiency: Empirical data and food web models. *Ecol. Monogr*, 72(2):251–270.

Ghosh, S.D.2017.All male *Oreochromis niloticus* in Sunderbans. (Person.comm.).

Ghosh, D. 1985a. Dhapa Report.From disposal ground to WAR (waste-as-resource) field. Submitted to Calcutta Municipal Corporation (Preliminary Draft). Government of West Bengal, India. 32p.

Ghosh, D. 1985b. Cleaner Rivers. The least cost approach. A village linked programme to recycle municipal sewage in fisheries and agriculture for food, employment and sanitation. Government of West Bengal, India. 32p.

Ghosh, D and S.Sen. 1987. Ecological history of Calcutta's wetland conversion. Environmental Conservation 14(3):219-226

Goldman, C.R. and A.J. Horne, 1983. Limnology. McGraw Hill, New York.

Goswami, C and V.S Zade.2015. Statistical Analysis of Fish Production in India. Int.J.ofInnov. Res. in Science, Engineering and Technology 4: 294-299.

Govind, B.V., S. Ayyappan, S.L. Raghavan and M.F. Rahman. 1988. Culture of catla (Catla catla) in floating net cages. Mysore Journal of Agricultural Science 22:517-522.

Gjedrem T, N. Robinson, M. Rye.2012.The importance of selective breeding in aquaculture to meet future demands for animal protein: a review.Aquaculture, 350–353, pp. 117-129pp.

Guerrero, R.D., 1983. Tilapia farming in the Philippines: practices, problems and prospects. Paper presented at PCARRD/ICLARM Workshop on Philippine Tilapia economics, Los Baños, Laguna, Philippines, August 10–13, 1983, 23 p. Abstr. in ICLARM Conf. Proc., (10):4.

Halver, J.E, 1979.Vitamin requirements of fin fish.In.Finfish nutrition and fish feed technology, Vol.I (Edt.by J.E.Halver and K.Tiews), pp.45-58.Schriften der Bundesforschungsanstalt fur Fisherei, Berlin.

Halver, J.E. 2002. The vitamins. In: Fish Nutrition. 3rd.Edn. (J.E. Halver and R.W. Hardy; Eds.). Academic Press.

Hargreaves, J. A.2006. Photosynthetic suspended growth systems in aquaculture. *Aquacult Eng*34: 344–363.

Hargreaves, J. A and C.S. Tucker, 2004. Managing Ammonia in Fish Ponds. SRAC Publication No. 4603.

Harpaz, S.2016. Enhancing profit through the use of Periphyton as partial replacement of commercial feeds for sustainable fish farming (culture). Inv. Lect. International Conference on Aquatic Resources and sustainable Management, Kolkata. West Bengal, India.

Hastings, W.H.1979. Fish nutrition and fish feed manufacture.In Advances in Aquaculture (Edt.by T.V.R. Pillay and W.A.Dill) pp.568-574.Fishing News Books.Oxford.

Hels O, Hassan N, Tetens H, Thilsted SH. 2002. Food consumption energy and nutrient intake and nutritional status in rural Bangladesh: Changes from 1981-82 to 1995- 96. European Journal of Clinical Nutrition 57: 586-594.

Helfrich. L.A and Libey. G, 1991. Fish farming in recirculating aquaculture systems (RAS). Published by Virginia Cooperative Extension. Department of Fisheries and Wildlife Sciences, Virginia Tech

Hewison, M. 2011. Vitamin D and innate and adaptive immunity. Vitam. Horm. 86: 23-62.

Hutchinson, G.E. 1957. A treatise on limnology, Vol-1. Geography; physics and chemistry. John Wiley and sons, New York. 1015 PP.

Hutchinson W, M. Jeffrey, D.O Sullivan, D. Casement, and S. Clark, 2004. *Recirculating Aquaculture System Minimum Standard for Design, Construction and Management.* South Australia Research and Development Institute.

Hickling, C.F. 1971. Fish Culture 2nd Edn. London: Faber and Faber 317 P.

Hickling, C.F., 1962. Fish culture. London. Faber and Faber. 295 p.

Hels O, Hassan N, Tetens H, Thilsted SH .2002. Food consumption energy and nutrient intake and nutritional status in rural Bangladesh: Changes from 1981-82 to 1995- 96. European Journal of Clinical Nutrition 57: 586-594.

Holm L.G., Plucknett D.L, Pancho J.V, J.P.Herberger.1977.The World's worst weed. Distribution and Biology.University of Hawaii Press: Honolulu.

Hora, S.L and Pillay T. V. R., 1962. Handbook of fish culture in the Indo-Pacific Region. FAO Fish.

Hu, Z., Lee, J.W., Chandran, K., Kim, S., Sharma, K., and Khanal, S.K. 2014. Influence of carbohydrate addition on nitrogen transformations and greenhouse gas emissions of intensive aquaculture system. *Sci Total Environ* 470: 193–200.Tech. Pap (14):204 p.

Hutchinson W, M.Jeffrey, D.O Sullivan, D.Casement, andS.Clark, 2004. *Recirculating Aquaculture System Minimum Standard for Design, Construction and Management.* South Australia Research and Development Institute.

Huong, D.T.T., W-J. Yang, A. Okuno, and M.N. Wilder, M.N. 2001. Changes in free amino acids in the hemolymph of giant freshwater prawn *Macrobrachium rosenbergii* exposed to varying salinities: Relationship to osmoregulatory ability. Comparative Biochemistry and Physiology Part A, 128: 317-326.

IDRC/Aquaculture Department SEAFDEC, 1979. International workshop on pen and cage culture of fish. 11–22 February 1979. Tigbauan, Iloilo, Philippines. Iloilo, Philippines, SEAFDEC, 164 p.

Idyll, C.P. 1978. The Sea against hunger. Thomas Y. Crowell Co., N.Y. 222 PP.

Ismail, A and Ikram, E.H.K. 2004. Effects of cooking practices (boiling and frying) on the protein and amino acids contents of four selected fishes. Nutr. Food Sci., Charlton.34 (2), : 54-59.

Jäger, R. Tand A. Kiwus, 1980. Aufzucht von Hechtsetzlingen in erleuchetennetzgehagen. Fisch.Teichwirt. 11:323–6.

Janes R.A, J.W. Eaton and K. Hardwick. 1996. The effects of floating mats of *Azolla filiculoides* Lam. and *Lemna minuta* on the growth of submerged macrophytes. *Hydrobiologia.* 340 (1/3):23-26.

Jana, B.B. 1998. Sewage-fed aquaculture: The Calcutta model. *Ecological Engineering,* 11: 73 – 85.

Jayachandran, K.V and N.I. Joseph. 1992. A key for the field identification of commercially important *Macrobrachium* spp. of India with a review of their bionomics. Fresh water prawns. Edt. by: E.G. Silus. Kerala Agricultural University 272p.

Jayalakshmy, B. and P. Natarajan, 1996. Influence of salinity on fertilization and hatching of *Macrobrachium idella* under laboratory condition. J. Aqua. Trop., 11: 33-38.

Jhingran, V.G. 1987. Introdsuction to aquaculture, UNDP, Nigerian Institute of Oceanography and marine research. PR7.

Jhingran, V.G. 2000. Fish and Fisheries of India. Hindusthan Publishing Corporation, New Delhi.

John Samuel, M., P. Soundarapandian and T. Kannupandi, 1997. In vitro embryo culture and effect of salinity on the embryonic development of the cultivable freshwater prawn *Macrobrachium malcolmsonii* (H. Milne Edwards). Curr. Sci., 73(3): 294-297.

Jhingran, V.G. 2000: Fish and Fisheries of India. Hindustan Publishing Corporation; New edition edition.

Jhingran, V.G. 1987. Introduction to aquaculture UNDP, Nigerian institute for Oceanography and marine research.PR7.

Ju, Z.Y., Forster, I., Conquest, L., Dominy, W., Kuo, W.C., and Horgen, F.D. 2008. Determination of microbial community structures of shrimp floc cultures by biomarkers and analysis of floc amino acid profiles. *Aquac Res* 39: 118–133.

Karim, M.R. and A.K.M. Harum-Al-Rashid Khan, 1982. Small-scale pen and cage culture for finfish in Bangladesh. Manila, Philippines, South China Sea Fisheries Development and Coordinating Programme, Manila, Philippines, SCS/GEN/82/34:157–60.

Kautsky N, Rönnbäck P, Tedengren M, Troell M. 2000 Ecosystem perspectives on management of disease in shrimp pond farming. *Aquaculture*.191 (1–3):145–161.

Kirchman, D. L.1994. The uptake of inorganic nutrients by heterotrophic bacteria. *MicrobEcol* 28: 255–271.

Kleiven E. 2013. Historical information on common carp (Cyprinuscarpio) in Norway. Fauna norvegica 33: 13-19.

Kohli, M.P.S., S. Ayyapan, S.N. Ogle, R.K. Langer, C. Prakash, K, Dube, A.K. Reddy, M.B. Patel and N. Saharan.2002. Observations on the performance of *Tor khudree* in floating cages in open waters. Applied Fisheries and Aquaculture 2(1):51-57.

Kumar Dilip and R. Sharma, 2012. Diversions in aquaculture Edt. A. Sinha, S. Datta and B.K. Mahapatra) 1-6p.Narendra Publ.

Kumar, J, A.S. K. Naik, V. Mahesh, and P. Nayana.1991.Stock manipulation and management. Aquafind. Aquatic fish database.

Kumar. J, A.K. Pandey, A.C. Dwivedi, A.S. Kumar Naik, V. Maheshand S. Benakappa. 2013. Ichthyofaunal diversity of District faizabad (U P), India, J. Exp. Zool. India.16 (1): 149-154.

Kohli, M.P.S., S. Ayyapan, S.N. Ogle, R.K. Langer, C. Prakash, K, Dube, A.K. Reddy, M.B. Patel and N. Saharan.2002. Observations on the performance of Tor khudree in floating cages in open waters. Applied Fisheries and Aquaculture 2(1):51-57.

Krebs, C. J. 1985. Ecology: The experimental Analysis of distribution and Abundance. Harper and Row Publishers, New York.

Kuhn D.D., Lawrence A.L., Boardman G.D., Patnaik S., Marsh L., and Flick G.J. 2010. Evaluation of two types of bioflocs derived from biological treatment of fish effluent as feed ingredients for Pacific white shrimp, *Litopenaeus vannamei*. Aquaculture 303: 28–33.

Kundu.N, M. Pal and S. Saha. 2008. East Kolkata Wetlands: A resource recovery system through productive activities.Proc.of Taal. Edt. by Sengupta, M & Dalwani R The 12th.world Lake Conference:861-881.

Kutty, M.N., Paviras, Sand P. Jayachandran., 1977. Comparative tolerance of freshwater fishes to selected pesticided. Seminar on "Environment impact on Development Activities. Proc. 16:20.

Lall, S.P. 1981. Minerals - a Review, In Biological aspects of aquaculture - nutrition. Proc. World Conf. Aquacult. Int. Trade Show. Sept. 20-23, Venice, Italy.

Leith, H. 1975. Historical survey of primary productivity research. In. H. Leith and R.H. Whittaker (eds.) primary productivity of the Biosphere Springer _ Verlag, Berlin, pp. 718.

Lucy Towers. 2014. Role of Vitamin C & Multivitamin Diets for Enhacement of Immunity, Growth & Biological Performance in Shrimps/Fish. (Mr Prakash Chandra Behera, Technical Manager (Aqua Division), PVS Group, India).Fish site.

Lagler, K. F. 1956. Freshwater Fishery Biology. (Second Ed.) W. C. Brown Co., Dubuque, Lowa, pp. 106-421.

Lam, T.J, 1982. Fish culture in Southeast Asia. Can.J.Fish.Aquat. Sci., 39(1):138–42.

Lander, T.R, S.M.C. Robinson, B.A. MacDonald, J.D, and Martin. 2013. Characterization of the suspended organic particles released from salmon farms and their potential as a food supply for the suspension feeder, *Mytilus edulis* in Integrated Multi-trophic Aquaculture (IMTA) systems. Aquaculture, 10.1016/J.aquaculture.2013.05.001.

Langeland.K.A.1996.*Hydrilla verticillata* (L.f.) Royle (Hydrocharitaceae), "The perfect aquatic weed". Castanea.61:293-304.

Largo D B, A D G Mario and S.Marabol, 2016. Development of an integrated multi-trophic aquaculture (IMTA) system for tropical marine species in southern Cebu, Central Philippines.3:67-76.

Lin, C.K, Y. Yi, M. K. Shrestha, R. B. Shivappa, and M.A. Kabir Chowdhury. 1999. Management of organic matterand nutrient generation in pond bottoms. Edt. By: K. McElwee, D. Burke, M. Niles, and H. Egna Sixteenth Annual Technical Report. Pond Dynamics/ Aquaculture CRSP, Oregon State University, Corvallis, Oregon.21-26.

Lindsay K and H.M.Hirt.1999. Use water Hyacinth. A practical Handbook of uses for the Water Hyacinth from across the world.Anamed.Winnenden.

Liu, R.H. 2003. Healthy benefits of fruits and vegetables are from additive and synergistic combinations of phytochemicals. Am. J. Clin. Nutr. 78: 517-520

Liao, I.C.2002..Roles and contributions of fisheries science in Asia in the 21st century.Fish. Sci., 68 (Suppl.1), pp. 3-13.

Little, D.C and P. Edwards. 2003. Integrated livestock-fish farming systems. Inland water resources and aquaculture service animal production service, FAO.

Losordo, T, M, Masser, P.MichaelandR.James.1998.*Recirculating Aquaculture Tanks Production System, AnOverview of Critical Considerations.* Southern Regional AquacultureCenter, SRAC Publication No. 451.

Luo. G, Gao. Q, W. C, Liu, W, Sun, D. Li, .2014.Growth, digestive activity, welfare, and partial cost effectiveness of genetically improved farmed tilapia (*Oreochromisniloticus*) cultured in a recirculating aquaculture system and an indoor biofloc system. *Aquaculture* 422: 1–7.

Macan, T.T., C.H. Mortima and E.B. Worthington. 1942. The production of fresh water fish for food. Freshw. Biol. Assoc. Sci. Publ., 6:36

Maier, P. G., 1998. Differential success of cyclopoid copepods in the pelagic zone of eutrophic lakes. *J. Mar. Systems*, 15 (1 – 4) : 135 – 138.

Mahapatra B.K 2004: Conservation of the Asiatic catfish, *Clarias batrachus* through artificial propagation and larval rearing technique in India. Sustainable Aquaculture, Vol. IX No. 4).: ***Clarias batrachus*** (Linnaeus, 1758)

Mohanta, K.N., S.Subramanian, N.Komarpant and S.Saurabh.2008. Alternate carp species for diversification in freshwater aquaculture in India. ICAR Research Complex for Goa,

Mahanta. P.C., Prem Kumar, N. N. Pandey, S.K.Srivastava, S.Ali and D Sarma.2010. Improved strains of Common Carp for Coldwater Aquaculture Champa -1 and Champa-2. Bulletin No: 16. Directorate of Coldwater Fisheries Research (Indian Council of Agricultural Research) Bhimtal - 263 136, Nainital (Uttara Khand).

Mair GC, Abucay JS, Beardmore JA, Skibinski.D, 1995. Growth performance trials of genetically male tilapia (GMT) derived from YY males in Oreochromisniloticus L.: On station comparisons with mixed sex and sex reversed male populations. Aquaculture 137:313-322.

Mallya Y. J. 2007. The Effects of Dissolved Oxygen on fish growth in aquaculture. Final Report.Kingolwira National Fish Farming Centre, Fisheries Division, Ministry of Natural Resources and Tourism, Tanzania.

Mandal, S. 2017. Disadvantages of culture of *Cyprinus carpio* in earthen ponds (Personnel communication).

Manorama.M, A.Gupta, K. K. Lal, R. K. Singh & V. Mohindra, 2017. Genetic divergence in natural and farm populations of Pengba fish, *Osteobrama belangeri* (Valenciennes, 1844), an endemic fish of North-East India derived from mtDNA *ATPase 6/8* gene, Mitochondrial DNA Part B, 2:2, 658-661, DOI: 10.1080/23802359.2017.1372700.

Masser, M.P, J.RakosyandT.M.Losordo, 1999. Recirculating Aquaculture Tanks Production *System, Management of Recirculating System.* Southern Regional Aquaculture Center, SRAC Publication No. 452.

Meyer, F.P., J.W. Warren and T.G.Carey(ed).1983. Aguide to integrated fish health management in the great lakes basin. Freat Lakes Fishery Commission. Ann. Arbor. Michigan, Spec.Pub.83-2:272p.

Miller, D. 1979. Recent developments in cage and enclosure aquaculture in Norway, in: Pillay, T.V.R. and W.A. Dill (eds). Advances in aquaculture. pp. 447–452. Fishing News Books Ltd.

Milne, P.H., 1979. Fish and shellfish farming in coastal waters. Farnham, Surrey, Fishing News Books Ltd., 208 p. 2nd ed.

Mishra S.S, Das R, Dhiman M, Choudhary P, Debbarma J, *et al.,* 2017. Present Status of Fish Disease Management in Freshwater Aquaculture in India: State-of-the-Art-Review. J Aquac Fisheries 1: 003.

MohantyB P, A.Mohanty and S.C.Parija.2018.Antimicrobial resistance and alternatives to antibiotics and drugs use in human and animal health.Nat.Workshop on Antimicrobial Resistance March.2018.129-139.

Mukhopadhyay, S.K., S. Biswas and A. Chatterjee. 1997. Effects of solar eclipse on zooplanktonic diel periodicity and primary productivity in fresh water ponds. *Kod. Obi.* Bull. 13:245–249.

Muller, F. and L.Varadi, 1980. The results of cage fish culture in Hungary. Aquacult. Hung., 2:154–67.

Murugesan, V.K., S. Manoharan and R. Palaniswamy. 2005. Pen fish culture in reservoir-An alternative to land based nurseries.Naga, World Fish Centre, newsletter, 28(1&2):49-52.

Nakajima, K. 1999. Prevention and Management of diseases of formed, sea caught and recreational caught fish in Asia, the far east and ocean. Conf. OIE. 227-240.

Nargis, A. 2006. Seasonal variation in the chemical composition of body flesh of koi fish, *Anabas testudineus* (Bloch), (Anabantidae: Perciformis). Bangladesh J. Sci. Industr. Res. 41: 219-226.

Nandeesha, M.C. 2002. Sewage fed aquaculture sustems of Kolkata- A century-old innovation of farmers. Aquaculture Asia.7 (2): 28-32.April-June 2002.

NFDB.2016. Guidelines for cage culture in inland water bodies of India.GOI, New Delhi.20p.

Nash, C. E., 1974. Crop Selection issues. In: Open sea mariculture. ..• E. J. Hanoon (Ed) p. 183-210.

Naskar K.R. 1985. A short history and the present trends of brackishwater fish culture in paddy fields at the Kulti Minakhan areas of Sundarbans in West Bengal. Journal of the Indian Society of Coastal Agricultural Research 3 (2): 115–124.

Nath. J. 2010. Hydrobiology of waste water fed fish ponds with special reference to fish production. Ph.D. thesis. Vidyasagar University, Midnapur. West Bengal.

Natarajan, P, V. Sundarajan and M. D. K. Kuthalingam. 1983. Review on cage and pen culture. Proc. Natl. Sem. Cage Pen Culture.6p.

Nutrition Research Council. 1977. Nutrient requirements of warmwater fishes. Natl. Acad. Sci., Washington, DC.78 p.

Nayak P. K., A. K. Pandey, B. N. Singh, J. Mishra, R. C. Das and S. Ayyappan 2000. Breeding, Larval Rearing and Seed Production of the Asian Catfish, Heleroptzeustesfossilis (Bloch). Central Institute of Freshwater Aquaculture, Kausalyaganga. Bhubaneswar-751002.

Nazar A.K.A, R. Jaykumar and G. Tamilmani. 1990. Recirculating aquaculture systems. CMFRI Manuel Customized training Book. Mandapam Regional Centre of CMFRI Mandapam Camp - 623520, Tamil Nadu, India.

Nevas I.F., Rocha O., Roche K.F. and Pinto A.A.2003. Zooplankton community structure of two marginal lakes of the river Cuiba (Mato Grasso, Brazil) with analysis of rotifera and cladocera diversity, Braz. J. Biol., 63(3), 329- 343.

Neuwinger HD. 2004. Plants used for poison fishing in tropical Africa. Toxicon. 44:417-430.

New, M.B.1987. Feed and feeding of fish and shrimp.UNDP, FAO, Rome.274p.

Nikolsky G.V. 1963. The ecology of fishes.Translated from Russian by L. Birkett. Academic press London and New York. 1-352.

Nunes, A.J.P., Sa, M.V.C., Andriola Neto, F.F., and Lemos, D. 2006. Behavioral response to selected feed attractants and stimulants in Pacific white shrimp, *Litopenaeus vannamei*. *Aquaculture* 260: 244–254.

Odum, E.P. 1971. Fundamentals of Ecology. 3rd Ed. W.B. Saunders co., Philadelphia: 574 PP.

Ogbeibu, A.E and R.Victor, 1995.Hydrological studies of water bodies in the okomu forest reserves (sanctuary) in Southern Nigeria, physico-chemical hydrology, Tropical Freshwater Biology, 4, pp 83-100.

Ozyurt, G., Polat, A. and Loker, G.B. 2009. Vitamin and mineral content of pike perch (Sander lucioperca), common carp (Cyprinuscarpio) and European catfish (Silurisglanis). Turk. J. Vet. Anim. Sci. 33: 351-356.

PCARRD (Philippine Council for Agriculture and Resources Research), 1981. State of the art: Lakes and reservoirs research. Fish.Ser. Philipp.Counc.Agric.Resour.Res.Dev., (1):70 p.

Pethon P. 1994. Aschehougs store Fiskebok. Allenorskefiskerifarger. Aschehoug, Oslo 447 p.

Primavera J.H.2005.Mangroves, fishponds, and the quest for sustainability.Science, 310. pp. 58-59.

Pal, R.N and A.K. Ghosh. 1990. Fish health protection in Aquaculture Operation. In: Technologies for Inland Fisheries Development.Edt.V.V. Sugunan and U. Bhowmik. CICFRI, ICAR, Barrackpore, West Bengal, pp.131-142.

Pal, M. and Ghosh, M. 2013. Assay of biochemical compositions of two Indian freshwater Eel with special emphasis on accumulation of toxic heavy metals. *J. Aquatic Food Prod. Technol.* 22: 27-35.

Paul, B.N, S. Chanda, N. Sridhar, G.S. Saha and S.S. Giri.2016. Proximate, Mineral and Vitamin Contents of Indian Major Carp. Indian J. Anim. Nutr. 2016. 33 (1): 102-107.

Peachey B. Environmental stewardship – what does it mean? 2008. Proc Saf Environ Protect. 86(4):227-236.

Piasecki, W., Andrew, E. G. Jorge, C. E and Barbara, F. N.2004. Importance of copepoda in fresh water aquaculture. Zoological Studies 43 (2): 193-205.

Pieterse, A.H, 1990. Introduction. In: Aquatic weeds the Ecology and Management of Nuisance Aquatic Vegetation.Eds.A H Pieterse & K J Murphy, pp.3-16.Oxford University Press: New York.

Pillay. T.V.R. 1993. Aquaculture: Principles and Practices. Wiley, 600 p.

Porto, H. L. R, A. C. Leal de Castro, V. E. M. Filho, and G. R.Baptista, 2016. Evaluation of the Chemical Composition of Fish Species Captured in the lower Stretch of Itapecuru River, Maranhão, Brazil. Int'l. Jour. Adv. in Agri& Environ. Engg. (IJAAEE) Vol. 3 (1): ISSN 2349-1523 EISSN 2349-1531.

Queiroz.J.F; G. Nicolella, C. W Wood and C.E. Boyd.2004. Lime application methods, water and bottom soil acidity in fresh water fish ponds. Sci. Agricola.61 (5): 469-475.

Rabanal.H.R. 1988. History of Aquaculture, Lecture contributed to the FAO/ UNDP Network of Aquaculture Centers in Asia (NACA) Training Programme for Senior Aquaculturists, SEAFDEC, Tigbau

Ramakrishniah N.1986. Studies on the fishery and biology of *Pangasius pangasius* (Hamilton) of the Nagarjunasagar reservoir in Andhra Pradesh. Indian Journal Fisheries 33: 320-335.an, Iloilo, Philippines, 24 March 1988 FAO.

Rana, S.S.2015. Recent advances in Integrated farming systems. Department of Agronomy. College of Agriculture. CSK Himachal Pradesh KrishiViswavidyalaya, Palampur, 204p.

Rao, K.J., 1986. Life history and behaviour of *Macrobrachium malcolmsonii*. Bull.Cent, Inland Fish, Res. Inst. Barrackpore, 47: 60-64.

128. Roberts, R.D. and T. Zohary, 1992. The influence of temperature and light on the upper limit of *Microcystis aeruginosa* production in a hypertrophic reservoir. *J. Plankton. Res*. 14(2):235–247.

Rosenberg, A., T. E. Bigford, S. Leathery, R. L. Hill, and K. Bickers. 2000. Ecosystem approaches to fishery management through essential fish habitat. Bulletin of marine science 66(3):535-542.

Reid, G.K. and R.D. Wood. 1976. Ecology of inland waters and Estuaries. D. Van Nostrand Company, N.Y. 485 PP.

Santra, S.C. ; S.C. Deb, 1996. Hydro-chemistry and hydrobiology of sewage fed jheels in the tropics: a case study, *Indian Hydrobiol*. 1, 5 – 17.

Santhosh, B. and Singh, N.P., (2007), Guidelines for water quality management for fish culture in Tripura, ICAR. Research Complex for NEH Region, Tripura Center, Publication no.29.

Sanyal, P. S K Chakraborty and P B Ghosh.2015. Phytoremediation of Sewage-Fed Wetlands of East-Kolkata, India - A Case Study.Int.Res.J.Env.Sci (ISCA).4(1):80-89.

Sasikumar.G and C. S. Viji.2016. Integrated Multi-Trophic Aquaculture Systems (IMTA). ICAR-CMFRI Research Centre, Mangalore ICAR-CIFE Mumbai.Winter School on Technological Advances in Mariculture for Production Enhancement and Sustainability: 47-55.

Schreiber, J.D. and Neumaier, E.E. 1987. Biochemical oxygen demand in agricultural run-off. J. Environ. Qual., 16 (1) : 6 – 10.

Schuster.W.H.1955. Fish culture in conjunction with rice cultivation. World crops. 7: 11–14 and 67–70.

Schaeperclaus W.1933. Test book of pond culture: Rearing and keeping of carp, trout and allied fishes. Berlin (English translation by F.H. Hund). Fishery Leaflet, Fish and Wildlife Service, U.S. Department of Interior, Washington.

Silva, E. I. L. 2007. Ecology of phytoplankton in tropical waters: Introduction to the tropic and ecosystem changes from Srilanka. *Asian J. Water. Env*. and *Poll.* 4(1):25–35.

Silva C F, Genaro M. Soto-Zarazúa, I T Pacheco and A F Rangel. 2013. Male tilapia production techniques: A mini-review. African Journal of Biotechnology Vol. 12(36), pp. 5496-5502.

Shina Babu, D. P., B. C. Ghosh, M. M. Panda and B. B. Reddy. 1983. Effect of fish on growth and yield of rice under rice-fish culture. Oryza, 20: 144–150.

Singh, P, S, Maqsood, M.H.Samoon, N.Verma, S. Singh and A. Saxena. 2017. Polyculture - A culture practice to utilize all ecological niches of pond ecosystem effectively. Aquafind.

Soto-Zarazúa, GM, Rico-García E, Ocampo R, Guevara-González RG, Herrera-Ruiz G 2010b. Fuzzy-logic-based feeder system for intensive tilapia production (Oreochromisniloticus). Aquacult. Int. 18:379-391.

Soundarapandian, P., M. John Samuel and T. Kannupandi, 1997. A simple method for the seed production of freshwater prawn Macrobrachium malcolmsonii (H. Milne Edwards). J. Aqua. Trop., 12: 261-266.

Soundarapandian, P., K.S. Prakash and G.K. Dinakaran. 2009. Simple Technology for the Hatchery Seed Production of Giant Palaemonid Prawn *Macrobrachium rosenbergii* (De Man). International Journal of Animal and Veterinary Advances 1(2): 49-53.

Sidorkewicj N.S, M.R. Sabbatini and J.H.Irigoyen. 2000. The spread of *Myriophyllum elatinoids* Gaudich and *M.aquaticum* (Vell.) Verdc.from stem fragments, Abstracts of the Third International Weed Congress.3:224.

Sishula J., Makasa M. L., Nkonde G. K., Kefi A. S. and Katongo C. 2011. Removal of ammonia from aquaculture water using maize COB activated carbon. Malawi j.aquac.fish. 1(2): 10-15.

Sikoki, F.D. and J.V. Veen. 2004. Aspects of Water Quality and the Potential for Fish Production of Shiroro Reservoir Nigeria, Living System Sustainable development, 2, pp 7.

Singh, P, S, Maqsood, M.H. Samoon, N.Verma, S. Singh and A. Saxena. 2017. Polyculture - A culture practice to utilize all ecological niches of pond ecosystem effectively. Aquafind.

Snieszko, S.E 1973. Recent advances in scientific knowledge and developments pertaining to diseases of fishes. Adv, Vet. Sci. Comp. Med. 17: 291-314.

Snieszko S.E, and G.L. Bullock. 1975. Fish furunculosis. U.S. Fish and Wildl. Serv., Fish Dis. Leafl. 43. Washington, DC. 10 p.

Stone, N. M. and Thomforde H. K.2004. Understanding Your Fish Pond Water Analysis Report. Cooperative Extension Program, University of Arkansas at Pine Bluff Aquaculture / Fisheries.

Summerfelt, ST, J. W. Davidson, T.Waldrop, T.B, Tsukuda, M.Scottand B.W, Julie. 2004. *Aquaculture Engineering* 31, p 157-181.

Svobodova, Z., Lloyd, R. and Vykusova, J. M. B. 1993. Water quality and fish health. EIFAC Technical Paper No. 54. FAO, Rome.

Talwar, P K and A G Jhingran, 1991. Inland fishes of India and adjacent countries.Vol.I.Oxford and IBH Publishing Co.(Pvt) Ltd. New Delhi-Calcutta.

Thilstead, S.H., Roos, N. and Hasan, N.1997. The role of small indigenous fish species in food and nutrition security in Bangladesh, NAGA the ICLARM Q, 20(3 and 4), 82-84.

Tiwari, K.K., 1949. On a new species of *Palaemon* from Banaras, with a note on *Palaemonlanchasteri* de Man. Rec.Indian Mus., 45: 333-345.

Troell, M, D. Robertson-Anderson, R.J Anderson, J J Bolton, G. Maneveldt, C.Halling. and T. Probyn, 2006.Abalone farming in South Africa: an overview with perspectives on kelp resources, abalone feed, potential for on-farm seaweed production and socio-economic importance. Aquaculture, 257 (1–4) pp. 266-281.

Tucker C.S and D'Abramo, 2008. Managing High pH in Freshwater Ponds. SRAC Publication No. 4604.

Tucker, C.S. 2006. Pond aeration. The Fish site.

Timmons M. B, Ebeling J, Wheaton F, Summerfelt S, Vinci B. 2002.*Recirculating Aquaculture Systems*. 2nd edition.N.R.A.C Publication, No.01-002. Cayuga Aqua Ventures.

Timmons, M.B, J.M. Ebeling, F.W Wheaton, S.TSummerfelt, andB.J.Vinci, B.2002. *Recirculating Aquaculture Systems*. 2nd Northeastern Regional Aquaculture Center Publication No. 01-002.

Thomaston, W. Wand Zeller, H.D.1961. Results of a six-year investigation of chemical soil and water analysis and lime treatment in Georgia fish ponds. In: Annual Conference of the South Eastern Association of Game and Fish Commissioners, 15:236-245.

Uttangi J.C.2001. Conservation and management strategy for the water fowls of minor irrigation tank habits and their importance as stopover site in Dharwad Dist. In: Hosetti and M. Venkateshwaralu (eds.) Trends in wild life and management, Daya Publ. House, New Delhi, India, 179- 221

Vallentyne, J.R. 1965. Net primary productivity and photosynthetic efficiency in the biosphere on C.R. Goldman (ed.). Primary production in aquatic environment PP. 309 – 312.

Vankara, A.P. and C. Vijayalakshmi. 2009. Metazoan parasites of Mysus vittatus (Bloch) of River Godavari with description of a new species of Acanthocephala, Raosentis godavarensis. Sp. nov. J. Paras. Dis.33 (1&2):77- 83.

Verdegem MCJ, Bosma RH, Verreth JAJ. 2006. Reducing water for animal production through aquaculture. Int J Water Resources and Development. 22:101-113.

Vila Nova C.M.V.M, Helena Teixeira Godoy, Mauro Luiz Aldrigue. 2005. Chemical composition, cholesterol content and characterization of total lipids of Niletilapia (*Oreochromis niloticus)* and pargo *(Lutjanuspurpureus).*Ciênc. ETecnol. Aliment, Campinas, v.25, n.3, p.430-436. (http://dx.doi.org/ 10.1590/ S 0101 -20).

Vimal. J.B and S. S. M.Das, 2015.Toxicity of Euphorbia antiquorum latex extract to fresh water fish *Poecilia reticulate*. Int. Jour. of Fisheries and Aquatic Studies. 2(3): 214-216

Vincke.M.M.J.1977. Integrated farmingon fish and livestock: Present status and future development.Proceedings of the FAO/IPT Workshop onIntegrated

Livestock-Fish Production Systems, 16–20 December 1991, Institute of Advanced Studies, University of Malaya, Kuala Lumpur, Malaysia.

Warren.J.W. 1983.The Nature of fish diseases. In. A Guide to integrated fish health management in the great lakes basin. Great lakes fishery commission.

Webber Mona, Myers, Elecia Edwards, Cambell C. and Webber D.2005. Phytoplankton and zooplankton as indicator of water quality in Discovery Bay Jamaica, Hydrobiologia, 545, 177-193.

Wetzel R.G 1983. Limnology, 2nd ed. Saunders Co, Philadelphia, 860p.

WHO. Antimicrobial resistance. Fact sheet No. 194. 2012, http://www. who. int/ media centre /factsheets / Nfs194/es/index.html

Whipple, G.C. 1927. The microscopy of drinking water. 4th ed. Revised by Fair and Whipple, New York. 586 PP.

Williams, M.J .1997. Aquaculture and Sustainable food security in the developing world. In.Susainable Aquaculture. Edt.J.E.Bardach john Wiley &Sons.15-51.